T0339525

AMBIENT ASSISTED LIVING AND ENHANCED LIVING ENVIRONMENTS

AMBIENT ASSISTED LIVING AND ENHANCED LIVING ENVIRONMENTS

Principles, Technologies and Control

Edited by

CIPRIAN DOBRE
University Politehnica of Bucharest, Romania

CONSTANDINOS X. MAVROMOUSTAKIS
University of Nicosia, Cyprus

NUNO M. GARCIA
Instituto de Telecomunicações, Universidade da Beira Interior, Portugal

ROSSITZA I. GOLEVA
Technical University of Sofia, Bulgaria

GEORGE MASTORAKIS
Technological Educational Institute of Crete, Greece

ELSEVIER

AMSTERDAM • BOSTON • HEIDELBERG • LONDON
NEW YORK • OXFORD • PARIS • SAN DIEGO
SAN FRANCISCO • SINGAPORE • SYDNEY • TOKYO

Butterworth-Heinemann is an imprint of Elsevier

Butterworth-Heinemann is an imprint of Elsevier
The Boulevard, Langford Lane, Kidlington, Oxford OX5 1GB, United Kingdom
50 Hampshire Street, 5th Floor, Cambridge, MA 02139, United States

Notices

Knowledge and best practice in this field are constantly changing. As new research and experience broaden our understanding, changes in research methods, professional practices, or medical treatment may become necessary.

Practitioners and researchers must always rely on their own experience and knowledge in evaluating and using any information, methods, compounds, or experiments described herein. In using such information or methods they should be mindful of their own safety and the safety of others, including parties for whom they have a professional responsibility.

To the fullest extent of the law, neither the Publisher nor the authors, contributors, or editors, assume any liability for any injury and/or damage to persons or property as a matter of products liability, negligence or otherwise, or from any use or operation of any methods, products, instructions, or ideas contained in the material herein.

Library of Congress Cataloging-in-Publication Data
A catalog record for this book is available from the Library of Congress

British Library Cataloguing-in-Publication Data
A catalogue record for this book is available from the British Library

ISBN: 978-0-12-805195-5

Publisher: Joe Hayton
Acquisition Editor: Sonnini R. Yura
Editorial Project Manager: Ana Claudia Abad Garcia
Production Project Manager: Kiruthika Govindaraju
Designer: Maria Inês Cruz

Typeset by VTeX

DEDICATIONS

Special thanks go to my family, Anamaria-Raluca and Iulia-Raluca, for their unconditional love and support, understanding and patience.

Ciprian Dobre

To my wife Aphrodite for her love, patience and caring.

Constandinos X. Mavromoustakis

To my friends and colleagues who team up to achieve higher goals and to my family, grateful for your love, wisdom and for being a permanent source of strength.

Nuno M. Garcia

To all who believe in me.

Rossitza I. Goleva

To my son Nikos, who makes me proud every day.

George Mastorakis

CONTENTS

CONTRIBUTORS

Ghufran Ahmed
COMSATS Institute of Information Technology, Park Road, 45550, Islamabad, Pakistan

Raluca Maria Aileni
Politehnica University of Bucharest, Faculty of Electronics, Telecommunication and Information Technology, Romania

Ionut Anghel
Technical University of Cluj-Napoca, Romania

Nauman Aslam
Department of Computer Science and Digital Technologies, Faculty of Engineering, Northumbria University, Newcastle, UK

Abdullah Balcı
Ege University, Department of Electrical & Electronics Engineering, Turkey

Liviu Breniuc
"Gheorghe Asachi" Technical University of Iasi, Faculty of Electrical Engineering, Bd. D. Mangeron, 23, 700050, Iasi, Romania

Ivan Chorbev
Faculty of Computer Science and Engineering, University Ss Cyril and Methodius, ul. Rugjer Boskovic 16, Karpos 2, 1000 Skopje, The former Yugoslav Republic of Macedonia

Chandreyee Chowdhury
Jadavpur University, Kolkota, India

Tudor Cioara
Technical University of Cluj-Napoca, Romania

Valeriu David
"Gheorghe Asachi" Technical University of Iasi, Faculty of Electrical Engineering, Bd. D. Mangeron, 23, 700050, Iasi, Romania

Antonio Del Campo
Dipartimento di Ingegneria dell'Informazione, Università Politecnica delle Marche, Via Brecce Bianche 12, I-60131 Ancona, Italy

Ciprian Dobre
University of Politehnica of Bucharest, Bucharest, Romania

Elisa Felici
INRCA, Lab. Bioinformatica, Bioingegneria e Domotica, Via Santa Margherita 5, I-60124 Ancona, Italy

Ennio Gambi
Dipartimento di Ingegneria dell'Informazione, Università Politecnica delle Marche, Via Brecce Bianche 12, I-60131 Ancona, Italy

Nuno M. Garcia
Instituto de Telecomunicações, Covilhã, Portugal
Universidade da Beira Interior, Covilhã, Portugal
Universidade Lusófona de Humanidades e Tecnologias, Lisbon, Portugal

Alejandro García Marchena
Ingeniería Y Soluciones Informáticas del Sur S.L., Spain

Rossitza I. Goleva
Technical University of Sofia, Kl. Ohridski blvd 8, Faculty of Telecommunications, Department of Communication Networks, 1756 Sofia, Bulgaria

Sérgio Guerreiro
Universidade da Beira Interior, Covilhã, Portugal
Universidade Lusófona de Humanidades e Tecnologias, Lisbon, Portugal

Cristian-Győző Haba
"Gheorghe Asachi" Technical University of Iasi, Faculty of Electrical Engineering, Bd. D. Mangeron, 23, 700050, Iasi, Romania

Vasos Hadjioannou
Department of Computer Science, University of Nicosia, 46 Makedonitissa Avenue, 1700 Nicosia, Cyprus

Najmul Hassan
Capital University of Science and Technology, Islamabad Expressway, Kahuta Road, Zone-V, Islamabad, Pakistan

Hilal Jan
COMSATS Institute of Information Technology, Park Road, 45550, Islamabad, Pakistan

Elisa Jimeno
Ingeniería Y Soluciones Informáticas del Sur S.L., Spain

Helmut Leopold
Austrian Institute of Technology, Austria

María Lindén
School of Innovation, Design & Engineering, Mälardalen University, Västerås, Sweden

Javier Malagón Hernández
Telematics and Electronics Engineering Department, ETSIS Telecommunication, Politécnica de Madrid University, Spain

María Luisa Martín Ruiz
Telematics and Electronics Engineering Department, ETSIS Telecommunication, Politécnica de Madrid University, Spain

George Mastorakis
Department of Business Administration, Technological Educational Institute of Crete, Heraklion, Crete, Greece

Constandinos X. Mavromoustakis
Department of Computer Science, University of Nicosia, 46 Makedonitissa Avenue, 1700 Nicosia, Cyprus

Laura Montanini
Dipartimento di Ingegneria dell'Informazione, Università Politecnica delle Marche, Via Brecce Bianche 12, I-60131 Ancona, Italy

Davide Perla
Dipartimento di Ingegneria dell'Informazione, Università Politecnica delle Marche, Via Brecce Bianche 12, I-60131 Ancona, Italy

Laura Raffaeli
Dipartimento di Ingegneria dell'Informazione, Università Politecnica delle Marche, Via Brecce Bianche 12, I-60131 Ancona, Italy

Muhammad Riaz
Capital University of Science and Technology, Islamabad Expressway, Kahuta Road, Zone-V, Islamabad, Pakistan

Lorena Rossi
INRCA, Lab. Bioinformatica, Bioingegneria e Domotica, Via Santa Margherita 5, I-60124 Ancona, Italy

Ioan Salomie
Technical University of Cluj-Napoca, Romania

Victor Sanchez Martin
Eindhoven University of Technology, Netherlands

Maham Shahid
COMSATS Institute of Information Technology, Park Road, 45550, Islamabad, Pakistan

Azfar Shakeel
COMSATS Institute of Information Technology, Park Road, 45550, Islamabad, Pakistan

Radosveta I. Sokullu
Ege University, Department of Electrical & Electronics Engineering, Turkey

Susanna Spinsante
Dipartimento di Ingegneria dell'Informazione, Università Politecnica delle Marche, Via Brecce Bianche 12, I-60131 Ancona, Italy

Rumen Stainov
Applied Computer Science Department, University of Applied Sciences, Leipziger Strasse 123, D-36037, Fulda, Germany

Vera Stara
INRCA, Lab. Bioinformatica, Bioingegneria e Domotica, Via Santa Margherita 5, I-60124 Ancona, Italy

Rodica Strungaru
Politehnica University of Bucharest, Faculty of Electronics, Telecommunication and Information Technology, Romania

Loizos Toumbas
University of Nicosia Medical School, St George's University of London Medical School, Nicosia, Cyprus

Vladimir Trajkovik
Faculty of Computer Science and Engineering, University Ss Cyril and Methodius, ul. Rugjer Boskovic 16, Karpos 2, 1000 Skopje, The former Yugoslav Republic of Macedonia

Saif ul Islam
COMSATS Institute of Information Technology, Park Road, 45550, Islamabad, Pakistan

Laura Vadillo Moreno
Telematics and Electronics Engineering Department, ETSIS Telecommunication, Politécnica de Madrid University, Spain

Alberto Carlos Valderrama
Polytechnic Faculty of Mons, Electronics and Microelectronics Department, Belgium

Dan Valea
Technical University of Cluj-Napoca, Romania

Miguel Ángel Valero Duboy
Telematics and Electronics Engineering Department, ETSIS Telecommunication, Politécnica de Madrid University, Spain
National Reference Centre of Personal Autonomy and Technical Aids (CEAPAT), Health, Social Services and Equality Ministry, Spain

Martijn Vastenburg
ConnectedCare R&D Office Arnhem, Netherlands

Florian Wamser
University of Würzburg, Germany

Irene Yu-Hua Gu
Department of Signals and Systems, Chalmers University of Technology, SE-412 96, Gothenburg, Sweden

Yixiao Yun
Department of Signals and Systems, Chalmers University of Technology, SE-412 96, Gothenburg, Sweden

Thomas Zinner
University of Würzburg, Germany

BIOGRAPHIES

Ghufran Ahmed is serving as Assistant Professor at the Department of Computer Science, COMSATS Institute of Information Technology, Islamabad, Pakistan. He received PostDoc in 2015 from Department of Computer Science and Digital Technology, Faculty of Engineering and Environment, Northumbria University, Newcastle Upon Tyne, UK. He has received Ph.D. in 2013 from Department of Computer Science, Capital University of Science and Technology, Islamabad. His area of research is Wireless Sensor Networks and Wireless Body Area Networks. He started Ph.D. in 2006 from Faculty of Computer Science and Engineering, GIK Institute, Topi, Swabi, KPK. He also worked as a visiting scholar at the CReWMaN Lab, Department of Computer Science and Engineering, University of Texas at Arlington in 2008–09.

Raluca Maria Aileni obtained in 2012 the Ph.D. degree in Industrial Engineering at Technical University "Gheorghe Asachi" of Iasi. She is a Ph.D. student at Faculty of Electronics, Telecommunication and Information Technology, Politehnica University of Bucharest. She graduated Faculty of Textile Leather and Industrial Engineering Management and Faculty of Computer Science. In 2010 during her Ph.D., she obtained a research fellowship for doctoral studies at ENSAIT — Lille University of Science and Technology, France, where she specialized in 3D modeling and simulation for textiles, using the Kawabata system, 2D–3D Design Concept for the design and simulation of technical textile articles. In 2015, she obtained the Excellence Fellowship Grant for doctoral studies in Belgium, Mons University.

Ionut Anghel received the Ph.D. degree in computer sciences at the Technical University of Cluj-Napoca in 2012. Currently he is a lecturer of computer science and also a member of the Distributed Systems Research Lab with main research areas ambient assisted living, artificial intelligence, semantic web and autonomic computing. His list of publication includes more than 30 peer-reviewed papers.

Nauman Aslam received his Ph.D. in Engineering Mathematics from Dalhousie University, Halifax, Nova Scotia, Canada in 2008. He is currently working as a Reader in the Department of Computer Science and

Digital Technologies at Northumbria University, UK. Prior to joining Northumbria University he worked as an Assistant Professor at Dalhousie University, Canada from 2008–2011. Dr. Nauman has extensive research experience in Wireless Ad hoc and Sensor Networks with focus on energy efficient protocols, quality of service, security and medium access control. He is currently leading two multi-partner collaborative projects (combined worth 5.5 million Euros) funded by EU's Erasmus Mundus programme. He has published over 50 refereed research articles in international journals and conference proceedings. He has also co-organized international workshops and conference. He also served on many technical program committees, edited journal issues and reviewed papers for several journals. Dr. Nauman is a member of IEEE and IAENG.

Abdullah Balcı received his B.Sc. degree in Electronics and Communication Engineering from Izmir University, Izmir, Turkey, in 2014. Currently he is a M.Sc. candidate in Electrical and Electronics Engineering at Ege University, Izmir, Turkey. Since 2015, he has been with the Department of Electrical and Electronics Engineering, Faculty of Engineering, Ege University, where he is currently a Research Assistant. His research interests are in the area of Wireless Sensor Networks, M2M communication over cellular networks, and energy efficient techniques.

Liviu Breniuc received Engineering Degree in Electronics and Telecommunications from the Faculty of Electrical Engineering, "Gheorghe Asachi" Technical University of Iasi (Romania) in 1979. He also received Ph.D. degree in Metrology from the same university in 1999. Between 1979–1988 he worked as a researcher in the Design Department for Professional Marine Radio Devices at Tehnoton Company in Iasi. From 1988 to present he is teaching at the "Gheorghe Asachi" Technical University of Iasi where he is Professor in the Department of Electrical Measurements and Materials. His research activity includes participation in several projects in the field of microcontroller and FPGA based systems with current research focused on developing medical, ambient assisted living and IoT systems with ZigBee or WiFi connectivity.

Ivan Chorbev earned his bachelor and master degrees in 2004 and 2006. He completed his Ph.D. studies at the Faculty of Electrical Engineering and Information Technologies in Skopje in 2009. His current position is Associate Professor and Dean of the Faculty of Computer Science and Engineering.

He gained his first professional experience at the Macedonian Telekom — T-Home, at the department for software development, during the period 2003–2005. He started working at the Institute for Computer Science and Engineering in the Faculty for Electrical Engineering in Skopje in 2005. In 2009 he became an assistant professor, and an associate professor in 2014. His teaching commitments included lectures in Structured Programming, Object-oriented programming, Programming practices, Web Design etc.

In his research he has participated in more than 70 scientific papers in journals and conference proceedings, book chapters, and has performed several research stays as visiting scientist. The fields of his research interests include combinatorial optimization, heuristic algorithms, constraint programming, web development technologies, application of computer science in medicine and telemedicine, medical expert systems, Assistive technologies, knowledge extraction, machine learning.

Chandreyee Chowdhury received Ph.D. in Engineering and Masters in Computer Science and Engineering from Jadavpur University in 2013 and 2005 respectively. She received Bachelor in Computer Science and Engineering from University of Kalyani in 2003. Currently she is Assistant Professor in the department of Computer Science and Engineering at Jadavpur University. Her research interests include routing issues of Wireless Sensor Networks and its variants, Applications of mobile crowd-sensing. She is a member of IEEE and IEEE Computer Society.

Tudor Cioara received the Ph.D. degree in computer sciences at the Technical University of Cluj-Napoca in 2012 where he is now a lecturer of computer science and member of Distributed Systems Research Lab (http://dsrl.coned.utcluj.ro/). His current research interest is focused on ambient assisted living, green IT, artificial intelligence, semantic web, context awareness and autonomic computing. He is reviewer at several international conferences and journals and is/was involved in eight national/international research projects. Also, he has published more than 40 peer-reviewed papers.

Valeriu David received the Engineering Degree in Electronics and Communications from the Faculty of Electrical Engineering, "Gheorghe Asachi" Technical University of Iaşi (Romania) in 1983 and Ph.D. degree in Electrical Measurement from the same university, in 1998.

He is Professor and Ph.D. supervisor in the Department of Electrical Measurements and Materials, Faculty of Electrical Engineering, "Gheorghe Asachi" Technical University of Iaşi.

Professor Valeriu David is author/coauthor of 10 books and about 150 scientific papers in the following domains: electric and electronic measurements; measurement in biomedicine and ecology; survey of the electromagnetic environment; biological effects of electromagnetic fields; new materials and techniques for electromagnetic shielding.

Antonio Del Campo received the Bachelor Degree in Biomedical Engineering in December 2010, and then he obtained the Master Degree in Electronic Engineering in October 2014 (both summa cum laude), from Università Politecnica delle Marche. In November 2014 he started his Ph.D. in Biomedical, Electronic and Telecomunications Engineering. Currently, he is working on low power wireless communication technologies for indoor applications, mainly related to AAL scenarios.

Ciprian Dobre is Professor within the Computer Science Department, University Politehnica of Bucharest (Habil. since 2014, Dr. since 2008 with Cum laudae). He currently leads the activities within Laboratory on Pervasive products and services, and MobyLab. Ciprian Dobre's research interests involve research subjects related to mobile wireless networks and computing applications, pervasive services, context-awareness, and people-centric or participatory sensing. He has scientific and scholarly contributions in the field of large scale distributed systems concerning mobile applications and smart technologies to reduce urban congestion and air pollution (MobiWay, TRANSYS), context-aware applications (CAPIM), opportunistic networks and mobile data offloading (SPRINT, SENSE), monitoring (MonALISA), high-speed networking (VINCI, FDT), Grid application development (EGEE, SEE-GRID), and evaluation using modeling and simulation (MONARC 2, VNSim). These contributions led to important results, demonstrating his qualifications and potential to go significantly beyond the state of the art. Ciprian Dobre was awarded a Ph.D. scholarship from California Institute of Technology (Caltech, USA), and another one from Oracle. His results received one IBM Faculty Award, two CENIC Awards, and three Best Paper Awards (in 2013, 2012, and 2010). The results were published in over 100 chapters in edited books, articles in major international peer-reviewed journals, and papers in well-established international conferences and workshops.

Elisa Felici graduated in Psychology and since 2010 she is a researcher at the National Institute of Health and Science on Aging — I.N.R.C.A. in the field of acceptance of technology among primary, secondary and tertiary end users of Ambient Assisted Living technologies. In the last years she has gained extensive experience in aging research and a long commitment to study the psycho-social impact of technology on health.

Ennio Gambi was born in Loreto (Ancona, Italy) in 1961. He received the Electronic Engineering graduate diploma at the Università Politecnica delle Marche in 1986 and the Microwave Engineering master degree in 1989. From 1984 to 1992 he worked with the Azienda di Stato per i Servizi Telefonici, while during 1988 he covered the role of Avionic Engineer as an Official of the Italian Military Air Force. Since 1992 he joined the Università Politecnica delle Marche in Ancona, where he is, currently, an Associate Professor. At present, on the basis of a long time interest on domotic system, he is working on the evolution of this technology toward the Ambient Assisted Living, with the aim to propose an integrated solution where the fusion of data provided by different sensor networks allows to offer the desired level of home assistance for disabled or elderly people.

Nuno M. Garcia holds a Ph.D. in Computer Science Engineering from the University of Beira Interior (UBI, Covilhã, Portugal) (2008) and he is a 5-year B.Sc. (Hons.) in Mathematics/Informatics also from UBI (1999–2004). He is Assistant Professor at UBI and Invited Associate Professor at the School of Communication, Architecture, Arts and Information Technologies of the Universidade Lusófona de Humanidades e Tecnologias (Lisbon, Portugal). He was founder and is coordinator of the Assisted Living Computing and Telecommunications Laboratory (ALLab), a research group within the Instituto de Telecomunicações at UBI. He was also co-founder and is Coordinator of the Executive Council of the BSAFE LAB — Law enforcement, Justice and Public Safety Research and Technology Transfer Laboratory, a multidisciplinar research laboratory in UBI. He is the coordinator of the Cisco Academy at UBI, Head of EyeSeeLab in EyeSee Lda. (Lisbon, Portugal), and member of the Consultative Council of Favvus IT HR SA (Lisbon). He is also Chair of the COST Action IC1303 AAPELE — Architectures, Algorithms and Platforms for Enhanced Living Environments. He is the main author of several international, European and Portuguese patents. He is member of the Non-Commercial Users Constituency, a group within GNSO in ICANN. His main interests include Next-Generation Networks, algorithms for bio-signal processing, distributed and cooperative protocols.

Alejandro García Marchena is a core member of the Innovation and Technology department of ISOIN, leading its technical team. Alejandro holds an engineering degree in Computer Engineering from the University of Seville. For the past 7 years he has been working as a software engineer on national and international research and consultancy projects focusing on interoperability, ambient assisted living, artificial intelligence and multimedia.

Rossitza I. Goleva received her Ph.D. in Communication Networks in 2016 and M.Sc. in Computer Science in 1982 at Technical University of Sofia, Bulgaria. She was part of the research staff of the research Institute of Bulgarian PTT between 1982 and 1987. Since 1987, she is with Department of Communication Networks at Technical University of Sofia. At present, Rossitza works on communication networks, communication protocols, and software engineering. Her research interests are in Quality of Service in communication networks, communication protocols, traffic engineering, cloud and fog computing, performance analyses. She is an IEEE Member, involved in IEEE Bulgaria section activities, has more than 85 research publications, was part of more than 30 research projects including and EU COST IC1303 AAPELE action.

Sérgio Guerreiro is an invited Assistant Professor at the Lusófona University, in Lisbon, Portugal, and at Universidade da Beira Interior, Covilhã, Portugal. He is a research member of CICANT/ULHT R&D unit. Concluded his Ph.D. degree in Information Systems and Computer Engineering at Instituto Superior Técnico/University of Lisbon (IST/UL) in 2012. Before that, concluded the M.Sc. degree in 2003 and the engineering degree in 1999. Both degrees also in IS and Computer Engineering at IST/UL. His research interests are Enterprise Ontology, Enterprise Architecture and Enterprise Governance, with a specific focus in developing and applying control systems that are able to support and deliver more information for the organizational actors involved in the decision-making processes. Moreover, he has professional expertise in the mobile telecommunications field.

Cristian-Győző Haba received his Ph.D. in Automatic Control in 2000 and Engineer Degree in Electrical Engineering in 1988 from "Gheorghe Asachi" Technical University of Iasi (Romania). Since 1990 he is teaching at Faculty of Electrical Engineering, "Gheorghe Asachi" Technical University of Iasi where he is now a Professor in the Department of

Electrical Engineering. At present he works on EDA, microcontroller and FPGA based embedded systems and e-learning. He is an IEEE, ACM and EAI member involved in several research projects on distributed measurement systems, home building automation, ambient assisted living and IoT.

Vasos Hadjioannou received his B.Sc. in Software Engineering from the Budapest University of Technology and Economics (BME) in 2014. There he chose the programming specialization, and his thesis project included the implementation of a location-based Android application. He continued his studies at the University of Nicosia (UNIC) where he enrolled in the M.Sc. course of Computer Science. At UNIC, he followed the Computer Networks specialization and expanded his knowledge on Android programming by choosing as a thesis project a context awareness location-based Android application. Currently, Vasos is working as a Software Developer at eBOS technologies where he improves his programming knowledge day by day.

Najmul Hassan did his M.S. in computer sciences from Capital University of Science and Technology, Islamabad, Pakistan in 2010. He is currently pursuing his Ph.D. research in air-to-ground communications in the Department of Electrical Engineering, Capital University of Science and Technology, Islamabad. His research areas are wireless sensor networks, ground-to-air and air-to-ground communication, handover in airborne Internet and Wireless Body Area Networks.

Hilal Jan received her M.S. degree in Software Engineering from Bahria University Islamabad. Currently, she is working as Lecturer in Computer Science Department at COMSATS Institute of Information Technology Islamabad, Pakistan. Her research interests include Wireless Sensor Networks and Wireless body Area Networks.

Elisa Jimeno holds a degree in Telecommunications Technology Engineering by the University of Cantabria (Spain). She is R&D Project Manager in the Innovation and Technology department of ISOIN and has wide experience in national and international research projects, with both management and technical skills. Her expertise lies on communication systems, especially in design and development of sensing infrastructures.

Helmut Leopold holds a Master degree (Dipl.-Ing.) in Computer Science from the Vienna University of Technology, 1989, and a degree in Electronics and Communications from the Technical College HTL Rankweil, Austria. After 10 years with Alcatel Austria he worked for 9 years at Telekom Austria, where he held the position as Managing Director Platform and Technology Management. In this role he was the driving force behind Telekom Austria's Next Generation Network (NGN) transformation programs. Since January 2009, Helmut has joined the AIT Austrian Institute of Technology, where he currently holds the position Head of Digital Safety & Security Department. The key research areas under his responsibility are Future Networks and Services with a special focus on Cyber Security, eHealth, Digital Insight based on big data analytics, Highly Reliable Systems and Intelligent Vision Systems. Helmut Leopold is President of the Austrian Organization for Information and Communication Technology (GIT) and Board Member of the Austrian Electrotechnical Association (OVE). He acts as evaluator for the R&D programs of the European Commission and is lecturer at various universities. Helmut Leopold was chairman of the San Francisco based Broadband Services Forum (BSF), Vice President of the Austrian Telecommunication Research Center (FTW) and President of the Austrian IPv6 Taskforce. He has served on the boards of many organizations providing strategic, technology and business leadership. He is IEEE and AOM member.

María Lindén is a Professor in Health Technology with specialization in biomedical sensor systems at Mälardalen University and she leads the research profile Embedded Sensor Systems for Health. She holds a Ph.D. in Biomedical Instrumentation Techniques from Linköping University (1998). Her research direction includes the development of sensor systems for monitoring physiological parameters. This includes wireless sensor systems monitoring vital parameters in the home environment and technologies to measure blood flow at different tissue depths combining the techniques laser Doppler flowmetry and photoplethysmography (PPG). It further includes research on sensor systems for movement analysis, based on gyroscopic and accelerometer principles and signal analysis.

Javier Malagón Hernández is an Operating Systems and Programming Professor at the Universidad Politécnica de Madrid. He received his Laurea (Graduate Diploma) Degree in Telecommunications Engineering in 1997 from the Universidad Politécnica de Madrid (Spain). At present, he works on intelligent systems, developing and validating in e-health scenarios. He is a Researcher Member of the T>SIC research group at the Universidad Politécnica de Madrid, where he is working on applied research projects in AAL and home building automation, supported by regional and national funds.

María Luisa Martín Ruiz received her Ph.D. from the Universidad Politécnica de Madrid (Spain) in 2014, and her Laurea (Graduate Diploma) Degree in Informatics Engineering in 2006 from the Universidad Carlos III, Madrid (Spain). Since December 2014, she has been a PostDoc in Telecommunications at the Universidad Politécnica de Madrid, where now she is Assistant Professor in Telecommunications. At present, María Luisa works on intelligent systems, developing and validating in e-health and telemedicine scenarios. She is a Researcher Member of an innovation research group at the Universidad Politécnica de Madrid (T>SIC), in which she is working in applied research projects on AAL and home building automation, supported by regional and national funds.

George Mastorakis received his B.Eng. (Hons) in Electronic Engineering from UMIST, UK in 2000, his M.Sc. in Telecommunications from UCL, UK in 2001 and his Ph.D. in Telecommunications from University of the Aegean, Greece in 2008. He is serving as an Associate Professor at Technological Educational Institute of Crete and as a Research Associate in Research & Development of Telecommunications Systems Laboratory at Centre for Technological Research of Crete, Greece. His research interests include cognitive radio networks, networking traffic analysis, radio resource management, energy efficient networks, Internet of Things and mobile computing. He has a more than 150 publications at various international conferences proceedings, workshops, scientific journals and book chapters.

Constandinos X. Mavromoustakis is currently a Professor at the Department of Computer Science at the University of Nicosia, Cyprus. He received a five-year dipl.Eng (B.Sc., B.Eng., M.Eng.) in Electronic and Computer Engineering from Technical University of Crete (2000), Greece, M.Sc. in Telecommunications from University College of London, UK (2001), and his Ph.D. from the department of Informatics at Aristotle University of Thessaloniki, Greece (2006). Dr. Mavromoustakis is leading the Mobile Systems Lab. (MOSys Lab., http://www.mosys.unic.ac.cy/) at the Department of Computer Science at the University of Nicosia, dealing with design and implementation of hybrid wireless testbed environments and MP2P systems, IoT configurations and smart applications, as well as high performance cloud and mobile cloud computing (MCC) systems, modeling and simulation of mobile computing environments and protocol development and deployment for large-scale heterogeneous networks and new 'green' mobility-based protocols. Dr. Mavromoustakis is an active member (vice-chair) of IEEE/R8 regional Cyprus section since Jan. 2016, and since May 2009 he serves as the Chair of C16 Computer Society Chapter of the Cyprus IEEE section. Dr. Mavromoustakis has a dense research work outcome (more than 200 papers) in Distributed Mobile Systems and spatio-temporal scheduling, consisting of numerous refereed publications including several Books (IDEA/IGI, Springer and Elsevier). He has served as a consultant to many industrial bodies (i.e. member of the Technical Experts for Internet of Things-IoT competition at Intel Corporation LLC (www.intel.com) for the ChallengeMe etc.), he is a management member of IEEE Communications Society (ComSoc) Radio Communications Committee (RCC) and a board member of the IEEE-SA Standards IEEE SCC42 WG2040 whereas he has served as track Chair and co-Chair of various IEEE International Conferences (including AINA, IWCMC, ICC, GlobeCom, IEEE Internet of Things etc.).

Laura Montanini was born in 1987 in Fermo. She obtained the High School diploma in Informatics from I.T.I.S. "Montani" in Fermo, on July 2006. She received the Bachelor's Degree in Informatics and Automation Engineering in 2010, and the Master's Degree in Electronic Engineering on July 2013, at the Università Politecnica delle Marche. Since November 2013 she is a Ph.D. student in Telecommunications. She is working on multimodal user interfaces in the Ambient Assisted Living context. She is also interested in technologies and tools for the Smart Home, and algorithms for the Activities of Daily Living recognition.

Davide Perla obtained his Bachelor's Degree in 2008 from Università Politecnica delle Marche with a thesis on Embedded Systems. He obtained his Master's Degree in 2013 in Electronics Engineering from the same University with a thesis in the area of Network on Chip. In 2013/2014 he collaborated with IDEA Soc. Coop. in Ancona developing a generic purpose platform for sensing, actuation and control. Since 2014 he is a Ph.D. student in Electronics. He is working on domotic communication systems, wearable devices and cloud-oriented communication devices and architectures for home and building automation and AAL.

Laura Raffaeli received the Bachelor's Degree in Telecommunication Engineering in 2010, and the Master's Degree in Electronic Engineering on February, 2013 at the Università Politecnica delle Marche, Ancona. She was involved in a project dealing with the design of a Smart TV application, to collect medical reports from remote labs and hospitals, in cooperation with the Regione Marche ICT Department. At the Department of Information Engineering she is working on interactive apps to interface users to remote healthcare services and Ambient Assisted Living solutions.

Muhammad Riaz was born in Pakistan. He did his Ph.D. degree in electronic engineering at the Capital University of Science and Technology, Islamabad, Pakistan. He received his M.Sc. from Quaid-i-Azam University, Islamabad, Pakistan, in 2002 and M.S. degree in electronic engineering from Mohammad Ali Jinnah University Islamabad, Pakistan in 2009. At present, his research interests include channel modeling, channel equalization, Mobile-to-Mobile Communications, Wireless Sensor Networks and Wireless Body Area Networks.

Lorena Rossi is the chief of Laboratory of Bioinformatics, Bioengineering and Domotics at the National Institute of Health and Science on Aging — I.N.R.C.A. She graduated in Electronic Engineering and since 1991 she has been worked at INRCA gaining a long experience in development and management of HIS (Health information system). Her main interest is about the use of ICT technologies for the assistance of elderly people.

Ioan Salomie (Ph.D.) is a professor of Computer Science at the Technical University of Cluj-Napoca, former invited professor at University of Limerick and Loyola College in Maryland and head of Distributed Systems Research Lab. His research areas of expertise include service oriented distributed computing and systems, context-aware and pervasive systems, autonomic systems, green computing and systems, self-adaptive bio-inspired systems, knowledge engineering.

Victor Sanchez Martin was Innovation and Technology Director at ISOIN (Ingeniería y Soluciones Informáticas, S.L.) where he managed and coordinated more than ten EU projects. He holds a Master in Telecommunication Engineering with a specialization in innovation management in high technological sectors. Victor has published several papers in the field of ICT applied to adaptive interfaces. His areas of interest include adaptive interfaces, data monitoring and integration, and ambient assisted living. He is now member of the Electronic Systems Department from Eindhoven University of Technology.

Maham Shahid received B.S. and M.S. in Computer Sciences from COMSATS university, Islamabad, Pakistan. She is currently a lecturer in

COMSATS. Her research interests include wireless sensor networks and body area networks.

Azfar Shakeel is working in COMSATS Institute of Information Technology, Computer Sciences Department as Lecturer since 2006. He did MS-IT from NUST SEECS in 2011, MBA from IIUI in 1997, MCS from Arid Agriculture University in 2001. His research interests include Computer Security, Application of Cellular Automata and Wireless Sensor Networks.

Radosveta I. Sokullu received her Ph.D. in Telecommunications in 1992 from the Technical University Sofia and in 2000 she joined Ege University as an Assistant Professor. She became Head of Telecommunications Studies Branch in 2004. Currently is Associate Professor at the same department she is also Head of the Wireless Communications Lab. Her current research covers different wireless communication networks and protocols (IEEE 802.15.4, Bluetooth, Wireless Sensor Networks, Cellular Networks) with a focus on MAC layer and PHY layer protocol design, resource sharing, resource allocation, and Machine-to-Machine Communications. She participated in FP 7 CRUISE project as a leader of the Turkish partners, and has lead a number of projects sponsored by the National Research Council in Turkey (TUBITAK), by the Ministry of Industry (SANTEZ) in cooperation with large electronic companies in Turkey as well as numerous projects sponsored by the University Research Fund. Radosveta Sokullu is a member of the International Federation of University Women (IAUW), an IEEE Member and a consultant of the Student IEEE Branch at the Department of Electrical and Electronics Engineering.

Susanna Spinsante received her Ph.D. in Electronic Engineering and Telecommunications in 2005, from Università Politecnica delle Marche (Italy), and her Laurea (Graduate Diploma) Degree in Electronic Engineering in 2002, from the same University. Since December 2005, she has been a PostDoc in Telecommunications, at the same University, where now she is Assistant Professor in Telecommunications. At present, Susanna works on radio communication systems exploiting spread spectrum techniques, for indoor and outdoor communications (Wi-Fi networks, Bluetooth, multiple access systems). She is an IEEE Senior Member, involved in several applied research projects on AAL and home building automation, supported by regional and national funds. From 2010 to 2013 she has been the CEO of ArieLAB Srl. In September 2012 she was co-founder of the academic spinoff DowSee Srl.

Rumen Stainov received his B.S. and M.S. from the Technical University Ilmenau, Germany and his Dr.-Ing. from the Dresden University of Technology, Germany. Before joining Fulda University he has been a Professor of time at the University of Aachen, Germany. In Fall 1997, in Fall 98, in Spring, and Summer 1999 he has been Visiting Associate Professor at Boston University (USA). From Fall 1999 through Summer 2002 he has been full-time Associate Professor of Computer Science at Boston University. He is currently Professor of Computer Science at Fulda University of Applied Sciences, Germany. His research interests are in the field of networking, distributed systems, mobile communications, and peer-to-peer networks. He has more than 75 research publications, he wrote 5 technical books, and was leading investigator in 12 research projects.

Vera Stara received a Ph.D. in Artificial Intelligence Systems from the University of Ancona in 2006 and a professional continuing education certificate as instructional designer for e-learning system from the University of Cagliari in 2005. Since 2010 she is a researcher at the National Institute of Health and Science on Aging — I.N.R.C.A. in the field of User Centered Design and Human Computer Interaction. Her main research focus are on human factors approaches to aging and technology use.

Rodica Strungaru is Ph.D. coordinator and founder of Medical Electronics and initiator of the first laboratory for biomedical research at "Politehnica" University of Bucharest — Faculty of Electronics, Telecommunications and Information Technology in Romania, 1973. Univ. Prof. Ph.D. Strungaru Rodica is an author of the first book on Medical Electronics in Romania and invited professor at Universitaet der Bundeswehr Muenchen, Institut fuer Physik, Medizinische Elektronik Abteilung, between 1992 and 1993. She is coordinator of 2 Erasmus (Socrates) agreements, with Technical University Eindhoven, The Netherlands and Universitaet der Bundeswehr, Munich, Germany.

Research and Teaching Areas:
- Medical Electronics and Informatics
- Neural Networks
- Signal Processing
- Feature Extraction and Pattern Recognition

Loizos Toumbas received his B.Sc. in Anatomy, Developmental and Human Biology in 2013, from King's College London (England). Since September, he has been studying Medicine at St George's, University of London Medical School at the University of Nicosia.

Vladimir Trajkovik received Ph.D. degrees 2003. He joined the Ss. Cyril and Methodius University, Skopje, R. Macedonia, in December 1997. His current position is the Full Professor at the Faculty of Computer Science and Engineering. He is currently responsible for several courses at undergraduate level, and "Mobile and Web Services", "Collaborative Systems" and "Innovative Technologies" at postgraduate level.

He realized multiple research stays with several European Universities as visiting scientist within the scope of different EU and international projects. He is an author of 4 books, 9 book chapters, and more than 120 journal and conference papers. Dr. Trajkovik has participated (as researcher or project leader) in 20 international projects sponsored by European Commission in the framework of TEMPUS, PHARE and FP programs.

Saif ul Islam received his Ph.D. in Computer Science at the University Toulouse III Paul Sabatier, France in 2015. Previously, he did his Master by Research in Computer Science from University of Limoges, France. He is Assistant Professor at the Department of Computer Science, COMSATS Institute of Information Technology, Islamabad, Pakistan. He has been part of the European Union funded research projects during his Ph.D. His research interests include Wireless Body Area Networks, Underwater Wireless Sensor Networks, resource and energy management in large scale distributed systems.

Laura Vadillo Moreno is a Telecommunications Engineer and a Master of Science from the Universidad Politécnica de Madrid (Spain). She has been a Researcher Member of the Telematics Systems for the Information and Knowledge Society (T>SIC) group of the Universidad Politécnica de Madrid. Since 2006, she is Associate Professor at the ESNE University School of Design, Innovation and Technology, and Professor of the Master of Accessibility at the Smart City of the Universidad de Jaen. Her research work includes the design, development and validation of distributed intelligent systems in AAL projects applied to telemedicine, telecare and e-health scenarios, smart home monitoring, and ICT accessibility solutions.

Alberto Carlos Valderrama obtained Ph.D. degree in Microelectronics at the INPG/TIMA lab in Grenoble, France, as member of the Brazilian government R&D program in 1998. In 1989, he graduated as electrical and electronics engineer from the UNC, Cordoba, Argentina. Since September 2004, he is leading the Electronics and Microelectronics Department of the Polytechnic Faculty of Mons FPMs, in Mons, Belgium. Between 1999 and 2004, he was leading the CoWare NV. Hardware Flow team located in Belgium. He was also invited professor in two Brazilian universities: a. in 2004, at the Federal University of Pernambuco UFPE; b. in 1998, at the Federal University of Rio Grande do Norte UFRN.

Dan Valea (Eng.) is a master student and former research assistant of the Distributed Systems Research Lab having as main research areas adaptive user interfaces for elders, matching techniques, optimization algorithms and in lab simulation prototype development.

Miguel Ángel Valero Duboy received his Ph.D. in Telecommunications Engineering from the Universidad Politécnica de Madrid (UPM) in 2001. He has been the Director of CEAPAT-IMSERSO, the Spanish Reference Centre of Personal Autonomy and Technical Aids, since 2014. He became an Associate Professor in Telematics Engineering at DIATEL-UPM in 2003 and is Visiting Professor at Mälardalen University, Sweden. He was Deputy Director of Research and Doctorates at UPM (2004/08) and co-founded in 2005 the Group of Telematic Systems for the Information Society and Knowledge, where he leads R&D in e-accessibility, telemedicine systems and e-health. His research activity started in 1995 and he has worked on 35 European and national projects related to accessible information systems, networks and telematic services for e-health by managing, modelling and deploying telemedicine and m-health solutions in hospitals, homes and rural scenarios. He is the author of three books, 11 book chapters, 12 journal papers and over 80 conference contributions on telematics, telemedicine, e-accessibility, telecare, e-health and human factors.

Martijn Vastenburg (Ph.D.) is the founder and CEO of ConnectedCare, an online care collaboration platform for formal and informal caregivers. Until 2012 he worked was Assistant Professor in Industrial Design Engineering. He started the spin-off ConnectedCare 2010 in order to accelerate the transition from academic research in ambient assisted living to commercial application. His core competencies are ICT (social networks and sensor systems), industrial design (user-centered design, design methodologies) and care (family care, professional care, independent living).

Florian Wamser received the diploma degree in computer science, in 2009, and the Ph.D. degree, in 2015. He studied at the University of Würzburg and at the Helsinki University of Technology, Espoo, Finland. He is a Research Associate with the Chair of Communication Networks, University of Würzburg, Würzburg, Germany. He leads the group on cloud networks and Internet applications at the Chair of Prof. Dr.-Ing. Phuoc Tran-Gia. The title of his dissertation is "Performance Assessment of Resource Management Strategies for Cellular and Wireless Mesh Networks." His research interests include analytical and simulative performance evaluation and optimization of cloud networks and related fields.

Irene Yu-Hua Gu received the Ph.D. degree in electrical engineering from the Eindhoven University of Technology, Eindhoven, The Netherlands, in 1992. From 1992 to 1996, she was Research Fellow at Philips Research Institute IPO, Eindhoven, The Netherlands, and post Dr. at Staffordshire University, Staffordshire, U.K., and Lecturer at the University of Birmingham, Birmingham, U.K. Since 1996, she has been with the Department of Signals and Systems, Chalmers University of Technology, Gothenburg, Sweden, where she is currently full Professor. Her research interests include statistical image and video processing, object tracking and video surveillance, pattern classification, and signal processing with applications to electric power systems.

Dr. Gu was an Associate Editor for the IEEE Transactions on Systems, Man, and Cybernetics, Part A: Systems and Humans, and Part B: Cybernetics from 2000 to 2005. She was the Chair-elect of the IEEE Swedish Signal Processing Chapter from 2002 to 2004. She has been an Associate Editor with the EURASIP Journal on Advances in Signal Processing since 2005, and with the Editorial board of the Journal of Ambient Intelligence and Smart Environments since 2011.

Yixiao Yun received B.Sc. degree in electronic information science and technology from the University of Electronic Science and Technology of China (UESTC), Chengdu, China in 2009, and M.Sc. degree in communication engineering and the degree of Licentiate of Engineering from Chalmers University of Technology, Gothenburg, Sweden in 2011 and 2013, respectively. He is currently pursuing the Ph.D. degree in image and video signal processing from Department of Signals and Systems, Chalmers University of Technology.

His current research interests include image processing, visual object tracking, pattern classification, and video activity analysis.

Thomas Zinner is heading the NGN research group "Next Generation Networks" at the Chair of Communication Networks in Würzburg. He finished his Ph.D. thesis on "Performance Modeling of QoE-Aware Multipath Video Transmission in the Future Internet" in 2012. His main research interests cover the performance assessment of novel networking technologies, in particular software-defined networking and network function virtualization, as well as network-application interaction mechanisms.

PREFACE

INTRODUCTION

The increase in medical expenses due to societal issues like demographic ageing puts strong pressure on the sustainability of health and social care systems, on labor participation, and on quality of life for older people or for persons with disabilities. Enhanced Living Environments (ELE) encompass all ICT technological achievements supporting true Ambient Assisted Living (AAL) environments. ELE promotes the provision of infrastructures and services for the independent or more autonomous living, via the seamless integration of information and communication technologies within homes and residences, thus increasing their quality of life and autonomy maintaining one's home the preferable living environment for as long as possible, therefore not causing disruption in the web of social and family interactions.

Different ELE technologies are used today to construct safe environments around assisted peoples and help them maintain independent living. Most efforts towards the realization of Ambient Assisted Living systems are based on developing pervasive devices and use Ambient Intelligence to integrate these devices together to construct a safe environment. The missing interaction of multiple stakeholders needing to collaborate for ELE environments supporting a multitude of AAL services, as well as barriers to innovation in the markets concerned, the governments, and health and care sector, these innovations do not yet take place at a relevant scale. *Many fundamental issues in ELE remain open.* Most of the current efforts still do not fully express the power of human being, and the importance of social connections and societal activities is less noticed. Effective ELE solutions require appropriate ICT algorithms, architectures and platforms, having in view the advance of science in this area and the development of new and innovative connected solutions (particularly in the area of pervasive and mobile systems). The book aims to provide, in this sense, a platform for the dissemination of research efforts and presentation of advances in the ELE area that explicitly aim at addressing these challenges. The book constitutes a flagship driver towards presenting and supporting advance research in the area of Enhanced Living Environments.

The book is intended to be used by different professionals from medical doctors, ICT engineers, mathematicians, programmers to the caregivers, third parties like insurance companies' personnel and end-users. It could be used as a notebook in the related curriculum. The chapters could be read in any order.

THE OVERALL OBJECTIVE OF THE BOOK

The overall objectives of this book are to:

- Offer a coherent and realistic image of today's architectures, techniques, protocols, components, orchestration, choreography and development related to the Ambient Assisted Living (AAL) and Enhanced Living Environment (ELE) areas.
- Explain state-of-the-art technological solutions for the main issues regarding AAL and ELE, as well as supporting systems: resource and data management, fault tolerance, security, monitoring and controlling.
- Present the benefits of AAL and ELE, and the development process of scientific and commercial applications and platforms to support them.

The book's mission is to make readers familiar with those concepts and technologies that are successfully used in the implementation of today's AAL/ELE systems, or have a good chance to be used in future developments. The approach is to not separate the theoretical concepts concerning the design of such systems from their real-world implementations. For each important topic that one should master, the book aims to play the role of bridge between theory and practice and of instrument needed by professionals in their activity. To this aim, the topics are presented in a logical sequence, and the introduction of each topic is motivated by the need to respond to claims and requirements coming from a wide range of AAL/ELE applications. The advantages and limitations of each model or technology in terms of capabilities and areas of applicability are presented through concrete case studies for AAL/ELE systems and applications.

The book presents also up-to-date technological solutions to the main aspects regarding AAL/ELE systems and applications, a highly dynamic scientific domain that gained much interest in the world of IT in the last decade. Such systems have matured to commercially viable business AAL computing and network infrastructures. The book discusses nowadays AAL/ELE technologies designed to solve some of the thorniest business problems affecting applications in areas such as health and medical supply, smart city and smart housing, Big Data and Internet of Things, and many more. Along with covering architectural components behind the ELE vision, the book introduces readers to technologies supporting the development of AAL applications. In this aspect, the book aims to present the actual AAL/ELE systems that are becoming more and more attractive in academia and industry for a wide-range of actual and next-generation applications. Most IT vendors and enterprise solutions adopters view such systems beyond developed today as foundations of the technology of the future AAL/ELE applications.

ORGANIZATION OF THE BOOK

The book consists of 17 chapters that are grouped logically depending on the topic:

- Chapter 1 is an introduction to the AAL and ELE systems. It explains how ELE promotes the provision of infrastructures and services for the independent or more autonomous living, via seamless integration of ICT within homes and residences. For this, ELE encompasses the latest developments associated with the Internet of Things, towards designing services to better help and support people, or as a general term, to better live their life and interact with their environment.
- Chapter 2 is dialing with body area networks and implanted sensors. One of the main concern is about energy saving and heating because heat dissipation is dangerous for human health.
- Chapter 3 presents energy efficient communication in Ambient Assisted Living technologies. The authors propose end-to-end communication architecture in a typical heterogeneous AAL system with key functional blocks of wireless body area and sensor network. They concern reliability, efficient use of spectrum and energy efficiency in multitier communication protocol.
- Chapter 4 demonstrates human factor in the design of successful Ambient Assisted Living technologies. It discusses the basic role of the human centric approach in the design of assistive technologies, and, by analyzing the outcomes of previous experiences, provides a set of guidelines that can help transforming a disruptive prototype into a successful product.
- Chapter 5 tries to match requirements for Ambient Assisted Living and Enhanced Living Environments with networking technologies. The contribution of this book chapter is to understand the technological barriers hindering the widespread real world usage of AAL/ELE systems and to identify technologies, which may help to overcome them.

- Chapter 6 is about recent advances in remote assisted medical operations. The advancement of medical robotics has ushered in new and exciting age for medicine, which could revolutionize surgery.
- Chapter 7 considers cloud based smart living system. It describes a model of a cloud based Ambient Assisted Living system. The logical and physical architecture of the model is illustrated with proof of concept services.
- Chapter 8 presents AAL and ELE platform architecture. It is an open distributed system for internetworking and defines a provision of services to the citizens. The work is describing the framework, main hierarchical layers, common and specific services and applications, important functionality, AAL as a Service (AALaaS) and ELE as a Service (ELEaaS).
- Chapter 9 develops embedded platforms for Ambient Assisted Living. It focuses on home automation and health monitoring systems with integrated specific functions, common support for elderly and disabled people with special attention for those living alone.
- Chapter 10 is dialing with wearable electronics for elderly health monitoring and active living. The objectives are ubiquitous monitoring, transmission and storage of data from wearable sensors to perform signal processing.
- Chapter 11 concerns cloud-oriented domain for AAL. It presents IoT and cloud computing used for remote monitoring of the physiological condition of a patient along with how IoT can be used in gathering the desired data. Additionally, Electronic Health Records (EHR) are presented, followed by how wearable technology, used in AAL, can benefit from an IoT cloud system.
- Chapter 12 is about adaptive workspace interface for facilitating the knowledge transfer from retired elders to start-up companies. The approach is based on monitoring elders' interaction with the workspace to adapt the interface layout and contents to their cognitive conditions. A genetic algorithm is used to automatically decide and select the best configuration (both layout and content) for presenting the information to the elders.
- Chapter 13 concerns telemonitoring as a core component to enforce remote biofeedback control systems. It presents a conceptualization of remote telemonitoring, crossrelated with the dynamic control systems concepts.

- Chapter 14 demonstrates the role of smart homes in intelligent home-care and healthcare environments. It includes the main building blocks and their interrelationship that is required to create a sustainable and replicable people-centered smart home.
- Chapter 15 is presenting visual information-based activity recognition and fall detection for assisted living and eHealthCare. It mainly focuses on describing visual information-based daily activity recognition and anomaly detection through using low resolution visual sensors. Detailed descriptions are given on three robust methods that exploit smooth manifolds.
- Chapter 16 explains details on end-users testing of enhanced living environment platform and services. It presents possible use-case scenarios for end-user testing, verification and validation of AALaaS and ELEaaS as well as the entire platforms.
- Chapter 17 is about machine-to-machine communications and their role in AAL. Based on M2M communications AAL will allow to intelligently monitor patients' physical conditions and intervene when necessary; to increase the quality of their life; to empower them to carry on a fulfilling social life even when physically restricted.

ACKNOWLEDGMENTS

This book would not have been possible without the help of many people. We would like to thank the reviewers, colleagues from Elsevier, authors, co-authors, and friends for their direct and indirect contribution.

We are responsible for all remaining errors. In case you find an error please, report to the Publisher.

Many special thanks to all colleagues from EU COST action IC1303 AAPELE for their encouragement, advices, proposals, reviews. Many of them are authors and had been cited in the book.

Special thanks to Sonnini Ruiz Yura, our acquisition editor, Ana Claudia Abad Garcia, Editorial Project Manager, Kiruthika Govindaraju, Production Project Manager and all people who made the project looking so well in Elsevier S&T Books for their patience, corrections, advices and valuable support.

GLOSSARY

Ambient Assisted Living a concept that combines the daily life activations with the information and communication technology to improve and increase the quality of life for people in a daily life.

Ambient Assisted Living (AAL) an emerging multi-disciplinary field at the intersection between information and communication technologies, sociological sciences, medical research, that aims to develop personal healthcare and telehealth systems for countering the effects of growing elderly population.

Ambient Assisted Living (AAL) enhances the independent living ability of the elderly through various intelligent services, reducing the need for direct care giving, indoors or outdoors.

Ambient Intelligence digital environment that is sensible and responsible supporting people in their life.

Ambient intelligence (or, AmI) a vision where people will be surrounded by intelligent and intuitive interfaces embedded in everyday objects and an environment recognizing and responding to the presence of individuals in an invisible way.

Application Server software framework in a distributed network that handles all application operations between users and databases.

Cloud computing virtualized environment for data storage and processing regardless of the place, time and mobility.

Communication Protocol system of rules that allow the devices in a network to transmit information in a coordinated way.

Cyber Physical Systems physical systems whose operations are monitored, controlled and coordinated by a communication core.

Dew computing small-scale cloud computing like home cloud.

End Device a simplest type of device in a network, which transmits or receives a data but often cannot route the data.

Enhanced Living Environments (ELE) a term coined to refer to where AAL meets and makes use of Information and Communication Technologies to sit at the intersection between AAL and AmI, and technology applying ICT for every day live and personal environment management.

Fog computing medium scale cloud computing like regional one.

Gateway   intermediate device to manage the other devices, which are related with it, and aggregate data from device to send the base station. Device that interconnects two different technologies.

Internet of Things   system of computing devices, mechanical machines, objects, animals or people that provide a communication between them without any human intervention.

M2M Communication   emerging technology that enables the communication between different devices and allows them to perform a variety of actions without or with only limited human intervention.

Network Architecture   layout of communication network that consists of the hardware and software components, their functional organization and configuration, their operational principles and procedures.

Specific Absorption Rate (SAR)   a measure of a rate of energy absorbed by a human body when exposed to electromagnetic waves, or the power absorbed per mass of tissue and has units of joules per second per kg or Watts per Kilogram (W/kg).

Telemedicine   sharing of (patients') information between Healthcare professionals, allowing for remote analysis of diagnostic images, audio signals and good resolution video in real time that help the "telehealth care professional" reach a diagnosis.

Telepresence   the ability to give surgeons the sense that they are operating on something directly in front of their eyes, which could in fact be on the other side of the room, or even thousands of miles away.

Ubiquitous computing   present computers, functioning invisibly and unobtrusively in the background able to serve people with everyday activities, at home and at work.

ACRONYMS

2PCS Personal Protection & Caring System
3GPP The 3rd Generation Partnership Project
6LoWPAN Ipv6 over Low Power Wireless Personal Area Networks
AAL Ambient Assisted Living
AALaaS AAL as a Service
AAPELE Algorithms, Architectures and Platforms for Enhanced Living Environments
ACE Ambient Cardiac Expert
ACL Access Control List
ADC Analog to Digital converter
ADL Activities of Daily Living
ADMR Demand-driven Multicast Routing
AIOTI Alliance for Internet of Things Innovation
ALIAS Adaptable Ambient LIving Assistant
ALIP Assisted Living Platform
AML Ambient Intelligence
AN Access Network
API Application Programming Interfaces
AS Application Servers
ATM Asynchronous Transfer Mode
BAN Body Area Network
BER Bit Error Rate
BLE Bluetooth Low Energy
BNC BAN Coordinator
BoW bag-of-words
BP blood pressure
BS Base Station
BSN Body Sensor Network
BTS body temperature sensor
BWT Burrows-Wheeler Transform
CAGR Compound Annual Growth Rate
CBR Constant Bit Rate
CCRC continuing care retirement communities
CCSA The China Communications Standards Association
CM Communication Module

CMA   Cloud-based Mobile Augmentation
CMaps   conceptual maps
CN   Core Network
CoAP   Constrained Application Protocol
COI   Community of Interests
CORBA   Common Object Request Broker Architecture
CSMA/CA   Code Division Multiple Access/Collision Avoidance
CSS   curvature scale space
DALLAS   Delivering Assisted Living Lifestyles at Scale
DARPA   Defence Advanced Research Project Administration
DICOM   Digital Imaging and Communications in Medicine
DOSGi   Distributed Open Service Gateway initiative
DPM   deformable part models
DSC   dynamic systems control
DSF   Dynamic Software Framework
DSR   design science research
DT   dense trajectories
DTN   Delay/Disruptive Tolerant Networks
DTW   dynamic time warping
E2E   end-to-end
ECG   Electrocardiography
ECP   Evolved Packet Core
EDA   electro dermal activity
EDAX   Energy Dispersive X-ray analysis
EEG   Electroencephalography
EHR   Electronic Health Recorder
ELE   Enhanced Living Environments
ELEaaS   ELE as a Service
ELEMENT   Enhanced Living EnvironMENTs
ELM   extreme learning machine
EMG   electromyography
EMR   Electronic Medical Record
eNB   Evolved Node B
EPC   Evolved Packet Core
EPR   Electronic Patient Records
ETSI   European Telecommunications Standards Institute
ETSI TC M2M   ETSI Technical Committee on Machine-to-Machine Communications
eUTRAN   Evolved UMTS Terrestrial Radio Access Network

FCP Fiber Channel Protocol
FEC Forward Error Correction
FFD Full Function Device
FIPA Foundation for Intelligent Physical Agents
FN Fixed Nodes
FTT finger touch test
GDT Graphical Design Tool
GISFI The Global ICT Standardization Forum for India
GMM Gaussian mixture model
GPS Global Positioning System
GSM Global System Mobile
GUI Graphical User Interface
GW gateway
H2H Human-to-Human
H2M human-to-machine
HA Home Automation System
HAN Home Area Network
HbA1c diabetes glycosylated hemoglobin
HCU Health Central Unit
HeNB Home eNB
HeNB Home Evolved Node B
HER Electronic Health Record
HER electronic history register
HFC Hybrid Fiber-Coaxial
HGI Home Gateway Initiative
HL7 Health Level 7
HMI Human Machine Interfaces
HMM hidden Markov model
HOF histogram of optical flow
HRS heart rate sensor
HRV heart rate variability
HS Health System
HSS Home Subscriber System
HTTP Hyper-Text Transfer Protocol
I/O Input/Output
IaaS Infrastructure as a Service
ICAW Interface Collaborative Adaptive Workspace
ICT Information and Communications Technologies
IDT improve dense trajectories

IEEE   The Institute of Electrical and Electronics Engineers
IFUS   Integrated Fusion Sensor
IIM-HEE   Information Integration Module for Health Education of the Elderly
IOPE   inputs, outputs, pre-conditions and effects
IoT   Internet of Things
IPSec   Internet protocol security
IR   infrared
IR-UWB   Impulse Radio Ultra Wide Band
ISP   Internet Service Provider
ISTAG   Information Society Technologies Advisory Group
ITK   Interoperability Toolkit
LAN   Local Area Network
LDL   low density lipoprotein
LDPC   Low Density Parity Check Code
LTE   Long Term Evolution
LZ77   Lempel-Ziv 77
LZW   Lempel-Ziv-Welch
M2M   Machine-to-Machine
M2M-NA   M2M Network Application
M2M-MF   M2M Management Functions
MBS   mobile body sensors
MCU   Micro Controller Unit
MC-UWB   Multi Carrier Ultra Wide Band
MF   Management Functions
MH   Message Handler
MME   Mobility Management Entity
MMM   Membership Management Module
MN   Mobile Nodes
MP2P   Mobile Peer-to-Peer
MSUS   Multi-Modal Sensing u-Healthcare System
MTC   Machine Type Communication
NASA   National Aeronautics and Space Administration
NAT   Network Address Translation
NEEMO   NASA Extreme Environment Mission Operations
NFV   Network Function Virtualization
NMEA   National Marine Electronics Association
NMF   Network Management Functions
NTP   Network Time Protocol

OFDM Orthogonal Frequency Division Multiplexing
OMA Open Mobile Alliance
OSGi Service Gateway Initiative
OSI RM Open System Interconnection Reference Model
OWL Web Ontology Language
P2P peer-to-peer
PA Applications
PaaS Platform as a Service
PAN Personal Area Network
PAN-IF Personal Area Network Interface
PCP primary care provider
PCS powerline carrier system
PD Parkinson's disease
pdf probability density function
PDF Probability Distribution Function
PFID Radio Frequency Identification Devices
P-GW Packet Date Network Gateway
PHMM Personal Health Management Module
PHR Personal Health Record
PLC power line carrier
PM Policy Manager
PP peer port
PP Publisher Point
PPG photoplethysmographic
PPM Pulse Position Modulation
PRP pressure pods
PUMA Programmable Universal Manipulation Arm
PVA Personal Virtual Assistant
QoE Quality of Experience
QoS Quality of Service
RAN Radio Access Network
REST Representational State Transfer Model
RF Radio Frequency
RFID Radio-Frequency Identification
RMD Remote M2M Device
RMD Remote Monitoring Device
RN Relay Node
ROI region of interest
RTC Real-time clock

SaaS Software as a Service
SAN Storage Area Network (SAN)
SAR Specific Absorption Rate
SC Service Capabilities
SCADA Supervisory Control and Data Acquisition
SCC Service Capability Client
SCE Service Capability Entity
SCL Service Capability Layer
SCS Service Capability Server
SDM Super Data Management
SDMA System for Developing Medical Applications
SDN Software-Defined Networking
SDS smoke detector sensor
SEP Sink Entry Point
S-GN Serving Gateway
S-GW Serving Gateway
SINR Signal-to-Noise plus iNterference Ratio
SLA Service Level Agreement
SN Social Networks
SNPS Sensor Node Plugin System
SPD symmetric positive definite
SPP Super Publisher Point
SSEP Super Sink Entry Point
SWRL semantic web language rules
TDMA Time Division Multiple Access
TR Type Resolver
TS temperature sensor
UAV Unmanned Aerial Vehicles
UCD User Centered Design
UE user entity
UE user equipment
UHF Ultra-High Frequency
UI user interface
UID unique identifier
UIML User Interface Description Language
UPDRS Unified Parkinson's Disease Rating Scale
URC Universal Remote Console
USB Universal Serial Bus
USDM ultra sound distance meters

UTRAN UMTS Terrestrial Radio Access Network
UWB Ultra Wideband
VCML Video Covariance Matrix Logarithm
VICG Virtual Caregiver
WAN Wide Area Networks
WBAN Wireless Body Area Network
WBN Wireless Body Network
WHO World Health Organization
Wi-Fi Wireless Fidelity
WiMAX Worldwide Interoperability for Microwave Access
WLAN Wireless Local Area Networks
WPAN Wireless Personal Area Network
WPN Wireless Personal Network
WS web services
WSAN Wireless Sensor and Actuator Networks
WSN Wireless Sensor Networks
xDSL Digital Subscriber Line

CHAPTER 1

Introduction to the AAL and ELE Systems

Ciprian Dobre*, Constandinos X. Mavromoustakis[†], Nuno M. Garcia[‡],
George Mastorakis[§], Rossitza I. Goleva[¶]
*University Politehnica of Bucharest, Romania
[†]University of Nicosia, Cyprus
[‡]University of Beira Interior, Portugal
[§]Technological Educational Institute of Crete, Greece
[¶]Technical University of Sofia, Bulgaria

1.1 INTRODUCTION

Today we witness the miniaturization of computer technology, which is best reflected in the processors and tiny sensors that are being integrated into everyday objects. Our future witnesses the integration of ICT into our clothes, appliances, our households — and soon enough, such technology-driven ordinary objects will start to behave "intelligently", to communicate between themselves and with large data centers privately holding our "digital life", running against such data smart algorithms, all for our benefit, to provide us with ever-more smart and self-aware decisions designed to improve our life. This corresponds to a future similar to what Mark Weiser foreseen some 20 years (Weiser, 1991), when he coined the term *ubiquitous computing*, a world where ever-more-present computers, functioning invisibly and unobtrusively in the background, will be able to serve people with everyday activities, at home and at work, to free us (to a large extent at least) from tedious routine tasks.

A similar vision was later coined by the European Union's Information Society Technologies Program Advisory Group (ISTAG) (Ahola, 2001), to define the term *ambient intelligence* (or, AmI for short) similarly to describe a vision where "people will be surrounded by intelligent and intuitive interfaces embedded in everyday objects around us and an environment recognizing and responding to the presence of individuals in an invisible way" (Ahola, 2001).

This vision of a future where technology allows everyday objects to interact and work together for mankind, offers fascinating possibilities. As

Ambient Assisted Living and Enhanced Living Environments.
DOI: http://dx.doi.org/10.1016/B978-0-12-805195-5.00001-6

1

Bohn et al. (2004) predict, parents will be able to keep track of their children in the busiest of crowds, when location sensors and communications modules are sewn into their clothes. Devices attached to timetables and signposts will guide blind people in unknown environments by "talking" to them via a wireless headset (Coroama and Röthenbacher, 2003). Tiny communicating computers, the size of dust particles, will act as sensors to detect dispersion of oil spills and forest fires, leading to a much more protected environment. And, faced with the problem of an increasingly older population, society will rely on smart objects to improve the quality-of-life towards old age, and provide ever-more protection and social inclusion, to its senior citizens.

The reality is that society is, even today, facing an increase ageing of our population. Health and well-being are driving engines to the realization of a technology-driven society that will be able to better deal with the projections of the future. In European Union, the ratio of people aged 65 years and above will increase to 30.0% by 2060 (to an alarming 151.5 million people) (Giannakouris, 2008). Something similar is estimated for USA, where people over 65 years will represent 20.2% of the population by 2050 (Vincent and Velkoff, 2010), and for Japan, with an estimate of 39.6% (Takahashi et al., 2003).

Senior people, in particular, have a higher incidence rate for physical and/or cognitive impairments. This rate increases with the passing of years. As we get older, we find ourselves affected by sensing problems, we face a constant decrease in the mental capabilities to process information, a reduction in mobility, and a negatively-affected precision of doing even simple tasks. With age, we experience difficulties in dealing with complex scenarios or with keeping a focus when solving problems over a longer period of time. As a consequence, we progressively lose the capability to perform autonomously even routine daily activities. Thus, instead of fostering independent living, household appliances become rather a burden that adds to ageing limitations.

Living assistance systems designed to support elderlies (but also other types of patients, affected in their daily life by various medical conditions) live a better and more comfortable life in their own environment, are generally referred to as *patient-care systems*. Ambient Assisted Living (AAL) is an emerging multi-disciplinary field at the intersection between information and communication technologies, sociological sciences, medical research, that aims to develop personal healthcare and telehealth systems for countering the effects of growing elderly population. One definition of AAL is given by Kung and Jean-Bart (2010): "intelligent systems that will assist elderly individuals for a better, healthier and safer life in the preferred living environment and covers concepts, products and services that interlink and improve new technologies and the social environment". Thus, AAL is a multi-disciplinary field dealing with personal healthcare and telehealth systems for countering the effects of growing elderly population (Belbachir et al., 2010). With AAL, technology and science work together to provide improvements in the quality of life for people in their homes, with goals such as reducing the financial burden on the budgets of healthcare providers and costs associated with the social inclusion of elderly people, worldwide.

AAL systems are developed with personalized (design for the patient), adaptive (adapt to the patient and running conditions), and anticipatory (anticipate the event before it can harm) requirements. The goal, and as we will see below, is to integrate high quality for th medical and comfort service being provided, to achieve interoperability (with technologies, and with similar other systems), usability (considering the patient capabilities), security (e.g., of the personal digital data being used), and accuracy (e.g., in the precision of a diagnosis). Such objectives are still posing challenges to the large-scale adoption of such systems, although solutions are being developed as we speak. To solve them, AAL use AmI as a tool to provide integral solutions for supporting the person (elderly) in his/her independent living in different contexts (Blasco et al., 2014): dwellings, transport, workplaces, etc. To describe the kind of applications AAL can support, the European Ambient Assisted Living Innovation Alliance proposes, in fact, three macro scenarios for AAL development (van den Broek et al., 2010):

1. *AAL4persons*, with the objective of "Ageing well for the person";
2. *AAL in the community*, where the focus is on applications designed to support the social inclusion of elderly people into society, to improve their communications and their participation in the community;
3. *AAL@work*, including applications supporting elderly and people with disabilities at work.

The general umbrella, for all three, as stated by O'Grady et al. (2010), is that "Ambient Assisted Living (AAL) is advocated as technological solutions that will enable the elderly population maintain their independence for a longer time than would otherwise be the case". To achieve such a goal, an AAL system should provide components at least for context-awareness, where the context includes the person being monitored, for providing help when needed, detecting abnormal situations and acting accordingly (Blasco et al., 2014).

In a response to such requirements for AAL, Enhanced Living Environments (ELE) is the term coined to refer to where AAL meets and makes use of Information and Communication Technologies — it sits at the intersection between AAL and AmI, and technology. To design, plan, deploy and operate, an AAL system often comprehends the integration of several scientific areas. Such systems need dedicated software, hardware and service architectures, specifically designed for AAL. They need efficient algorithms for AAL, that can deal with processing of large amounts of data and of biosignals in lossy environments. Finally, they rely on communication and data transmission protocols for AAL.

Different ELE technologies are today aiming to construct safe environments around assisted peoples and help them maintain independent living. Most efforts towards the realization of ambient assisted living systems are based on developing pervasive devices and use Ambient Intelligence to integrate these devices together to construct a safety environment. The missing interaction of multiple stakeholders needing to collaborate for ELE environments supporting a multitude of AAL services, as well as barriers to innovation in the markets concerned, the governments, and health and care sector, these innovations do not yet take place at a relevant scale.

1.2 AAL/ELE SYSTEMS AND APPLICATIONS

We believe the reasons for the increasing interest in ambient assisted living applications are twofold; the well-known demographic shift with growing share of elderly in the population, and the rapid development of wirelessly connected embedded sensor devices in combination with efficient IT and storage infrastructures. Focus for the research on ambient assisted living has been technical systems, infrastructures and services to support elderly people in their daily routine, to allow an independent and safe lifestyle as long as possible, via the seamless integration of information and communication technologies within homes and residences.

The European initiative Active and Assisted Living Joint Programme (AAL JP) supports applied research on ICT-enhanced services for ageing (AAL-Europe, 2016). EIT Health is another European project for active ageing (EIT, 2016). The European COST action IC1303 Algorithms, Architectures and Platforms for Enhanced Living Environments (AAPELE) is a third example of research in this field (AAPELE, 2016).

A classification of the AAL/ELE application subdomains is presented by Nehmer et al. (2006):

For **Indoor Assistance**, applications can be classified as:
- **Emergency Treatment Services**: emergency prediction, emergency detection, and emergency prevention;
- **Autonomy Enhancement Services**: assistance for activities such as cooking, eating, drinking, cleaning, dressing, and medication;
- **Comfort Services**: logistic services, services for finding things, and infotainment services.

For **Outdoor Assistance**, applications can be classified as:
- **Emergency Treatment Services**: emergency prediction, emergency detection, and emergency prevention;
- **Autonomy Enhancement Services**: assistance for activities such as shopping, traveling, and banking;
- **Comfort Services**: transportation services, and orientation services.

To illustrate better the different aspects of such applications, we briefly mention below several AAL/ELE projects.

For prevention and management of chronic conditions of elderly people, projects such as ALADDIN[1] developed an integrated solution for the self-management of dementia patients, together with the tools to support such a procedure. The result is an integrated platform enabling distant monitoring of patient status and facilitating personalized intervention and adaptive care.

Similarly, DOMEO[2] developed assisted robotic technology for the integration and adaptation of personalized homecare services, as well as cognitive and physical assistance.

To promote social interaction of elderly people, the project HELAS-COL[3] developed the ICT tools to provide the elderly people with the means of maintaining social relations by developing an easy to use and easy to understand communication platform with social and entertainment capabilities that can be easily upgraded with security and medical features.

To help and support older persons live a more independent life, the project GOLDUI[4] developed a Cloud-based platform, allowing older persons access online "self-serve" services and benefit from the modern digital world (thus, integration into society) by using the familiar home technologies of domestic radio, TV and telephone augmented by a mobile smartphone interface when away from home.

---

[1] http://www.aladdin-project.eu/.
[2] http://www.aal-domeo.eu/.
[3] http://www.aal-europe.eu/projects/helascol/.
[4] http://www.goldui.eu/.

To support mobility for older persons, projects such as IMAGO[5] develop navigation and positioning technology for the blind and visually impaired. On a different note, the project GameUp[6] developed a platform for social and exercise games to reduce physical and motivational barriers to elderly people's mobility.

To deal with (self-)management of daily life activities for older adults, while being at home, the project DALIA[7] developed an integrated home system that hides any technical complexity behind a Personal Virtual Assistant (PVA), a human-looking avatar endowed with speech recognition and speech capabilities, designed to support the elderly with daily routine activities. Similarly, DIET4Elders[8] developed adults diet support services to support long-term healthy feeding habits for older adults. In the same category, the Necessity system (Muñoz et al., 2012) can represent and validate alerts in a domestic environment.

The i2Home project tries to make appliances and devices easier to understand for people with mild cognitive impairment and the elderly using a new mainstream user interface standard: the Universal Remote Console (URC) (Frey et al., 2011). The project creates support services integrating several technologies and devices: appliances (hood, oven, fridge, freezer, dishwasher and air conditioning), touch screen, RFID antenna which implement sensitive surfaces for products equipped with smart labels and lighting equipment (Neßelrath et al., 2011).

Patient monitoring, another AAL/ELE direction, involves periodically transmitting routine vital signs (and, in some cases, alerting signals when vital signs cross a threshold; in other cases, the data processing of vital signs is moved on a specialized in-the-Cloud platform) of patients, across certain boundaries. Still, there are many challenges in (wireless generally) monitoring of patients, including the coverage, reliability and quality of monitoring (Varshney, 2008). The current work done in patient monitoring includes, among others, home monitoring (Lee et al., 2000), wireless systems for digitized ECG (Khoor et al., 2001), hospital-wide mobile monitoring systems (Pollard et al., 2001), mobile telemedicine (Hung and Zhang, 2003), and/or real-time home monitoring of patients (Mendoza and Tran, 2002).

---

[5] http://www.aal-imago.eu/.
[6] http://www.gameupproject.eu/.
[7] http://www.dalia-aal.eu/.
[8] http://www.diet4elders.eu/en.

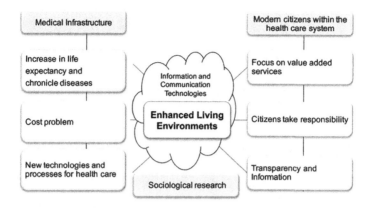

Figure 1.1 Driving factors for Enhanced Living Environments.

Generally, AAL/ELE systems offer support for elderly people, under some form of:

- **Prevention** — deals with tasks supporting and maintaining cognitive and motor abilities before severe diseases (monitoring, fall prevention, etc.)
- **Compensation and support** — deal with tasks supporting cognitive and physical abilities (smart walker, pedestrian GPS, etc.)
- **Independent and active ageing** — include tasks supporting independence of elderly (i.e., social inclusion, work, leisure and entertainment, etc.).

1.3 A VISION FOR AMBIENT ASSISTED LIVING

As stated in the introduction, Enhanced Living Environments (ELE) is the scientific area where Ambient Assisted Living intersects with several other inter-disciplinary domains, like information science, sociological and medical research, and with Information and Communication Technologies (see Figure 1.1).

The major problem of current medical infrastructure is its dependence on traditional outdated procedures. The costs of providing care for an aging population increases with a decreasing age-dependency ratio defined by the number of working individuals divided by the number of handicapped people of a country (Nehmer et al., 2006). This ratio will dramatically approach 1 in the next 10 to 20 years (Giannakouris, 2008). And, correspondingly, health system expenses increase more rapidly than GDP!

Thus, it is obvious that society has to react somehow to this dramatic process. Automated patient-care systems based on ambient intelligence technology are a promising approach towards this. They aim at the prolongation of a self-conducted life of assisted persons, reducing the dependency on intensive personal care to a minimum and thereby increasing the quality of life for the affected group while substantially decreasing the costs for society.

Governmental bodies such as hospitals, healthcare institutions, and social care institutions have expressed their concern about this development, which (i) creates enormous costs for the public and the individuals, caused by the need for intensive care in elderly people care homes, and (ii) dramatically decreases the quality of life of elderly people with the morbidity, which results in limitations to activities of everyday life.

On the other hand, there are a number of technologies that today act as enablers for AAL/ELE, and they can be grouped into sensing, reasoning, acting, interacting (interfaces) and communicatings (van den Broek et al., 2010). The ultimate goals of AAL/ELE systems, designed for the prolongation of a self-conducted life of assisted persons, reducing the dependency on intensive personal care to a minimum and thereby increasing the quality of life for the affected group while substantially decreasing the costs for society, are: (a) high recall in detecting every real emergency immediately; and (b) high precision, to prevent invalid emergency detections and alerts as a consequence of misinterpretations. Requirement (a) is mandatory to provide a trustworthy service quality to the affected persons in case of emergency situations that should be much safer than anything else they experienced before. Requirement (b) is essential for economic reasons, since invalid emergency alerts may unacceptably increase care costs and decrease trustworthiness. It is highly desirable to extend a pure emergency detection service by an emergency prediction service, which attempts to recognize a critical health condition before it escalates into an emergency. As a reaction to the detection of such critical situations, the service may assist the person in preventing the emergency, e.g., by suggesting appropriate medication.

Sensors (e.g., measuring motion, blood pressure) are at the heart of many of today's projects (see eMotion ECG (Mega, 2016), or Tru-Vue (Systems, 2016)), and they are used to monitor an elderly's well-being and detect critical situations that prompt caregivers take actions. For privacy issues, only health- and mobility-related sensors need to be used — no cameras or microphones are assumed and/or used. Interoperability plays also an important part, as some of these systems need to support connectivity with sensors using standards and certifications such as ETSI, ITU, Continua Alliance, mHealth Alliance, and HIMSS.

The data is usually sent from the patient, using a combination of point-to-point and point-to-multipoint networking technology, to a central repository. The communication layer needs to combine cross-layer management with underlying transfer mechanisms, over transmission protocols. Among requirements at this level we mention secure and reliable communication between patient and caregivers, by following a-priori established privacy policies (that need to be defined by the patient, who needs to remain in complete control over how the sensed data is used), and reliability and trustfulness capabilities from the supporting network. Reliability can be provided by creating parallel network paths, when needed, in order to switch from one to another if any problem is discovered in the first path. Trustfulness can be achieved by two ways: (1) at the high level, by providing clear interfaces of private policies to the end users, and (2) at the low level, by differentiating the traffic at the network level in order to integrate private policies.

Apart from technology considerations, AAL/ELE systems generally involve society and social actors (right side of Figure 1.1):

- **Patient**: They use monitoring apps for actual self-check. Advantages and benefits brought to patients generally include the ability to enjoy independent everyday activities. Through the integration of technology, patients are able to be constantly cared for and supported through electronic means, and their well-being telemonitored. Patient monitoring can be accomplished using indoor external sensors or wearable sensors. For the monitoring process, the patient places a large accent on usability — in some cases, the developer cannot assume elderlies would use a smartphone simply because they don't, and in this case the patient has to be monitored with simpler devices (i.e., wearable devices are currently used in several projects as activity trackers, and memory reminders are being constructed with simpler led-based indicators). In most cases, patients are getting as service specific information about his current conditions, e.g., after running; or maybe some time series showing the cardiogram in the last 30 minutes. Accuracy in such cases is not so important, as the user/patient only wants some rough trend of the result (e.g., he is not interested in the exact value, but in a specific interpretation like condition fine, or the raise of a problem, he wants just to see if regularly training enhances his pulse frequency). On the other side, cardiogram for physicians or caregivers (see below) has completely different meaning ok.

- **Physicians**: They are more interested in the specific monitoring of time series of sensor information, possibly in high accuracy to be able to track problems, and even identify specific illness. According to studies, professional care accounts only for around 10% of the total care provided (Nedopil et al., 2013). The employment of professional caregivers is more likely for seniors with a higher level of care dependency or intensity, and for single seniors without permanent informal care. Professional caregivers are influenced by many different and changing factors and requirements: societal issues such as the increasing need for mobility in the modern work environment, institutional motivations to reduce seniors' length of stay in hospitals, and changes in ethics towards technology use in order to remain independent. Professional caregivers are expected to meet the requirements resulting from these influencing dimensions: they have to be friendly, empathetic and active with elderly recipients of care, have to take the latest scientific findings into consideration, while complying with a facility's current quality standards, rules and routines. No wonder that working in the care sector is considered extremely stressful and burdensome, and the absence rate owing to illness is comparatively high. With living assistance systems, the professional caregiver is included in the monitoring loop, providing them with remote access to the patient data in near-real time, providing enhanced communication, coordination and monitoring of care and assistance by linking together healthcare providers, the elderly and their informal carers.

- **Care domain** (which includes long-term illness analysis by hospitals/government, and family): they are more interested in periodical information, to be able to react if something is happening. Sensor accuracy did not be so high since it does not aim at identifying specific illnesses, but to somehow be able to "only" react if real problems occur. Seniors aged 50 and above are mostly cared for by their children or spouses (Nedopil et al., 2013). In central Europe, for example, a spouse provides 42.3% of intensive care (Nedopil et al., 2013). In southern Europe, children and other relatives provide intensive care more frequently. Caring for extra-residential activities can actually affect the work of caregivers — providing personal care can be a very demanding task that is not compatible with a full-time or part-time job. To date, available jobs are often not flexible enough in terms of working hours, or they do not leave enough options to accommodate caring responsibilities. Caring duties can also be unpredictable at times regarding their intensity, which can lead to short-term absences from work. Hence, caregiving is associated with a higher probability of experiencing poverty across countries, especially for women. On the one hand, taking care of another person can be a source of satisfaction, fulfillment and personal growth. But intensive care can also be especially stressful and strainful, potentially leading to burnout and stress. Drug consumption increases with caring activities, especially the intake of sleeping pills, tranquilizers and painkillers. Overall, the prevalence of mental health problems among carers is 20% higher than among noncarers (Nedopil et al., 2013). Relatives are often torn between their responsibilities towards the cared person and their own needs. Thus, telemonitoring for example, brings the advantage of reducing the stress associated with caregiving, because the patient will be technologically and independently monitored, with the caregivers only being kept in the loop for informed consent about potential problems (it is good that the caregiver, even if not physical present and accompanying the elderly, will still know at all-time his whereabouts, having as such an increase sense of support and safety for knowing everything that is happening with the cared-for person).

1.4 CHALLENGES AND RESEARCH OPPORTUNITIES

From the customer perspective it is of key importance that a relatively expensive AAL product is "future-proof", meaning that the system can be maintained over several years, and can be extended to grow and adapt to the changing needs of the user. Most likely many of the houses that will be used in the next 50 years have already been built, so as developer of an AAL/ELE system, one needs to plan for systems that can be installed in already-existing homes. This poses a challenge particularly because technology behind such systems must integrate with local infrastructure (e.g. home automation) and local service providers (e.g., care providers, taxi service, meals on wheels, delivery services, and many others).

The adaptability and focus correspond to an important challenge. Applications up-to-now can be categorized, as seen, in several categories, and this is only natural as there is certainly a very large variety of user needs among older people. Thus, we cannot think of AAL/ELE systems can could "fit all sizes". But even with a well-articulated focus of an application, a challenge remains as to whether it could still deliver the same quality over the following years. This is particularly challenging, as technology of tomorrow was most-likely not been invented yet. Think of how different the technology was several years ago, before society was so dependent on mobile smartphones and other miniaturized devices.

Such challenges can only be addressed with Modularity, an idea where sensors, actors, and subsystems could be combined in multiple forms, much like "lego building blocks". Such an AAL/ELE system could combine various pieces in different ways, according to the needed functionality. But for such an idea to work, we need standardized interfaces between systems and system components. This leads to the next challenge: interoperability. Interoperability is definitely a key requirement for the success of AAL solutions!

Standardization activities, next, are necessary for interoperability to work. Standards offer support for many topics of relevance for AAL system development, not only interoperability, but also for usability and ergonomics, product safety and risk management, data protection and privacy, processes and services (e.g. quality criteria), and certification.

Interoperability and integration of medical devices in healthcare systems processing citizens' vital signs are the most frequently addressed attributes in AAL systems and platforms. This is very important if a telemonitoring system is to be integrated, for example, with various medical support systems already in use today (for example, with the medical systems available on hospital grounds). For monitoring of daily living activities, such systems could be integrated with the Open Service Gateway Initiative (OSGi) platform. The Continua Alliance has also proposed its own reference architecture to construct an ecosystem of interoperable and connected medical devices for personal healthcare and well-being. A solution to integrate with this could be through the Peripheral Area Network Interface (PAN-IF). Similarly, the Assisted Living Platform (ALIP) in the DALLAS (Delivering Assisted Living Lifestyles at Scale) program addresses interoperability as the major concern to achieve scalability in medical device communication. The Interoperability Toolkit (ITK) in DALLAS helps integrate systems based on standardizing technologies and interoperability specifications. More research in the ALIP i-focus reference architecture solves interoperability issues by using the DALLAS Interoperability Layer, which extends the ALIP platform with security features for user and component authentication. And there are several other standardization efforts that we aim to consider further.

However, what many authors do not recognize when they advertise standardization as the golden goose: standards are only one building block in the larger picture, and they have their own problems, too. We refer here particularly to the addition of technological complexity (standards are often designed to cover many use cases, and thus very complex). One side effect of complexity is that standards can be quite expensive to implement.

Among other problems we recognize: competition between standards (standardized products can more easily be exchanged by cheaper competitors), interoperability (standards can be ambiguous in description, and their implementation can often be faulty — see the WiFi Direct standard, implemented differently by various phone manufacturers, making phone-to-phone communication rather difficult to achieve across-vendors). Innovation can also be hindered by standardization activities, as consenting and publishing standards takes time (i.e., it is common to see standards often years behind the state of the art). Finally, access to official standards can be expensive, and free abstracts are often not sufficient to clarify if one standard is useful or not for a given scope.

An interesting challenge relates to security, privacy and data protection for the monitored patient. Security and data protection are critical issues in healthcare systems, which exchange and store citizens' medical data. All monitored medical data is, of course, security sensitive. Therefore, legislation provides guidelines to design authorization policies regarding access and usage of the medical data to avoid intentional misuse or accidental disclosure. Security is an important aspect to consider, and this need advances solutions that could use Public Key Infrastructure-based data encryption and digital certificate infrastructure for ensuring confidentiality and integrity of the medical data. A solution might be to collect medical data from mobile/wearable sensors, and catalog them depending on the level of privacy/protection needed. For example, the most critical data will be securely transmitted for access only to caregivers and family members of the citizens. A photo of the patient (without references to the actual illness) might be disseminated to trustful users only when needed to support (track, add) the patient (as in the park scenario presented below).

Another challenge relates to the quality of the sensing data. Quality attributes have a greater impact on the usability of AAL systems. Sensors might read imprecise data, or the sensing data might be completely unavailable (e.g., the GPS sensing might be wrongly affected by the weather conditions, or it might not work at all indoors). In such cases, mechanisms to augment the sensing with external sensors (e.g., for indoors, sensing data from the BAN mote could be superimpose with the data collected from the sensors within the house premises; for outdoor conditions, the quality of the sensing data can be re-enforced against a set of sensing data obtained opportunistically whenever available for nearby mobile devices) need to be explored.

1.5 CONCLUSIONS

Many fundamental issues in ELE remain open. Most of the current efforts still do not fully express the power of human being, and the importance of social connections and societal activities is less noticed. And effective ELE solutions requires appropriate ICT algorithms, architectures and platforms, having in view the advance of science in this area and the development of new and innovative connected solutions (particularly in the area of pervasive and mobile systems).

For the remaining of the following book, the reader will find a platform for the dissemination of research efforts and presentation of advances in the ELE area that explicitly aim at addressing these challenges. The book constitutes a flagship driver towards presenting and supporting advance research in the area of Enhanced Living Environments, as it offers a path-breaking collection of research and theory about this emerging scientific field. Its form provides a material artifact that intervenes in the digital realms of health, ICT and sociological sciences, presenting advances in ubiquitous computing, social media, and mobile. The book contributes to better understanding the opportunities as well as challenges behind the technologies for Enhanced Living Environments, to better understand what is need to construct tools, interfaces, methods, and practices of social and mobile technology that enable participation and engagement of society towards constructing safer environments around assisted peoples and help them maintain independent living.

As a material artifact, the book signifies the coming together of an international group of academics and practitioners from a diverse range of disciplines such as computing and engineering, social sciences, design, digital media, and human — computer interaction. In some cases, contributors' connection to the book has developed over the course of a few months through electronic exchanges; for others, it is the result of participation in one of two workshops, such as the International Workshop on Enhanced Living EnvironMENTs (ELEMENT) that preceded the editorial work on this book.

REFERENCES

AAL-Europe, 2016. European AAL joint programme. http://www.aal-europe.eu/. Last access on May 30, 2016. http://www.aal-europe.eu/. Last access on May 30, 2016.

AAPELE, 2016. ICT COST action IC1303 — algorithms, architectures and platforms for enhanced living environments (AAPELE). http://aapele.eu/. Last access on May 30, 2016. http://aapele.eu/. Last access on May 30, 2016.

Ahola, J., 2001. Ambient intelligence. ERCIM News 47 (10), 8–9.

Belbachir, A.N., Drobics, M., Marschitz, W., 2010. Ambient assisted living for ageing well — an overview. E&I, Elektrotech. Inf.tech. 127 (7–8), 200–205.

Blasco, R., Marco, Á., Casas, R., Cirujano, D., Picking, R., 2014. A smart kitchen for ambient assisted living. Sensors 14 (1), 1629–1653.

Bohn, J., Coroamă, V., Langheinrich, M., Mattern, F., Rohs, M., 2004. Living in a world of smart everyday objects — social, economic, and ethical implications. Hum. Ecol. Risk Assess. 10 (5), 763–785.

Coroama, V., Röthenbacher, F., 2003. The chatty environment – providing everyday independence to the visually impaired. Workshop on ubiquitous computing for pervasive healthcare applications at UbiComp.

EIT, 2016. EIT health. https://eithealth.eu/. Last access on May 30, 2016. https://eithealth.eu/. Last access on May 30, 2016.

Frey, J., Neßelrath, R., Stahl, C., 2011. An open standardized platform for dual reality applications. In: Proceedings of the International Conference on Intelligent User Interfaces (IUI) Workshop on Location Awareness for Mixed and Dual Reality. LAMDa11, Palo Alto, CA, USA, vol. 1316.

Giannakouris, K., 2008. Ageing characterises the demographic perspectives of the European societies. Stat. Focus 72, 2008.

Hung, K., Zhang, Y.-T., 2003. Implementation of a WAP-based telemedicine system for patient monitoring. IEEE Trans. Inf. Technol. Biomed. 7 (2), 101–107.

Khoor, S., Nieberl, J., Fügedi, K., Kail, E., 2001. Telemedicine ECG-telemetry with Bluetooth technology. In: Computers in Cardiology 2001. IEEE, pp. 585–588.

Kung, A., Jean-Bart, B., 2010. Making AAL platforms a reality. In: Ambient Intelligence. Springer, pp. 187–196.

Lee, R.-G., Chen, H.-S., Lin, C.-C., Chang, K.-C., Chen, J.-H., 2000. Home telecare system using cable television plants — an experimental field trial. IEEE Trans. Inf. Technol. Biomed. 4 (1), 37–44.

Mega, 2016. eMotion ECG. http://www.megaemg.com/products/emotion-ecg/. Last access on May 30, 2016. http://www.megaemg.com/products/emotion-ecg/. Last access on May 30, 2016.

Mendoza, G., Tran, B., 2002. In-home wireless monitoring of physiological data for heart failure patients. In: Engineering in Medicine and Biology, 2002. Proceedings of the Second Joint 24th Annual Conference and the Annual Fall Meeting of the Biomedical Engineering Society EMBS/BMES Conference, 2002, vol. 3. IEEE, pp. 1849–1850.

Muñoz, A., Serrano, E., Villa, A., Valdés, M., Botía, J.A., 2012. An approach for representing sensor data to validate alerts in ambient assisted living. Sensors 12 (5), 6282–6306.

Nedopil, C., Schauber, C., Glende, S., 2013. Knowledge Base: AAL Stakeholders and Their Requirements. Ambient Assisted Living Association.

Nehmer, J., Becker, M., Karshmer, A., Lamm, R., 2006. Living assistance systems: an ambient intelligence approach. In: Proceedings of the 28th International Conference on Software Engineering. ACM, pp. 43–50.

Neßelrath, R., Haupert, J., Frey, J., Brandherm, B., 2011. Supporting persons with special needs in their daily life in a smart home. In: 2011 7th International Conference on Intelligent Environments (IE). IEEE, pp. 370–373.

O'Grady, M.J., Muldoon, C., Dragone, M., Tynan, R., O'Hare, G.M., 2010. Towards evolutionary ambient assisted living systems. J. Ambient Intell. Humaniz. Comput. 1 (1), 15–29.

Pollard, J., Rohman, S., Fry, M., 2001. A web-based mobile medical monitoring system. In: International Workshop on Intelligent Data Acquisition and Advanced Computing Systems: Technology and Applications 2001. IEEE, pp. 32–35.

Systems, B., 2016. TruVue. http://www.biomedsys.com/truvue/. Last access on May 30, 2016. http://www.biomedsys.com/truvue/. Last access on May 30, 2016.

Takahashi, S., et al., 2003. Population projections for Japan 2001–2050. J. Popul. Social Secur. (Population) 1 (1), 1–43.

van den Broek, G., Cavallo, F., Wehrmann, C., 2010. AALIANCE Ambient Assisted Living Roadmap, vol. 6. IOS Press.

Varshney, U., 2008. A framework for supporting emergency messages in wireless patient monitoring. Decis. Support Syst. 45 (4), 981–996.

Vincent, G.K., Velkoff, V.A., 2010. The next four decades: the older population in the United States: 2010 to 2050. 1138. US Department of Commerce, Economics and Statistics Administration, US Census Bureau.

Weiser, M., 1991. The computer for the 21st century. Sci. Am. 265 (3), 94–104.

CHAPTER 2

Implanted Wireless Body Area Networks: Energy Management, Specific Absorption Rate and Safety Aspects

Saif ul Islam*, Ghufran Ahmed*, Maham Shahid*, Najmul Hassan†,
Muhammad Riaz†, Hilal Jan*, Azfar Shakeel*

*COMSATS Institute of Information Technology, Park Road, 45550, Islamabad, Pakistan
†Capital University of Science and Technology, Islamabad Expressway, Kahuta Road, Zone-V,
Islamabad, Pakistan

2.1 INTRODUCTION TO WBAN

2.1.1 Overview

WBAN is a network of either wearable or implantable devices in close proximity to a person's body. These nodes cooperate with each other to perform health monitoring. It covers human body with resident sensor nodes. Manufacturers may construct WBAN sensor nodes for a range of applications; for example, collecting and examining a person's vital signs (heart rhythm, blood pressure, oxygen and temperature), fitness information or human activities for medical purpose. WBAN may communicate to Internet and other wireless technologies like ZigBee, Wireless Sensor Networks (WSNs), Bluetooth, Wireless Local Area Networks (WLAN), Wireless Personal Area Network (WPAN), video surveillance systems and cellular networks (Movassaghi et al., 2014). WBAN is compatible with current technologies either wired or wireless such as Wireless Personal Area Network (WPAN), Wireless Local Area Networks (WLAN), mobile telecommunication technologies and video capturing (Movassaghi et al., 2014). Here, sensor nodes collect information from person's body and forward it to Internet through base unit on the body. Sensor nodes in WBAN have the capability to perform different tasks, e.g. to process, monitor and communicate vital signs to a remote control center like hospital. Finally, personnel at remote location can provide feedback without caus-

Ambient Assisted Living and Enhanced Living Environments.
DOI: http://dx.doi.org/10.1016/B978-0-12-805195-5.00002-8

ing any discomfort to user (Movassaghi et al., 2014; Otto et al., 2006; Ullah et al., 2010; Yang, 2006). According to IEEE 802.15.6, we can formally define BAN as "communication standard optimized for low power devices and operations on, in or around human body (but not limited to humans) to serve a variety of applications including medical, consumer electronics/entertainment and others" (Astrin et al., 2012). For practical implementation of WBAN, there are different social as well as technical issues which require to be handled. The user-oriented requirements such as user friendly environment, secure, safe and compatible solutions are the challenges that must be ensured in WBAN. These challenges offer different prospects of designing and implementing systems to achieve maximum network life time, reducing delay, minimizing unimportant communication and maximizing throughput (Movassaghi et al., 2014).

2.1.2 Background

BAN technology has a short history. It emerged in the form of natural byproduct of current WSN and biomedical technologies. Initially, Professor Guang-Zhong Yang formally defines the sentence "Body Sensor Network (BSN)" in his book titled Body Sensor Networks which was published in 2006 (Yang, 2006). The IEEE 802 working group performed a key role in international standardization of WBAN. However, in February 2012 IEEE 802.15.6 standard was ratified (Movassaghi et al., 2014; Astrin et al., 2012).

2.2 APPLICATIONS OF WBAN

WBAN has a wide range of applications such as remote health monitoring, defense, sport, entertainment and various other fields. According to IEEE 802.15.6 standard, WBAN applications are divided into mainly two categories: medical and non-medical (Consumer Electronics) as presented in Figure 2.1. The aim of WBAN applications is to improve the user's quality of life, however, for a specific application the technological requirements of WBANs may be diverse (Latré et al., 2011).

WBAN has the capacity to revolutionize the future of medical care. Many life threatening diseases can be monitored and diagnosed through WBANs (Kwak et al., 2010a). The medical applications of WBANs let enduring to monitor physiological characteristics of a person, for example, blood pressure, pulse and temperature of body. When anomalous situation is detected, data collected from sensor nodes can be communicated to a

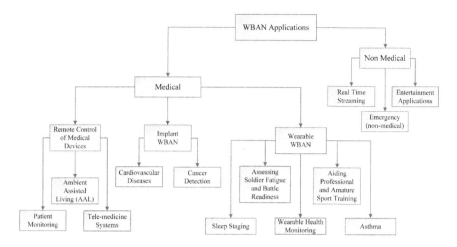

Figure 2.1 WBAN applications.

sink like a mobile phone. The sink sends this data to a faraway location, for example, a help desk in hospital or a doctor's clinic who can take necessary action. A cellular network or the Internet can be used for delivery of data from gateway to destination. Furthermore, WBANs have a role for taking preemptive health measures, remote monitoring and cure of patients with mortal diseases such as cardiovascular diseases, paralysis, hypertension and diabetes (Kwak et al., 2010a). WBAN also has non-medical applications; further categorization has been shown in Figure 2.1.

2.3 USE OF ULTRA WIDEBAND (UWB) IN WBAN APPLICATIONS

Any device or gadget that falls into wearable or implantable category requires certain standards to be followed so that not to make it harmful for the body in medical terms. The device must not interfere with traditional radio network as well as from any other devices like pace maker in the heart or cochlear implant in internal ear. The essence is that we need a network-enabled implant that can operate using a very small amount of battery, has no chances of interference within short range and has high bandwidth properties; the solution comes from Orthogonal Frequency Division Multiplexing (OFDM) based systems, formerly known as pulse radio and currently called as Ultra Wide Band or simply UWB. UWB operates in a frequency range greater than 500 MHz on a very low level of energy for a short range supporting high bandwidth needs (Reed, 2005).

2.3.1 Technology

UWB offers advantages by varying the radio wave energy at different instances of time. It is caused by using pulse positioning or time-based modulation. If one needs to share multiple channels of UWB, one can modulate on pulse polarity, pulse amplitude or its phase in order to make it either orthogonal division multiplexing or orthogonal time hopping. In UWB-based systems, Forward Error Correction (FEC) is performed using Low Density Parity Check Code (LDPC) based algorithms. Multipath fading can be resolved using channel equalization or raking-receiver techniques. The official IEEE standard that defines parameters for the use of UWB is 802.15.3a personal area networks that was improved by WiMedia Alliance and USB Implementer forum. The latest standard for WBAN is IEEE 802.15.6 which defines the standards for Quality of Service (QoS), non-interference, type of antennas, radiation pattern and Specific Absorption Rate (SAR). On the security side, the standard defines a three tier security approach to address all the classical security threats namely confidentiality, authenticity, integrity, privacy protection and replay attacks. Strict MAC level associations and temporal keys per sessions are used to maintain security (Kwak et al., 2010b). An optional mode known as FM-UWB is also available in the IEEE 802.15.6 standard that is based on double frequency shift keying by combining low index digital modulation with high index analog modulation. This mode is useful for simultaneous multiple users that require medium to low level bit rates (10 to 250 Kbits/sec) (Gerrits et al., 2005). When OFDM is used, the advantage of multiple carriers can be obtained. In this case, the communication system is known as Multi Carrier Ultra Wide Band (MC-UWB). This configuration makes the system more complex and increases the requirement of power consumption making it unsuitable for WBANs. Instead the preferred approach is to use Impulse Radio Ultra Wide Band (IR-UWB) for WBANs as Pulse Position Modulation (PPM) combined with on-off keying significantly reduces the power consumption.

2.3.2 Effects of IR-UWB on Human Body

Different case studies have been made to check the effects of electromagnetism on human tissues. This includes simulation of effect of heat produced and absorption rate models, variations in SAR and temperature variations, SAR calculations, device positioning effects, antenna models. A detailed study has been given in Thotahewa et al. (2014). In another experiment, researchers applied different common positions on human body for sensors and receivers. They proved that under static channel conditions, power can be reduced by 26 dB if the receiver is placed at optimal location on body. Moreover, if a body is moving then an increase of 7 dB in power is required to maintain same Bit Error Rate (BER) (Ho et al., 2012). A comprehensive survey article (Ragesh and Baskaran, 2012a) has been written that discusses the application and benefits of UWB in WBAN. The article (Ragesh and Baskaran, 2012a) also explains some of the existing WBAN projects.

2.4 DESIGN OF IMPLANTED SENSOR NODES

The sensor nodes in WBAN system have to bear harsh and inflexible environment (Cavallari et al., 2014). In applications where we have a choice for using either wearable or implantable sensor nodes, implanted nodes are more preferable. Implantable sensor nodes have an ability to measure chemical processes in human body regardless of requirement for patient intervention and without patient's physiological state. The major design requirement for these nodes are to be extremely small, lightweight and takes as less power as possible (Antonescu and Basagni, 2013).

2.4.1 Components of Sensor Node

WBAN and WSN nodes have similar functional architecture. As mentioned earlier, WBAN sensor nodes operate in extreme stringent surroundings. They have different operational characteristics in terms of event sensing, signal processing and communication as opposed to WSN nodes (Khan and Yuce, 2010). The sensor nodes must be smaller in size and consume very low energy thus presenting unique challenges and opportunities for WBAN nodes. Figure 2.2 represent the basic components of a sensor node.

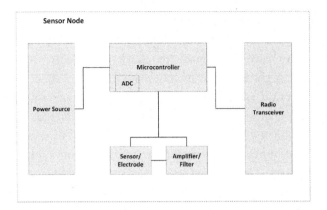

Figure 2.2 Major components of a sensor node.

2.4.1.1 Sensor Interface

In order to capture data from environment, sensors or electrodes are used by wireless sensor nodes. They collect analog signals which corresponds to physiological activities like change in temperature or pressure. Moreover, in order to detect various physiological signals, electrodes are designed to recognize different signal frequencies (Keong and Yuce, 2011).

2.4.1.2 Amplifier

In human body environment, the physiological signals are usually weak and contains noise. These signals are passed through an amplifier to increase the signal strength. The noise in these signals are further removed by going through a filtering process. The filtered analog signals pass through Analog to Digital converter (ADC) to be converted into digital signals for further processing.

2.4.1.3 Micro-controller

The micro-controller oversees the overall operation of a sensor node and is one of the most important component. It controls the functionality of other components in a sensor node. Moreover, it performs tasks and processes data. The digitized signal is processed and stored by the controller. The sensor node contains radio transceiver which transmits data of human body to an external device for monitoring (Keong and Yuce, 2011).

2.4.1.4 Power Management Module

The implantable nodes are usually placed in difficult to access location. Changing the energy source of sensor node is not an applicable solution. In addition to data acquisition and processing, the micro-controller also maintains a power management module. The goal is to provide as much energy as possible at smallest cost. The micro-controller is programmed to control the distribution of energy from battery in an optimized manner. In some cases, it is programmed to turn the battery connections off when sensor node is not functioning, i.e. during sleep mode (Khan and Yuce, 2010).

2.4.2 Design Challenges

As mentioned earlier, implantable sensor nodes have to operate under stringent environment. Therefore, to achieve unobtrusive monitoring of patient's health, these nodes must be lightweight and very small in size (Ragesh and Baskaran, 2012b). Ideally its battery should last many years as it would be impractical to replace it, however, the small size of sensor node itself limits the size of the battery, thus, affecting the power resources. The sensor nodes need to be reliable and sensitive especially when the users wear them in harsh environment: it should not cause any kind of discomfort or damage to human tissues through overheating (Darwish and Hassanien, 2011). The main challenge for the developers of implantable sensor node is to achieve advances in node miniaturization with low power circuitry (Ragesh and Baskaran, 2012b). In such a small system, optimizing every component based on its requirement is crucial as it can save development time, space and money.

2.5 IMPLANT POWER CONSTRAINTS AND BATTERY CONSIDERATIONS

In WBAN, sensor nodes inside or around the body communicate through electromagnetic waves. It is the power of the electromagnetic waves that decides the coverage area around the human body, i.e. the signals that are transmitted with low power have small coverage range and vice versa. The signal transmission with high power leads to high battery consumption that in turn makes more chances of battery replacement. It is essential to monitor transmit power control among these tiny nodes and need to be reliable too (Chepkwony et al., 2013).

An IEEE 802.15.4 standard is proposed specially for the systems with low power requirements that supports low data rate where latency is not so critical. Such a standard is also applicable to WBAN. In WBAN, battery life time of implanted sensor nodes is about 10 to 15 years (Timmons and Scanlon, 2004; Howitt and Gutierrez, 2003). To monitor patients in aged-care hospitals, it is necessary to continuously transmit sensed data from the patient's body to the health-care center that requires long life of batteries (Xiao et al., 2009). Storage capacity of batteries depends upon their sizes, large size of batteries have enough charging capacity while small sized batteries have low charging capacity (Mehmet and Khan, 2012). Various researchers have worked on technical issues of batteries like drain of batteries which is critical in continuous communication among the devices and the remote network (González-Valenzuela et al., 2012). The drain problem of battery requires special attention of software and manufacturers of batteries. It is necessary to manufacture a low sized battery with huge life time that can operate up to 15 years within human body to monitor various parameters. With recent advancement in solid state physics, hardware manufacturers are able to manufacture very sophisticated radio chips and microprocessors that consume current in nano-ampere range and uses very low power. On the software side, it is the duty of computer scientists and software engineers to design and create such efficient algorithms to process data transmission and reception that utilize very low battery and prolong the battery life (González-Valenzuela et al., 2012).

2.5.1 Energy Harvesting

Improvement in the battery technology is gradual. Hence, it is necessary to utilize energy resources efficiently. To achieve high performance of monitoring of patients, more investment is needed in battery industry. The lithium-based batteries are suitable for portable handheld equipments, however, life time of such batteries is limited in WBAN. Frequent replacement or recharging of batteries make the WBAN impractical. Carbon-nanotube-based batteries have much storage capacity and suitable for WBAN but have not been commercialized so far (Hanson et al., 2009).

Energy harvesting means collecting or extracting energy from natural sources like sunlight or vibration. This is not only a good solution to improve battery life time by recharging them with harvested energy but also feasible for WBAN, however, exposure of human body to electromagnetic radiation is a big constraint (Chalasani and Conrad, 2008;

Hanson et al., 2009). In Hanson et al. (2009), average harvested energy by WBAN user per hour per day is estimated. According to the estimated results, it is the sunlight that contributes more in harvested energy than other sources. It can also be seen that maximum harvested energy can be taken 1200 hrs to 1600 hrs.

Consequently, harvesting sources can increase battery life time and can bring revolution in the WBAN. It is highly recommended to do more research for efficient and hybrid design to store enough energy to prolong battery life time so that battery replacement and recharging be avoided.

2.6 ENERGY MANAGEMENT

Energy management is one of the basic requirements of WBAN deployment. The size of the sensor nodes should be small as they have to be implanted within the body. The battery size is one of the major contributors for device dimension and weight (Latré et al., 2011). Therefore, the sensor nodes need stringent restriction on energy utilization as the battery size has to be kept small.

Mainly, WBAN consumes energy during event sensing, communication and data processing (Chepkwony et al., 2013). This opens up many challenging issues to maximize the network life by achieving energy efficiency.

2.6.1 Issues

As discussed earlier, the sensor nodes are implanted within body through surgery to detect various physiological processes. These nodes consume energy mainly during data transfer (Chepkwony et al., 2013). It is impractical to replace the batteries of these nodes as it results in discomfort to the patient and causes health risks.

The human body is complex and its tissues have electrical attributes which affect the communication of electromagnetic signals emitted from these nodes (Ragesh and Baskaran, 2012b). Propagation of waves in a lossy medium causes signal attenuation (Ragesh and Baskaran, 2012b). In order to optimize energy consumption, WBAN needs to employ low power transmission, however, these signals are absorbed by human tissues. As a result, the process of data re-transmission occurs and eventually affects the energy consumption.

Different kinds of implantable nodes are placed in body to measure different physiological traits. Some of them do not require constant listening of event. These nodes consume energy during idle listening (Hughes et al., 2012; Ye et al., 2002). Furthermore, if a patient has more than one sensor implanted on his body, the issue of packet collision will also occur as different sensors will send the packet simultaneously. Collisions are the major source of energy inefficiency. When the sensor nodes consume energy, it dissipates heat which can affect the surrounding tissues and can cause damage to organs system in body such as cataract and bone system (Latré et al., 2011).

2.6.2 Solutions

As discussed earlier, replacement of batteries is not feasible in the case of WBAN nodes. It is due to the reason that it could damage a node and put at risk the patient's health. "Harvesting energy" is a method where the idea is to harvest energy during the working of system (Barakah and Ammaduddin, 2012). The energy collected depends on various parameters, for example, sensor node location in body and harvester's dimension and characteristics. The energy sources in a body of human can be categorized into predictable (heart contractions and chest movement from breathing etc.) and non-predictable sources (walking movements and body temperature etc.) (Ibarra et al., 2013; Pawar et al., 2011). The energy harvester uses some mechanism to produce electricity to charge the nodes (Ibarra et al., 2016; Tan and Panda, 2011).

The MAC layer is considered to be most suitable layer to address the energy efficiency. Its fundamental task is to avoid collision while maximizing throughput and energy efficiency. Multiple channel assessment protocols like TDMA (Time Division Multiple Access) and CSMA/CA (Code Division Multiple Access/Collision Avoidance) checks the status of transmission channel before sending the data packets (Zois et al., 2013). Node can send the data when there is no traffic in channel that results in avoiding collision and re-transmission of packet.

Depending on application, WBAN nodes can transmit data on–demand basis. Some of them require a node to monitor only when they are asked to do so, thus preventing it from idle listening. Some authors proposed a mechanism where the state of a node is turned to sleep when they are not sensing, thus, achieving energy efficiency (Al Ameen et al., 2012).

In contrast to WSN, WBAN nodes are mobile and therefore do not require extensive multi hop routing requirement (Joseph et al., 2011). Some routing techniques have been developed which will be discussed in detail in further section that achieve some level of energy efficiency. The hardware design of the node also has some impact on energy usage. The radio transceiver uses large portion of energy among other components when it is idle or in sending/receiving mode. While designing the node, choosing transceiver which requires small current consumption will extend battery life.

2.7 SPECIFIC ABSORPTION RATE (SAR) AND SAFETY ASPECTS

Definition. "Specific Absorption Rate (SAR) is a measure of the rate of energy absorbed by a human body when exposed to electromagnetic waves" (Wang et al., 2016).

OR

"It is also defined as the power absorbed per mass of tissue and has units of joules per second per kg or Watts per kilogram (W/kg)" (Wang et al., 2016).

OR

"Specific absorption rate (SAR) is the time derivative of dissipated energy per unit mass within the exposed body caused by the incident of electromagnetic waves" (Wang et al., 2016).

SAR of electromagnetic radiations within a human body tissue can be found using following mathematical relation (ICNIRP Guideline, 2010),

$$\text{SAR} = \frac{1}{V} \int_{\text{sample}} \frac{\sigma(r)|E(r)|^2}{\rho(r)} dr \tag{2.1}$$

where,
V is the volume of tissue sample
σ represents conductivity of tissue sample
ρ is the sample density, i.e. mass per unit volume of tissue
E represents the value of root mean square electric field strength

Various application-specific sensor nodes are deployed within or around the human body to monitor wide range of body parameters. These sensor nodes share information with one another or with the home station, from where the information is transmitted in the form of signals (De Santis et al., 2010; Huang et al., 2009). The ultra-high frequency (UHF) band of radio frequency spectrum is very fascinating for implanted wireless communications that ensure small antenna size for short communication range (Scanlon et al., 2000). The advantage to use UHF band is to keep the signal–to–noise plus interference ratio (SINR) in a limit because man–made signals of low frequencies never interfere the signals in UHF band. The exposure of human body to these high frequency electromagnetic radiations greatly effect the internal tissues of human body.

It is a usual practice to limit the absorption of RF electromagnetic radiations. Measurement of these absorbed radiations is known as absorption dose and the rate at which these radiations are absorbed in human body per unit mass of tissues is referred to as SAR. The SAR values are kept at minimum level according to the national and international regulations. The regulations for transmitting of electromagnetic signals in WBAN in proximity of human body is similar to cellular mobile phones (O.N.-I.R.H. IEEE Standards Coordinating Committee 28, 1992).

In the frequency range from 100 kHz to 10 GHz in the microwave window, the current safety SAR values reach the electromagnetic field strength and power density limits (ICNIRP Guideline, 2010). These limits must be established when exposure of human body to radiations is within a range of 20 cm from source to avoid any biological effect. For the above mentioned frequency band, the mean value of SAR for the whole human body is 0.08 Watts per kg; in the head and trunk region of the body, its value is 2 Watts per kg and for the arms and legs regions it is 4 Watts per kg. The biological effects are dependent on numerous aspects that involve physical geometry of tissues, its conductivity, permittivity, the orientation of a body and the SAR distribution which is either uniform or non–uniform.

Two types of biological effects are associated due to data transmission using wireless links from a sensing device inside or around a body to a remote network:

- **Short Term Effects:** These are immediate effects caused due to exposure to electromagnetic radiations like thermal effects.
- **Long Term Effects:** Such effects are not so dominant, however, it causes some degenerative diseases after exposure of body tissues to electromagnetic radiations for many years (De Santis et al., 2010).

2.8 ENERGY EFFICIENT ROUTING IN WBAN

There is need to consider limitations of each type of network while designing routing algorithm for different networks. In wired networks, usually energy is not a big problem as compared to wireless networks. In a wireless environment, considering battery limitations of wireless nodes, it is required to consider energy constraint. This limitation gets worst when implanted nodes are considered. Hence, there is a need to build energy-efficient routing algorithm that consume least amount of energy to send data from source to destination node (Singh et al., 2010).

2.8.1 How BAN Routing Is Different from Conventional Routing Algorithm

In traditional routing algorithms of wireless networks, the aim is to reduce congestion at intermediate nodes while at the same time, preserving connectivity among devices. Energy is not the primary concern in such routing protocols. However, in WBAN where nodes are deployed on the human body, conventional routing protocols are not feasible due to many reasons (Latré et al., 2011). Therefore, there is a need of a routing mechanism which is not only reliable and robust, but also energy efficient. Nevertheless, designing such routing protocol is not an easy job. It is a very challenging task to design such energy-efficient routing algorithm (Movassaghi et al., 2013; Khan et al., 2012).

2.8.2 Challenges in Designing Energy Efficient Routing Algorithm for WBAN

There are a number of challenges in designing routing protocols for WBAN (Movassaghi et al., 2013; Dangi and Panda, 2014; Schiller, 2015) due to its specific nature and limitations, some of these are listed as follows:

- **Body Movement:** One of the challenges of WBAN is the link quality degradation due to frequent movement of a body. Normally, sensor nodes are deployed on the human body that moves dynamically and frequently. This results in frequent and unexpected variations in network topology. This is due to the reason that the link health among nodes in WBANs changes due to the change of posture and movement of body. Therefore, routing protocol should be flexible in order to adapt dynamic topology changes.
- **Temperature:** In case of implantable nodes, a routing algorithm must be biocompatible and must not overheat the body tissues. Therefore, extremely low energy is used while transmitting and relaying the data.

- **Local Energy Awareness:** Local energy awareness refers to the fact that the network nodes should know the energy level of other nodes in the network. It can be achieved when routing protocol for BAN communicate important parameters like status of energy to other nodes of the network. This process balances the load to avoid disconnectivity of network.
- **Network Lifetime:** Network lifetime is usually referred to as the time till the death of the first node. It means when significant damage in terms of energy takes place, nodes start to disconnect with neighboring nodes due to energy depletion of nodes in the network. It is very difficult to design a routing algorithm which gracefully depletes equal amount of energy of all collaborative nodes in a uniform manner.
- **Transmission Range:** Energy of a node in WBAN is limited, therefore ultra-low transmission power is used. This may result in less transmission range and low coverage area. A routing algorithm is vital under such circumstances. It should be robust enough to deal with transmission range problem effectively. Especially, when the transmission range of nodes is less than a particular threshold value, the only option available for routing is to forward data to a small number of neighboring nodes located at a closer proximity. This leads to a high number of transmissions which in turn rise the overall temperature.
- **Non-Homogeneous Environment:** Different applications of WBAN may require non-homogeneous information gathering from various sensing devices with different data rates. Therefore, Quality of Service (QoS) provision in WBAN is not an easy task to achieve.
- **Limitation of Resources:** There is a large number of resource limitations including energy, processing power, device lifetime and data capacity. Due to these constraints, it is very difficult to make a routing algorithm which can fulfill the desired goals and may help in QoS and other important parameters.

2.9 ADAPTIVE THERMAL-AWARE ENERGY EFFICIENT ROUTING

There are various reasons of heat dissipation by sensor nodes in WBAN. WBAN often uses Radio Frequency (RF) to establish communication among sensor nodes. These waves may penetrate into tissues that may result in tissue damage. Other reasons of heat generation may be power dissipation of node circuitry and radiation from sensor nodes. Moreover, continuous usage of a node can heat it up. Thus, presence of heat for a longer period of time may damage the adjacent tissues. This can also provide ideal environment for bacterial growth (Alemdar and Ersoy, 2010). Therefore, it is important to consider heat factor while proposing routing solutions in WBAN. Routing algorithm should be smart enough to produce as low heat as possible. Hence, communication protocols must aim at reducing the temperature of implanted sensor nodes.

2.9.1 WBAN Routing Constraints

One of the most important areas of concern in WSNs is energy limitations. Hence, most of the routing solutions in WSN are designed to be energy efficient. As these networks are not designed for Medical solutions (Biosensors), so it does not focus on restricting the communication of devices due to the continuous propagation of communicating waves. Sensor nodes used in WBAN are very light weight, small in size and limited in energy, hence, should use ultra low power in communication. WBAN sensor nodes are not homogeneous in their attributes, so they may have different energy levels and can generate data of different sizes. In contrast, nodes in WSN are homogeneous, i.e. almost same devices are used in terms of initial energy levels and other parameters. Use of routing solutions of WSN may not be feasible for WBAN. The state-of-the-art routing solutions designed for WSNs are not suitable for WBAN. In addition, the biosensors in WBAN are used in sensitive procedures like organ monitoring. Therefore, heat generated due to radiation in communication may cause serious health hazards. In order to avoid such problems, thermal aware routing protocols are used for implanted nodes in WBAN.

2.9.2 Issues with Adaptive Thermal Aware Routing Protocols

As discussed earlier that thermal aware routing protocols are very important in implanted sensor nodes, however, there are some consequences of these algorithms (Galluccio et al., 2012; Dangi and Panda, 2014).

- **Lack of Energy Efficiency:** The primary importance in thermal aware routing algorithms is how to reduce heat generations as much as possible. The trade-off of this factor is in the form of poor energy utilization of sensor nodes.

- **Network Dis-connectivity:** Network may be partitioned when use of the only available relay node between two sub-networks is avoided due to its thermal vulnerability.
- **Poor Selection of Optimum Route:** A non-optimal or expensive route can be selected using thermal-aware routing algorithm, as the optimum path may be heated. Moreover, there are risks to heat up the relay nodes if the optimal path is used continuously.
- **Longer End-to-end Delays:** In adaptive thermal-aware routing, it is possible that overall end-to-end (E2E) delay may increase from a specific threshold due to adaptive selection of thermal invulnerable routes (Monowar et al., 2014).
- **Not suitable for Real Time Traffic:** Real time data is time sensitive in nature. WBAN biomedical sensors have to sense a number of real time attributes from human body that have to be delivered through the quickest available route (Hassan et al., 2013). While using thermal aware routing protocol, it is not possible all the time.

2.10 CONCLUSION

This chapter has provided a comprehensive overview of WBAN specially energy management, SAR and safety aspects in implanted wireless body area networks. Due to the energy constraints and the level of difficulty to replace battery of implanted sensor nodes, it is important to propose energy efficient WBAN solutions. Moreover, heating issues caused by sensor nodes implanted in human body have a significant impact on tissues. Human tissues are sensitive to temperature. Sensor heat dissipation is caused by communication of signals among sensor nodes. An increase of heat around tissues may cause malfunctioning and even can damage the tissues. For safety of human body, it is crucial to propose and implement thermal aware mechanisms. Hence, this chapter has covered the techniques used to cope with previously mentioned issues and provided a global overview of energy efficient and thermal aware routing solutions to avoid tissue damage and to maximize network life time.

List of acronyms
SAR   Specific Absorption Rate
WBAN   Wireless Body Area Network
WSN   Wireless Sensor Network
ECG   Electrocardiography

EEG   Electroencephalography
UHF   Ultra-High Frequency
UWB   Ultra Wide Band

REFERENCES

Al Ameen, M., Ullah, N., Chowdhury, M.S., Islam, S.R., Kwak, K., 2012. A power efficient MAC protocol for wireless body area networks. EURASIP J. Wirel. Commun. Netw. 2012 (1), 1–17.

Alemdar, H., Ersoy, C., 2010. Wireless sensor networks for healthcare: a survey. Comput. Netw. 54 (15), 2688–2710.

Antonescu, B., Basagni, S., 2013. Wireless body area networks: challenges, trends and emerging technologies. In: Proceedings of the 8th International Conference on Body Area Networks. ICST (Institute for Computer Sciences, Social-Informatics and Telecommunications Engineering), pp. 1–7.

Astrin, A., et al., 2012. IEEE standard for local and metropolitan area networks part 15.6: wireless body area networks: IEEE std 802.15.6-2012. The document is available at IEEE Xplore.

Barakah, D.M., Ammad-uddin, M., 2012. A survey of challenges and applications of wireless body area network (WBAN) and role of a virtual doctor server in existing architecture. In: 2012 Third International Conference on Intelligent Systems, Modelling and Simulation (ISMS). IEEE, pp. 214–219.

Cavallari, R., Martelli, F., Rosini, R., Buratti, C., Verdone, R., 2014. A survey on wireless body area networks: technologies and design challenges. IEEE Commun. Surv. Tutor. 16 (3), 1635–1657.

Chalasani, S., Conrad, J.M., 2008. A survey of energy harvesting sources for embedded systems. In: IEEE, Southeastcon, 2008. IEEE, pp. 442–447.

Chepkwony, R.C., Gwendo, J.O., Kemei, P.K. 2013. Energy efficient model for deploying wireless body area networks using multi-hop network topology.

Dangi, K.G., Panda, S.P., 2014. Challenges in wireless body area network—a survey. In: 2014 International Conference on Optimization, Reliability, and Information Technology (ICROIT). IEEE, pp. 204–207.

Darwish, A., Hassanien, A.E., 2011. Wearable and implantable wireless sensor network solutions for healthcare monitoring. Sensors 11 (6), 5561–5595.

De Santis, V., Feliziani, M., Maradei, F., 2010. Safety assessment of UWB radio systems for body area network by the method. IEEE Trans. Magn. 46 (8), 3245–3248.

Galluccio, L., Melodia, T., Palazzo, S., Santagati, G.E., 2012. Challenges and implications of using ultrasonic communications in intra-body area networks. In: 2012 9th Annual Conference on Wireless On-demand Network Systems and Services (WONS). IEEE, pp. 182–189.

Gerrits, J.F., Kouwenhoven, M.H., van der Meer, P.R., Farserotu, J.R., Long, J.R., 2005. Principles and limitations of ultra-wideband FM communications systems. EURASIP J. Adv. Signal Process. 2005 (3), 1–15. http://dx.doi.org/10.1155/ASP.2005.382. Available [Online].

González-Valenzuela, S., Liang, X., Cao, H., Chen, M., Leung, V.C., 2012. Body area networks. In: Autonomous Sensor Networks. Springer, pp. 17–37.

Hanson, M.A., Powell Jr, H.C., Barth, A.T., Ringgenberg, K., Calhoun, B.H., Aylor, J.H., Lach, J., 2009. Body area sensor networks: challenges and opportunities. Computer 1, 58–65.

Hassan, N., Khan, N.M., Ahmed, G., Ramer, R., 2013. Real-time gradient cost establishment (RT-GRACE) for an energy-aware routing in wireless sensor networks. In: 2013 IEEE Eighth International Conference on Intelligent Sensors, Sensor Networks and Information Processing. IEEE, pp. 54–59.

Ho, C.K., See, T.S., Yuce, M.R., 2012. An ultra-wideband wireless body area network: evaluation in static and dynamic channel conditions. Sens. Actuators A, Phys. 180 (Complete), 137–147.

Howitt, I., Gutierrez, J.A., 2003. IEEE 802.15.4 low rate-wireless personal area network coexistence issues. In: Wireless Communications and Networking, 2003, vol. 3. WCNC 2003. 2003 IEEE. IEEE, pp. 1481–1486.

Huang, L., Ashouei, M., Yazicioglu, R.F., Penders, J., Vullers, R.J., Dolmans, G., Merken, P., Huisken, J., de Groot, H., Van Hoof, C., et al., 2009. Ultra-low power sensor design for wireless body area networks-challenges, potential solutions, and applications. Int. J. Digit. Content Technol. Appl. 3 (3), 136–148.

Hughes, L., Wang, X., Chen, T., 2012. A review of protocol implementations and energy efficient cross-layer design for wireless body area networks. Sensors 12 (11), 14730–14773.

Ibarra, E., Antonopoulos, A., Kartsakli, E., Verikoukis, C., 2013. Energy harvesting aware hybrid MAC protocol for WBANs. In: 2013 IEEE 15th International Conference on e-Health Networking, Applications & Services (Healthcom). IEEE, pp. 120–124.

Ibarra, E., Antonopoulos, A., Kartsakli, E., Rodrigues, J.J., Verikoukis, C., 2016. QoS-aware energy management in body sensor nodes powered by human energy harvesting. IEEE Sens. J. 16 (2), 542–549.

ICNIRP Guideline, 2010. Guidelines for limiting exposure to time-varying electric and magnetic fields (1 Hz to 100 kHz). Health Phys. 99 (6), 818–836.

Joseph, W., Braem, B., Reusens, E., Latré, B., Martens, L., Moerman, I., Blondia, C., 2011. Design of energy efficient topologies for wireless on-body channel. In: 11th European Wireless Conference 2011—Sustainable Wireless Technologies (European Wireless). VDE, pp. 1–7.

Keong, H.C., Yuce, M., 2011. UWB-WBAN sensor node design. In: 2011 Annual International Conference of the IEEE, Engineering in Medicine and Biology Society, EMBC. IEEE, pp. 2176–2179.

Khan, J.Y., Yuce, M.R., 2010. Wireless body area network (WBAN) for medical applications. In: New Developments in Biomedical Engineering. INTECH.

Khan, Z., Aslam, N., Sivakumar, S., Phillips, W., 2012. Energy-aware peering routing protocol for indoor hospital body area network communication. Proc. Comput. Sci. 10, 188–196.

Kwak, K.S., Ullah, S., Ullah, N., 2010a. An overview of IEEE 802.15.6 standard. In: 2010 3rd International Symposium on Applied Sciences in Biomedical and Communication Technologies (ISABEL). IEEE, pp. 1–6.

Kwak, K.S., Ullah, S., Ullah, N., 2010b. An overview of IEEE 802.15.6 standard. In: 2010 3rd International Symposium on Applied Sciences in Biomedical and Communication Technologies. ISABEL 2010, pp. 1–6.

Latré, B., Braem, B., Moerman, I., Blondia, C., Demeester, P., 2011. A survey on wireless body area networks. Wirel. Netw. 17 (1), 1–18.

Mehmet, R.Y., Khan, J., 2012. Wireless Body Area Network: Technology, Implementation, and Applications. Pan Stanford Publishing Pte. Ltd.

Monowar, M.M., Mehedi Hassan, M., Bajaber, F., Hamid, M.A., Alamri, A., 2014. Thermal-aware multiconstrained intrabody QoS routing for wireless body area networks. Int. J. Distrib. Sens. Netw. 2014.

Movassaghi, S., Abolhasan, M., Lipman, J., 2013. A review of routing protocols in wireless body area networks. J. Netw. 8 (3), 559–575.

Movassaghi, S., Abolhasan, M., Lipman, J., Smith, D., Jamalipour, A., 2014. Wireless body area networks: a survey. IEEE Commun. Surv. Tutor. 16 (3), 1658–1686.

O.N.-I.R.H. IEEE Standards Coordinating Committee 28, 1992. IEEE standard for safety levels with respect to human exposure to radio frequency electromagnetic fields, 3 kHz to 300 GHz. IEEE.

Otto, C., Milenkovic, A., Sanders, C., Jovanov, E., 2006. System architecture of a wireless body area sensor network for ubiquitous health monitoring. J. Mob. Multimed. 1 (4), 307–326.

Pawar, P., Nielsen, R., Prasad, N., Ohmori, S., Prasad, R., 2011. Hybrid mechanisms: towards an efficient wireless sensor network medium access control. In: 2011 14th International Symposium on Wireless Personal Multimedia Communications (WPMC). IEEE, pp. 1–5.

Ragesh, G.K., Baskaran, K., 2012a. A survey on futuristic health care system: {WBANs}. In: International Conference on Communication Technology and System Design 2011. Proc. Eng. 30, 889–896. http://www.sciencedirect.com/science/article/pii/S1877705812009526. Available [Online].

Ragesh, G.K., Baskaran, K., 2012b. An overview of applications, standards and challenges in futuristic wireless body area networks. IJCSI Int. J. Comput. Sci. Issues 9 (1), 180–186. http://www.ijcsi.org/papers/IJCSI-9-1-2-180-186.pdf. Available [Online].

Reed, J., 2005. An Introduction to Ultra Wideband Communication Systems, 1st ed. Prentice Hall Press, Upper Saddle River, NJ, USA.

Scanlon, W.G., Burns, J.B., Evans, N.E., 2000. Radiowave propagation from a tissue-implanted source at 418 MHz and 916.5 MHz. IEEE Trans. Biomed. Eng. 47 (4), 527–534.

Schiller, J., 2015. Towards motion characterization and assessment within a wireless body area network. In: Internet and Distributed Computing Systems: 8th International Conference. IDCS 2015, Windsor, UK, September 2–4, 2015. Proceedings, vol. 9258. Springer, p. 63.

Singh, S.K., Singh, M., Singh, D., 2010. A survey of energy-efficient hierarchical cluster-based routing in wireless sensor networks. Int. J. Adv. Netw. Appl. 2 (02), 570–580.

Tan, Y.K., Panda, S.K., 2011. Energy harvesting from hybrid indoor ambient light and thermal energy sources for enhanced performance of wireless sensor nodes. IEEE Trans. Ind. Electron. 58 (9), 4424–4435.

Thotahewa, K.M.S., Redout, J.-M., Yuce, M.R., 2014. Ultra Wideband Wireless Body Area Networks. Springer Publishing Company, Incorporated.

Timmons, N.F., Scanlon, W.G., 2004. Analysis of the performance of IEEE 802.15.4 for medical sensor body area networking. In: 2004 First Annual IEEE Communications Society Conference on Sensor and ad hoc Communications and Networks, 2004. IEEE SECON 2004. IEEE, pp. 16–24.

Ullah, S., Shen, B., Islam, S., Khan, P., Saleem, S., Kwak, K.S., 2010. A study of medium access control protocols for wireless body area networks. arXiv preprint arXiv:1004.3890.

Wang, J., Lim, E., Leach, M., Wang, Z., Man, K., Huang, Y., 2016. Two methods of SAR measurement for wearable electronic devices. In: Proceedings of the International MultiConference of Engineers and Computer Scientists, vol. 2.

Xiao, S., Dhamdhere, A., Sivaraman, V., Burdett, A., 2009. Transmission power control in body area sensor networks for healthcare monitoring. IEEE J. Sel. Areas Commun. 27 (1), 37–48.

Yang, G.-Z., 2006. Body Sensor Networks. Springer-Verlag New York, Inc., Secaucus, NJ, USA.

Ye, W., Heidemann, J., Estrin, D., 2002. An energy-efficient MAC protocol for wireless sensor networks. In: IEEE, INFOCOM 2002. Twenty-First Annual Joint Conference of the IEEE Computer and Communications Societies. Proceedings, vol. 3. IEEE, pp. 1567–1576.

Zois, D.-S., Levorato, M., Mitra, U., 2013. Energy-efficient, heterogeneous sensor selection for physical activity detection in wireless body area networks. IEEE Trans. Signal Process. 61 (7), 1581–1594.

CHAPTER 3

Energy Efficient Communication in Ambient Assisted Living

Chandreyee Chowdhury*, Nauman Aslam[†], Susanna Spinsante[‡],
Davide Perla[‡], Antonio Del Campo[‡], Ennio Gambi[‡]

*Jadavpur University, Kolkota, India
[†]Department of Computer Science and Digital Technologies, Faculty of Engineering, Northumbria University, Newcastle, UK
[‡]Dipartimento di Ingegneria dell'Informazione, Università Politecnica delle Marche, Via Brecce Bianche 12, I-60131 Ancona, Italy

3.1 INTRODUCTION/MOTIVATION

With the advancement of medical science, the average longevity of human is increasing day-by-day resulting in more elderly population in many countries (United Nations Department of Economic and Social Affairs, 2013). This trend demands new ways of medical care that acts proactively rather than being reactive. Currently, treatment is provided to handle emergencies. However with more elderly population, day-to-day monitoring is needed for early detection of any anomaly in vital signs. Hospitals cannot handle this huge load. Rather, the living environment at home can be made smart enough to monitor and detect any vital signs and communicate it to the proper place, like a nursing center or the hospital. As mentioned in Wang et al. (2015), "Ambient Assisted Living (AAL) enhances the independent living ability of the elderly through various intelligent services, reducing the need for direct care giving, indoors or outdoors". The use of wireless technology, in particular medical applications of wireless sensor networks and wireless body area networks in AAL, improves the existing health-care and monitoring services especially for the elderly and chronically ill. A recent report (Prescot, 2013), forecasts that the number of home monitoring systems with integrated communication capabilities will grow at a compound annual growth rate (CAGR) of 26.9% between 2011 and 2017, to reach 9.4 million connections worldwide. These figures imply that the demand for health-care devices and AAL systems is increasing to involve citizens' in personal health-care, support independent living and economize the health-care expenses. These solutions provide a number of benefits includ-

Ambient Assisted Living and Enhanced Living Environments.
DOI: http://dx.doi.org/10.1016/B978-0-12-805195-5.00003-X

ing, remote monitoring and reduced costs for managing the health-care systems. With remote monitoring, the identification of emergency conditions for at risk patients will become easy and the people with different degrees of cognitive and physical disabilities will receive proper assistance to have a more independent and easy life. First responders could receive immediate notifications from such smart environments on any changes in patient status, such as respiratory failure or cardiac arrest (Shnayder et al., 2005).

A typical AAL system consists of diverse technologies and systems incorporating medical sensors, wireless sensor and actuator networks (WSANs), RFID tags and readers, computer systems, computer-networks, software applications, and databases, which are interconnected to exchange data and provide services in an ambient assisted environment (Lloret et al., 2015; Gambi et al., 2015). Wireless communication serves as a key enabling technology to support connectivity among different components of the system including, medical sensors, ambient sensors, home gateways and wireless routers that enable the logging/monitoring applications to report data in soft real-time to health-care professionals. Most of the existing solutions include one or more types of sensors carried by the patient, forming a Body Area Network (BAN), and one or more types of sensors deployed in the environment forming a Personal Area Network (PAN). These two are connected to a backbone network via a gateway node. In general, two classes of sensors can be used for monitoring in AAL: sensors to monitor the environment and wearable sensors to sense a person's context and activity. Data from these two networks of sensors should be analyzed to monitor any vital sign. For instance, the heartbeat sensor may give higher reading when a person is jogging, so does the temperature sensor. Thus, to filter out false positives, data from the accelerometer and location of the sensor are very important to appropriately contextualize the information.

In AAL system comprising BAN and PAN sub-components, one of the most precious resource is energy. Since sensor nodes rely on battery power and can only operate as long as their batteries maintain power, their hardware systems (micro-controller and radio communication subsystems e.g. ZigBee or 802.15.4) should be designed for low power consumption. Generally speaking, the amount of energy consumed in radio communication is significantly higher than the energy consumed by micro-controller. Since, the AAL system depends on communication activities to relay the sensor data across BAN and PAN subsystems, the energy efficiency remains a key challenge for end-to-end communication. These challenges necessitate that energy efficient communication protocols are designed. In

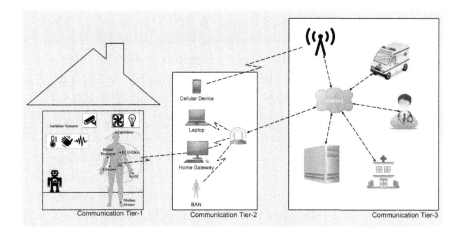

Figure 3.1 Typical network architecture of AAL with BAN.

addition to the energy efficiency, AAL systems also need to address many more complex challenges including, reliability of communication, security, privacy, user mobility, interference and quality of service (QoS) (Porambage et al., 2015). A brief description of these challenges is provided below;

1. Reliable communication: Interfacing different communication technologies should be reliable enough to transport the important medical data (physiological sensor readings, patient's vital signs etc.). Any loss is not acceptable.

2. Security and privacy: Handling massive private data–Any violation of privacy and hence data integrity is extremely important. The data may consists of sensitive and private medical or personal data which should be adequately protected.

3. Multiple receivers: Data could be sent to doctors or a display unit at the room. So multi-cast rather than uni-cast is more suitable.

4. In-network aggregation cannot be applied as it is not meaningful to combine data from multiple patients.

5. User mobility: Mobility of the patient results in the change of topology with time as BAN connects to different point of attachment.

An example architecture of AAL system is shown in Figure 3.1. As can be seen, the BAN constitutes tier-1 where body sensors communicate directly or indirectly to the BAN coordinator. In tier-2, information is received from the BAN coordinator to local access point or a home gateway, smart devices acting as a gateway of a simple workstation that is connected to the Internet. The information received by any tier-2 device is finally transmitted to tier-3 via the Internet or cellular network. Such a system can operate in a centralized or distributed fashion. Wearable sensors carried by a person constitute a BAN configuration consisting of a BAN coordinator and multiple sensors like the one shown in Figure 3.1. The sensors communicate with the BAN coordinator following IEEE 802.15 standard, or other communication protocols (Spinsante et al., 2015). These BAN coordinators may have a direct or indirect connection to some central device (PDA or nursing station coordinator) in tier 2 in the building (Aquino-Santos et al., 2013), the main coordinator (say) following IEEE 802.11 LAN standard. This point of attachment can be distributed as well. For instance, whenever a patient moves, the BAN coordinator registers itself to the nearest tier 2 device (Khan et al., 2013). Tier 2 devices further connect to Internet or wireless mobile network in tier 3 providing global access to sampled information from AAL BAN and PAN.

The rest of this chapter is organized as follows. Section 3.2 explores the related work focusing on system and network architecture, as well as inter-BAN communication. Section 3.3 articulates the problem formulation for the energy efficient route calculation. In Section 3.4 we present the evaluation results for the proposed protocol. Finally, the summary and conclusions are provided in Section 3.5.

3.2 BACKGROUND AND RELATED WORK

Many AAL proposals could be found in literature. According to Lloret et al. (2015), most of these applications focus on elderly people, however a few of them also address people with special needs, babies and children. The system must be less expensive than a service based on care-givers. Many works on AAL systems aim at investigating suitable architecture for AAL. These can be grouped into two categories, both addressing the target of energy efficient architectures.

3.2.1 Energy Efficient Architectures

The energy efficient design of AAL architectures typically address two targets: networks, or systems. The available literature on energy efficient networks design, and systems design, is accordingly reviewed and presented.

3.2.1.1 Network Architecture

In Kim and Cho (2009), a network architecture for inter-BAN communication is proposed with the mechanism of combining or splitting a BAN. Though the proposed mechanism manages relative mobility of BANs in hospital environment, yet it does not provide support for real time display of BAN data. According to Rashidi and Mihailidis (2013), most AAL systems having wearable sensors rely on network architecture similar to that of Wireless Body Area Networks (WBAN) as discussed by Khan et al. (2013), Ganti et al. (2006) and Zhou et al. (2007), which is shown in Figure 3.1. Communication within tier 1 devices that is, intra-BAN communication can be done using Bluetooth (IEEE 802.15.1), ZigBee (IEEE 802.15.4) or WiMedia (IEEE 802.15.3). For communicating with implantable sensor nodes IEEE 15.6 standard can be used. Generally, star topology is used where all BAN sensors are connected to a BAN coordinator. However, multi-hop communication (upto 2 hops) with relay nodes is also supported by IEEE 802.15 standard. Multi-hop communication is more energy efficient as the nodes do not need to transmit with the maximum transmission power. In CodeBlue project by Shnayder et al. (2005), a publish/subscribe routing framework is proposed based on Adaptive Demand-driven Multicast Routing (ADMR) protocol for communication between sensors in tier 1 and display devices for doctors and/or nurses in tier 2 or tier 3.

Communication between tier 1 and tier 2 that is, inter-BAN communication happens through the BAN Coordinator (BNC) using Bluetooth or WLAN (IEEE 802.11), cellular, 3G/4G etc. as mentioned by Rashidi and Mihailidis (2013). The authors also predicted that technologies like "Bluetooth Low Energy" (BLE) can be used here to reduce communication overhead. Inter-BAN communication can take place in infrastructure mode or in ad-hoc mode, as is supported by WLAN. This communication could be point-to-point (p-p) or point-to-multipoint (p-mp) as mentioned by Khan et al. (2013). In p-p mode, the BNC sends data for a single destination while in p-mp mode, the BNC sends data packets for multiple destinations. The p-mp mode of communication can be particularly useful in hospital environments where the doctors from their respective office(s) may need the data from BNC. Mobility should also be considered in inter-BAN communication. CodeBlue handles mobility using controlled flooding. This approach, despite exploiting the robustness of flooding, is a bit expensive in terms of energy. To save energy, data is opportunistically sent upon detection of user activity in SATIRE project proposed by Ganti et al. (2006). Mobility is not effectively handled in many proposed AAL architectures, like the one by Aquino-Santos et al. (2013). Although it is mentioned by the authors that PANDORA protocol proposed by Santos et al. (2009) could be effectively utilized to handle the effects of mobility in inter BAN communication. Due to mobility, energy consumption also increases as the nodes try to re-route packets. Mobility should be carefully handled in order to save energy like by treating some nodes as forwarders. Moreover, both in p-p and p-mp it is important ensure that data is accessed by authorized users only.

The communication between tier 2 and tier 3 is often through Internet. Most of the existing works assume a UDP traffic on standardized IP or IPv6 based communication infrastructure for this part, as mentioned by Wu et al. (2015).

3.2.1.2 System Architecture

Designing system architecture is one of the main investigated issues of AAL according to Memon et al. (2014). These systems take inputs from not only BANs but also from other devices like cameras and thermographic devices. These inputs are collected through wireless medium and analyzed for activity recognition, context modeling, anomaly detection, location and identification according to Rashidi and Mihailidis (2013). This can be extended to incorporate actuators for generating alerts (Lloret et al., 2015) as well. Typical system architecture is shown in Figure 3.2. It is a three tier architecture where the first layer (input subsystem) and third layer (patient status system) have separate physical existence. The Service system is implemented in software. Here samples collected from input subsystem are included in a corpus depending on the application for which it is used. For example, as shown by Lloret et al. (2015), samples collected for identifying pain using face image recognition are different from the ones used to identify falls. Semi-supervised or supervised learning is used on these data. For supervised learning, all corpus samples should be labeled by caregiver or doctors. Then, data gathered are pre-processed (or normalized if needed) to filter out unnecessary components. Finally these data are fed for training and/or feature extraction and feature selection. Finally decision algorithms may be used to detect or predict user behavior. This process has to be iterated to guarantee a certain level of accuracy. The accuracy of the decision algorithms should be assessed by doctor and/or caregiver and fed back to training algorithms for learning according to Lloret et al. (2015). In deployment phase, the output from service system (that of the decision

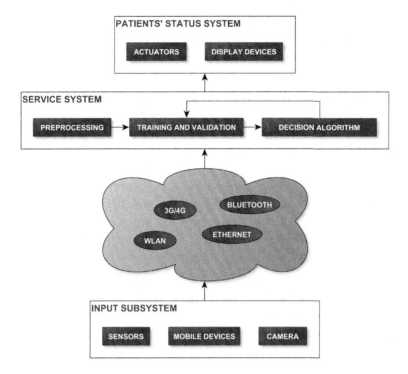

Figure 3.2 Typical system architecture of AAL.

algorithm) may be fed to actuators occasionally to further generate alarms or movement of the devices etc.

For the successful operation of this architecture, it has to be energy efficient. To ensure energy efficiency the following issues need to be considered;

- all collected data should be aggregated at the BNCs for BANs and/or at the sink for sensor networks
- transmitted data from each device has a large proportion of replication with itself
- data should be properly prioritized
- motion prediction algorithms based on data from accelerometer, gyro etc. could be effectively utilized to further reduce energy consumption.

Aquino-Santos et al. (2013) proposed a novel system architecture for detecting arrhythmia. Three nodes are used as coordinators to communicate to a static central coordinator. Fixed routing strategy is used here. ACK based MAC protocol as supported by IEEE 802.11.4 is used. The architecture is simulated in TOSSIM and implemented using TelOSB and Shimmer2R. Star topology is assumed for intra–BAN communication and the central coordinator is assigned a fixed position. QoS parameters are also considered in this work. The local coordinators are assigned tasks like routing and synchronization between sub-networks, distributing QoS indices of nodes etc. However, movement of the nodes relative to the coordinator is not handled here. The problem of placing the local coordinators in such a way to cover the nodes in the monitored area is also an issue to be dealt with.

Lloret et al. (2015) proposed a smart system architecture for AAL applications that can decide on specific user action and send alarms. The architecture is smart as it not only uses sensor data but also takes into account human behavior for decision making. It uses supervised learning with tagging of data done by doctor and/or care-giver. It works in p-p mode only, thus saving energy. Remaining energy of the devices seem to affect the performance of the training algorithms as it can control user movement too.

Wu et al. (2015) proposed a system architecture comprising of service system and communication platform. The work presented a summary of challenges in AAL and proposed an asynchronous flow scheduling scheme that takes care of QoS requirements of data. The need for proper authorization is also emphasized.

Amoretti et al. (2013) proposed a functional architecture for activity monitoring that is non-invasive of the user's life. An activity monitoring sub system named PERSONA is designed that exploits contextual information of the end users like location and posture of the user along with environmental data (state of the devices surrounding the users). These two information are combined for predicting activity.

Software agents can also be effectively utilized in AAL system architecture to make it highly reconfigurable as mentioned by Ayala et al. (2013). The authors proposed a system architecture based on MalacaTiny agent architecture. Here agents can be executed in different devices providing autonomic management tasks.

3.2.2 Energy Efficient Protocols

Based on network architecture described in Section 3.2.1.1, few protocols are proposed in AAL environment for activity recognition (Ugolotti et al., 2013) and routing. Most of these protocols are centralized and only a few protocols address the challenges imposed by mobility. Energy efficiency with respect to static nodes are handled by these protocols.

Activity recognition using four calibrated cameras and a body-mounted wireless accelerometer is discussed in works by Ugolotti et al. (2013). Several instances of a hybrid classifier based on Support Vector Machines and Hierarchical Temporal Memories, and bio-inspired computational paradigm, are used to classify movements (like walking, jumping etc.) mainly aiming at detecting falls.

Though there are many routing protocols for BAN (Bangash et al., 2014) that aims to find a route from any node in the BAN to the BAN coordinator, only a few protocols include communication from BNC to a tier 2 device. In Quwaider et al. (2011) a stochastic modeling framework for store-and-forward packet routing is proposed with postural partitioning. Delay modeling techniques for evaluating single-copy on-body Delay Tolerant Network (DTN) routing protocols are then developed. End-to-end routing delay for a series of protocols including opportunistic, randomized, and two other mechanisms that capture multiscale topological localities in human postural movements have been evaluated. Finally, a mechanism for evaluating the topological importance of individual on-body sensor nodes is developed. It is shown that information can be used for selectively reducing the on-body sensor-count without substantially sacrificing the packet delivery delay.

In Movassaghi et al. (2012), authors propose a routing protocol for communication between any sensor node to the BAN coordinator. *Hello* message is broadcast in the network. After receiving *Hello* packet from all neighbors, a node decides its minimum cost neighbor depending on temperature, received power and remaining energy. Otherwise the packet is stored at the node itself. Packets are given a maximum hop count assuming that the packets will be reaching the destination within the predefined threshold value. However, minimum cost neighbor may not always be the one nearest to the sink (BNC in this context). So it does not ensure shortest routing path. Multihop routing protocol is proposed in Ortiz et al. (2012) where nodes nearer to the sink join the sink first, then broadcast *Hello* packets. This ensures that a parent node is always closer to the sink than itself. This protocol allows a node to change its parent during deployment to be able to send data to the sink even when a parent runs out of battery. However, significant energy is consumed in communicating many control packets. Setup time can also be longer. In Lu and Wong (2007) a multipath routing protocol is proposed where a node gets a level ring number during topology formation. In the setup phase, the sink broadcasts topology setup packet. Nodes receiving this packet are assigned ring number 0 and they rebroadcast it. Nodes getting this packet are assigned ring number 1 and the process continues until all nodes are connected at some level. However, for small networks, multihop communication particularly more than 2 hop is found to incur significant overhead in terms of both time and energy. Secondly, many broadcast messages may overload the network. In Razzaque et al. (2011), a routing protocol is proposed that routes packets having different QoS requirements to BAN coordinators. Here also *Hello* packets are broadcast periodically. After receiving these packets, the routing table of the nodes are updated. However broadcasting Hello packets consume significant energy. The communication between tier 1 and tier 2 devices is covered in Khan et al. (2013) that assumes location of the nodes to be known. Then, depending on distance and type of device, communication cost is calculated. Packets are sent to the least cost neighbor. *Hello* packets are also used to update the routing tables. *Hello* packet is again an overhead here. Secondly, two data structures, neighbor table and routing table are needed to be maintained.

3.3 PROBLEM DESCRIPTION

We focus on the problem of finding out an effective route from BNC to any device in tier 2 (network architecture shown in Figure 3.1) connected by WLAN. Consequently, a mechanism for energy efficient routing is designed here that aims at optimizing energy efficiency of the BAN coordinators subject to user mobility and multicast communication. The communication architecture of WBAN is assumed to be based on star topology, where any node in the BAN may send data to its BNC directly, using multi-path routing (MPR) as described by Lu and Wong (2007).

Figure 3.3 shows the routing framework considered here. The main problem is how to find nearest point of attachment so that BNC can effectively communicate data collected from BAN sensors to the devices in tier 2. WiFi based indoor localization techniques could be employed to get the nearest point of attachment of a BAN coordinator thus making the architecture fully distributed. But indoor localization needs considerable effort in fingerprinting each reference point beforehand as mentioned by Farid et al. (2013). However, a BAN coordinator may directly or indirectly connect to a display unit (in tier 2) by using other nearby BAN coordinators

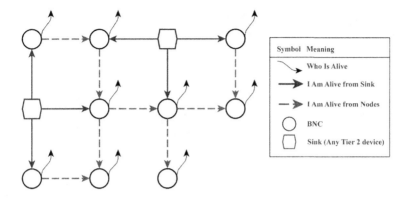

Figure 3.3 Routing framework for inter BAN communication in AAL.

as forwarder. This would save considerable energy as multi-hop communication is mostly energy efficient than single-hop. For example, in a hospital environment, patients may move along the corridor or from one ward to the other where all the wards may not be covered by the WLAN providing similar bandwidth. Thus it might happen that a patient at some point moves away from his nearest point of attachment such that the BNC may not get good RSSI from its nearest tier 2 device. In such a case, a nearby BNC of another patient may help to communicate the data by acting as a forwarder. Patient mobility can be handled in this way.

However to save energy, a BAN coordinator may behave selfishly and not act as a forwarder. The decision of forwarding or not forwarding depends on factors like:

1. amount of remaining energy of the BAN coordinator
2. current data rate
3. priority of data to be forwarded

The BAN coordinators may transmit using the power level that is just sufficient to connect to a nearby device. If it does not find any, it will search for one by increasing its transmission power level. Thus transmission power control is employed to further improve energy efficiency.

To address the issues mentioned above, a routing algorithm for inter BAN communication, *RouteToSink()* is proposed here (Algorithm 3.1). This algorithm results in a topology where a BNC is maximum 2-hop away from its nearest tier 2 device while any sensor node in BAN is maximum 3 hop away from the nearest tier 2 point of attachment. Possible collision of frames is assumed to be handled by the MAC protocol in place like IEEE 802.11 standard. It is assumed that the data is sent from BNCs to the sink.

(a)   wholsAlive Packet Format

Source$_{id}$	ResidualEnergy

(b)   iAmAlive Packet Format

Dest$_{id}$	Source$_{id}$	ResidualEnergy	edge(this,sink)

Figure 3.4 Routing Packet Format used in *RouteToSink*() (Algorithm 3.1): (a) Packet format of wholsAlive; (b) Packet format of iAmAlive.

RoutingTable

edge(this,sink)	Forwarder$_{id}$	ResidualEnergy$_{forwarder}$	TrPower	Timestamp

Figure 3.5 Routing Table Format used in *RouteToSink*() (Algorithm 3.1).

It is a proactive routing protocol where the BNCs periodically broadcast *whoIsAlive* packet in the network. A typical structure of the packet is shown in Figure 3.4(a). Since the packet is broadcast, destination id is not needed. Remaining energy is mentioned so that the receivers can act promptly if the *RemainingEnergy* is below a threshold. This packet is initially broadcast with maximum power level but the power is adjusted to the current power level. After sending the packet, the BNC switches to the receiving mode and accepts any *iAmAlive* packet for the next x seconds. Here the value of x depends on network bandwidth. If sink (a tier 2 device) receives the packet, it responds with *iAmAlive* packet. The structure of *iAmAlive* packet is shown in Figure 3.4(b). The destination id of this packet is the source id of its corresponding *whoIsAlive* packet. For the sink, the other fields need not contain valid data as these would not be processed by the receiver of the packet. Upon receiving *iAmAlive* from sink, a node makes an entry in its *RoutingTable* as shown in Figure 3.5. *edge(this,sink)* will be 1 as there is an edge from this node to sink. *Forwarder$_{id}$* and *ResidualEnergy* will not be needed. *TrPower* indicates the current transmission power with which it sent the *whoIsAlive* packet. The *Timestamp* field denotes when this entry was added so that the node can predict the link existence probability when it sends data. Thus this node is designated as a direct neighbor of sink.

However, if a node receives *whoIsAlive* packet, it checks its *RoutingTable* for an entry about sink. If there is an edge to sink, the node will respond to the sender with an *iAmAlive* packet. Here the two fields, *ResidualEnergy* and *edge(this,sink)* are significant, and *edge(this,sink)* = *1*. Residual energy of the device is also sent to indicate how long it can act as a rational forwarder. If the energy falls below a certain threshold ($REnergy_{th}$), the device may act

Algorithm 3.1 Algorithm for calculating route by BNC.

1: **procedure** ROUTETOSINK(
)
2:     **if** t%timer == 0 **then**
3:         broadcast *whoIsAlive* packet
4:         receive packets for x seconds
5:         **if** no packets received **then**
6:             **if** $TrPower == TrPower_{max}$ **then**
7:                 wait for the next period
8:             **else**
9:                 increase *TrPower* to the next higher level
10:             **end if**
11:         **end if**
12:     **end if**
13:     **if** *iAmAlive* packet is received from sink **then**
14:         update routing table
15:         Assign edge(this,sink) = 1
16:     **else if** *iAmAlive* packet is received from another BNC **then**
17:         update routing table
18:     **end if**
19:     **if** *data* packet is received from a neighbor **then**
20:         forward data to sink
21:     **end if**
22:     **if** *whoIsAlive* packet is received **then**
23:         **if** $REnergy > REnergy_{th}$ AND $edge(this, sink) == 1$ **then**
24:             reply with *iAmAlive*
25:         **end if**
26:     **end if**
27:     **if** *data* packet is ready to be sent **then**
28:         **if** $edge(this, sink) == 1$ in routing table **then**
29:             Send data to sink
30:         **else if** $Forwarder_{id} \neq null$ AND $aging(REnergy) > REnergy_{th}$ **then**
31:             Send data to $Forwarder_{id}$
32:         **end if**
33:     **end if**
34:     **if** RSSI value from sink $> RSSI_{hi}$ **then**
35:         reduce *TrPower* to the next lower level
36:     **else if** highest RSSI value in the routing table $< RSSI_{lo}$ **then**
37:         increase *TrPower* to the next higher level
38:     **end if**
39: **end procedure**

selfishly to save energy. A *Timestamp* is associated with each entry so that *ResidualEnergy* at current time can be predicted from the routing table entry by applying an aging function. Any node, after receiving *iAmAlive* packet, checks if it is from sink or not. If it is not from sink then *Forwarder_{id}* is kept as the sender of *iAmAlive* packet. Periodically, RSSI values are adjusted by tuning to appropriate *trPower*. The working of the protocol is depicted in Figure 3.3 with 9 BNCs and 2 sinks. As shown, whenever a node receives *iAmAlive* from a sink, it updates its routing table and sends data directly to the sink. Otherwise, data is forwarded to a forwarder having maximum residual energy according to step 30 in Algorithm 3.1.

3.4 EXPERIMENTAL RESULTS

The proposed protocol is simulated and compared with the state-of-the-art methods. The simulation setup is discussed first followed by the discussion of the results obtained through the simulation experiments.

3.4.1 Simulation Setup

The protocol is simulated in Castalia, a simulator for Wireless Networks, especially networks with low power wireless devices. It is based on OM-NeT++ simulator (Boulis, 2015; Varga and Hornig, 2008). It is considered one of the most realistic simulators for wireless networks and in particular wireless sensor and body area networks. This reputation is based on the fact that the simulator incorporates detailed models for wireless channels, fading, mobility and path loss. Further, it has a number of built-in protocols for fast prototyping. Here, the programmers can get information about and can also tune parameters like bandwidth, RSSI, Tx output power level, consumed power, modulation etc. The transceiver specifications at the nodes are taken from CC2420 data sheet. Here 7 different transmission power levels of the nodes are assumed from -25 dBm to -1 dBm. The simulation parameters are given in Table 3.1. The nodes are assumed to move with some speed according to line mobility manager supported in Castalia. However, the sink node is fixed and does not have mobility as a tier 2 device is generally assumed to be static. Three threshold values are used in the simulation of the algorithm RouteToSink() summarized in Algorithm 3.1. Two of these threshold values are temporal.

- The time period of sending event of whoIsAlive packet ($timeout_1$)
- The timeout period after getting whoIsAlive packet and before sending iAmAlive ($timeout_2$) (according to step 3)

Table 3.1 Simulation parameter

Parameter	Default value
Mobilty model	LineMobility
Time	50 sec (results are averaged after 5 rotations)
No. of mobile nodes (N)	15
Deployment area	30 m × 30 m
Deployment pattern	3 × 5 grid
Packet size	32 bytes
Sink node	Node 0
Initial node energy	18720 J (= 2 AA batteries)
Traffic type	CBR (Constant Bit Rate)
netSetupTimeout	5000 ms

Both of these threshold values are taken in comparison to the parameter netSetupTimeout(ms) in Castalia, that determines the period between sending data packets ($timeout_3$). The parameters are taken in a way such that $timeout_3 < timeout_2 < timeout_1$. All of these parameters are factors of netSetupTimeout parameter.

The other threshold used in the simulation is RSSI that is used to monitor signal quality. The RSSI value in the range of −50 dBm to −80 dBm is considered for the simulation experiments. This is the range that we normally get for smart devices when connected to WLAN. When the values are higher, the Tx output power level is reduced to save power according to step 35, and vice versa.

3.4.2 Simulation Results and Discussion

The metric used for simulation results include Consumed Power at the nodes, Ratio of Time Spent at Tx Power level, Number of Received Packets at network layer and at application layer.

Figure 3.6 shows variation of received packets by the nodes subject to varying data rate for $netSetupTimeout = 2000$ ms. At the network layer the sink (node 0) ultimately receives all the packets. However due to multihop communication ability, a distant node may use one of its neighbors as forwarder if it is connected to sink in one hop. Thus in this protocol any node can be maximum 2-hop away from the sink. So the graph effectively shows the nature of forwarding traffic subject to increasing data rate. As more packets are generated, the chances of sending forwarded packets decreases as a node has more packets to send. However, when the traffic is moder-

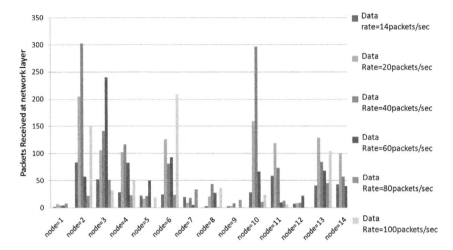

Figure 3.6 Variation of data packets received by the nodes at the network layer for different data rates.

ate like 20–80 packets per second, the forwarding traffic can be effectively handled by the network.

Performance of the application layer is shown in Figure 3.7 in context of netSetupTimeout interval for two data different data rates. Here, the performance of our proposed protocol (marked as ALIVING in Figure 3.7) is also compared with Multipath Rings routing protocol and the routing protocol described in Khan et al. (2013). It can be seen from the results that the proposed protocol offers improved results when data requirements increase. For cases with lower data rates, the performance is observed as comparable with the other two protocols.

Energy efficiency of the protocol is shown in terms of consumed energy in mJ and ratio of time spent by the nodes at a given Tx power level. The variation of consumed energy of the nodes is shown in Figure 3.8. Performance is compared with the protocol in Khan et al. (2013), as this protocol also considers AAL environment beyond a single BAN. It can be observed that the consumed energy of the nodes across the network is comparable. The improvement is even more noticeable when it is taken into account that more packets are being received by our protocol, as shown in Figure 3.7.

Energy efficiency of the nodes is measured in terms of:

$$\text{Time Spent at Tx Level} = 100 * \frac{\text{total time spent at Tx level}}{\text{Total simulation time}} \tag{3.1}$$

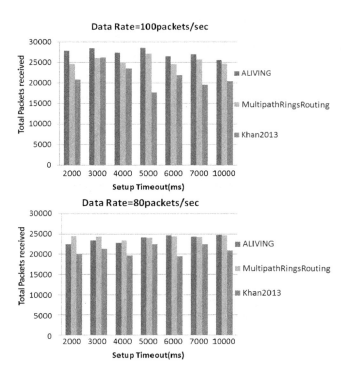

Figure 3.7 Comparison of ALIVING protocol with Multipath Rings routing and Khan et al. (2013) in terms of total packets received at the application layer, for different data rates. The performance subject to varying netSetupTimeout interval is shown.

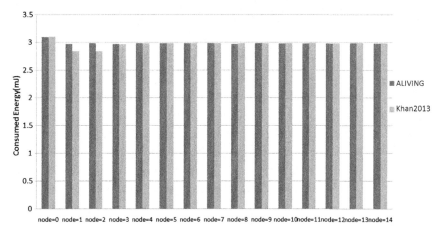

Figure 3.8 Comparison of ALIVING protocol with Khan et al. (2013) in terms of energy consumed by the nodes for total simulation time.

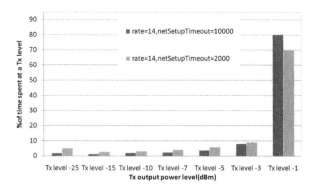

Figure 3.9 Energy efficiency of the network in terms of ratio of time spent at a Tx power level subject to varying netSetupTimeout interval.

Energy efficiency of the network is measured in terms of average of Time Spent at Tx Level for each Tx level by the nodes in the network. The energy efficiency of the network is shown in Figure 3.9. It shows that the network spends most of the time at the highest power level (−1 dBm) used in the simulation. However, whenever the channel quality improves (in terms of RSSI) the nodes change to lower power levels thus saving energy. It can also be observed that we get better energy efficiency when the value of netSetupTimeout is small. This is because the network can effectively handle frequent topology changes by frequently sending setup packets. The corresponding energy efficiency of the nodes is shown in Figure 3.10. These results strongly supports the observations made in Figure 3.9, which also indicates almost equal distribution of consumed energy across the nodes supporting results in Figure 3.8.

The median of the ratio defined in Equation (3.1) at Tx level −1 dBm is plotted in Figure 3.11 for varying data rate and netSetupTimeout interval. The figure shows that the performance is better when the data rate is low as the network spends least time sending at −1 dBm. However, higher data rate calls for more packets being sent at −1 dBm, as the network cannot get enough time to handle topology changes, and sends packets at the highest energy level according to step 9 of RouteToSink() algorithm summarized in Algorithm 3.1. The netSetupTimeout seems to have little influence on the energy efficiency of the network, when the data rate is high. However, for lower data rate, the effect of varying netSetupTimeout is more visible.

The percentage of energy consumed by a node at each power level is plotted in Figure 3.12. The figure is consistent with Figure 3.11 and

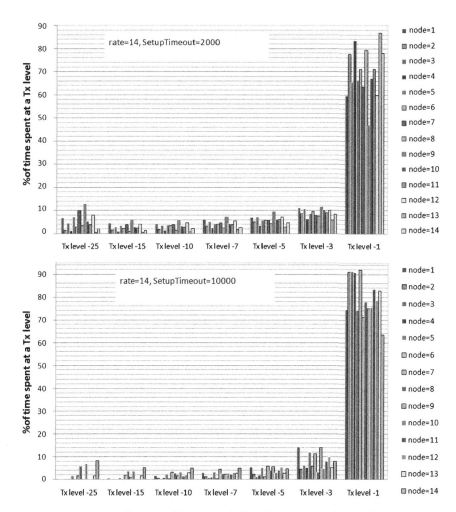

Figure 3.10 Energy efficiency of the individual nodes in terms of ratio of time spent at a Tx power level subject to varying netSetupTimeout interval.

Figure 3.10. The nodes spend maximum power when at the highest power level, however, at −10 dBm, there is a little peak as the nodes start with this power and switches only when they do not receive any *iAmAlive* message for some time. For clarity, performance level of some of the nodes is not shown here.

The next experiment is done with completely different setup where among 15 nodes, more than one node may act as sink. The result is shown in Figure 3.13 and Figure 3.14. It is observed that the forwarding traffic

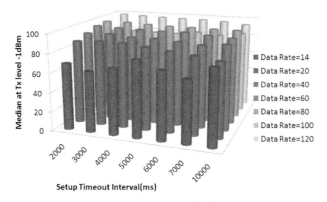

Figure 3.11 Energy efficiency of the network for varying netSetupTimeout interval subject to different data rates.

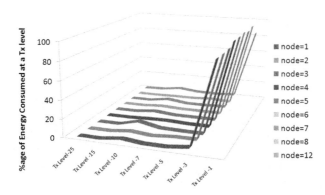

Figure 3.12 Energy efficiency of the network with respect to ratio of consumed energy at each power level.

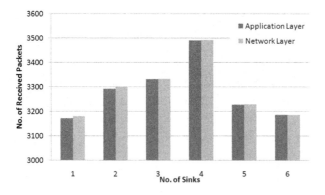

Figure 3.13 Variation of received packets with varying number of sinks.

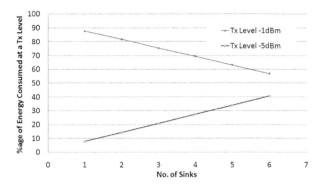

Figure 3.14 Variation of energy consumed at different Tx level with varying number of sinks.

decreases with more sinks as nodes get direct path to send their packets (Figure 3.13). The number of packets received increases initially with more sinks, however, if more nodes out of 15 become tier 2 device, the data packet generated would decrease as well. Moreover mobility does not allow to maintain a consistent topology for long. As a result packet reception rate decreases after an initial increase.

The performance in terms of energy consumption is shown in Figure 3.14. With increased number of sinks, the nodes in the network tends to stay at lower power levels for longer duration. However due to mobility the topology frequently changes resulting in the need of raising Tx level to transmit data. However the overall performance is better with more tier 2 devices in vicinity.

3.5 SUMMARY

This chapter explored important challenges concerning energy efficient communication in a multi-tier AAL systems. We presented detailed discussion on requirements for energy efficiency focusing on popular architectures and communication protocols was also presented. We also discussed recent research outputs highlighting design and development of a multi-tier communication protocol. A 3-tier architecture was discussed focusing on inter and intra–BAN communication in first two tiers. Since the BAN communication (both intra- and inter-BAN) constitutes a fundamental building block in realizing an AAL systems, it is imperative that the communication strategies are reliable and energy efficient. We also presented a new protocol aimed at optimizing the energy efficiency in inter-BAN communications. The developed protocol optimizes energy consumption between multiple BAN coordinators by exchanging residual energy information and adaptively changing the transmit power to suit the current transmission. The protocol was implemented in OMNeT++ Castalia simulator and thoroughly evaluated against diverse performance metrics. The simulation studies showed promising performance with respect to energy efficiency and other network related performance metrics, to justify its viability.

ACKNOWLEDGEMENT

The authors would like to acknowledge that this work was supported in part by the COST Action IC1303 AAPELE project.

REFERENCES

Amoretti, M., Copelli, S., Wientapper, F., Furfari, F., Lenzi, S., Chessa, S., 2013. Sensor data fusion for activity monitoring in the persona ambient assisted living project. J. Ambient Intell. Humaniz. Comput., 67–84.

Aquino-Santos, R., Martinez-Castro, D., Edwards-Block, A., Murillo-Piedrahita, A., 2013. Wireless sensor networks for ambient assisted living. Sensors 13 (12), 16384–16405.

Ayala, I., Amor, M., Fuentes, L., 2013. Self-configuring agents for ambient assisted living applications. Pers. Ubiquitous Comput. 17 (6), 1159–1169.

Bangash, J.I., Abdullah, A.H., Anisi, M.H., Khan, A.W., 2014. A survey of routing protocols in wireless body sensor networks. Sensors 14 (1), 1322–1357.

Boulis, A., 2015. Castalia user manual. https://castalia.forge.nicta.com.au/index.php/en/index.html. Last accessed on October 2015.

Farid, Z., Nordin, R., Ismail, M., 2013. Recent advances in wireless indoor localization techniques and system. J. Comput. Netw. Commun. 2013, 1–12.

Gambi, E., Perla, D., Spinsante, S., Pelliccioni, G., Del Campo, A., Montanini, L., De Santis, A., 2015. Unobtrusive monitoring of physical activity in AAL: a simple wearable device designed for older adults. In: Proc. ICT4 Ageing Well 2015 International Conference, vol. 3. Multidisciplinary Digital Publishing Institute, pp. 200–205.

Ganti, R., Jayachandran, P., Abdelzaher, T., Stankovic, J., 2006. A software architecture for smart attire. In: Proc. of the 4th Intl. Conf. on Mobile Systems and Applications Services, pp. 19–22.

Khan, Z., Sivakumar, S., Phillips, W., Aslam, N., 2013. A new patient monitoring framework and energy-aware peering routing protocol (EPR) for body area network communication. J. Ambient Intell. Humaniz. Comput., 1–15.

Kim, D.-Y., Cho, J., 2009. WBAN meets WBAN: smart mobile space over wireless body area networks. In: IEEE 70th Vehicular Technology Conference Fall. VTC 2009-Fall, pp. 1–5.

Lloret, J., Canovas, A., Sendra, S., Parra, L., 2015. A smart communication architecture for ambient assisted living. IEEE Commun. Mag. 53 (1), 26–33.

Lu, Y., Wong, V., 2007. An energy-efficient multipath routing protocol for wireless sensor networks. Int. J. Commun. Syst. 20, 746–766.

Memon, M., Wagner, S., Pedersen, C., Beevi, F., Hansen, F.O., 2014. Ambient assisted living healthcare frameworks, platforms, standards, and quality attributes. Sensors 14 (3), 4312–4341.

Movassaghi, S., Abolhasan, M., Lipman, J., 2012. Energy efficient thermal and power aware (ETPA) routing in body area networks. In: 2012 IEEE 23rd International Symposium on Personal Indoor and Mobile Radio Communications (PIMRC). IEEE, pp. 1108–1113.

Ortiz, A.M., Ababneh, N., Timmons, N., Morrison, J., 2012. Adaptive routing for multihop IEEE 802.15.6 wireless body area networks. In: 2012 20th International Conference on Software, Telecommunications and Computer Networks (SoftCOM). IEEE, pp. 1–5.

Porambage, P., Braeken, A., Gurtov, A., Ylianttila, M., Spinsante, S., 2015. Secure end-to-end communication for constrained devices in IoT-enabled ambient assisted living systems. In: IEEE 2nd World Forum on Internet of Things (WF-IoT), pp. 711–714.

Prescot, R., 2013. M-health, 2.8 million patients remotely monitored worldwide in 2012. http://www.rcrwireless.com/20130116/wireless/2-8-m-patients-remotely-monitored-worldwide-2012#prettyPhoto.

Quwaider, M., Taghizadeh, M., Biswas, S., 2011. Modeling on-body DTN packet routing delay in the presence of postural disconnections. EURASIP J. Wirel. Commun. Netw. 2011, 3.

Rashidi, P., Mihailidis, A., 2013. A survey on ambient-assisted living tools for older adults. IEEE J. Biomed. Health Inform. 17 (3), 579–590.

Razzaque, M., Hong, C., Lee, S., 2011. Data-centric multiobjective QoS-aware routing protocol for body sensor networks. Sensors 11 (1), 917–937.

Santos, R., Gonzalez-Potes, A., Garcia-Ruiz, M., Edwards-Block, A., Rangel-Licea, V., Villasenor-Gonzalez, L., 2009. Hybrid routing algorithm for emergency and rural wireless networks. Elektron. Elektrotech. 1 (89), 3–8.

Shnayder, V., Chen, B., Lorincz, K., Fulford-Jones, T., Welsh, M., 2005. Sensor networks for medical care. In: SenSys, vol. 5, p. 314.

Spinsante, S., Gambi, E., Montanini, L., Raffaeli, L., 2015. Data management in ambient assisted living platforms approaching IoT: a case study. In: IEEE Globecom Workshops (GC Wkshps), pp. 1–7.

Ugolotti, R., Sassi, F., Mordonini, M., Cagnoni, S., 2013. Multi-sensor system for detection and classification of human activities. J. Ambient Intell. Humaniz. Comput. 4 (1), 27–41.

United Nations Department of Economic and Social Affairs, 2013. World population ageing 2013. UN Report, ST/ESA/SER.A/348.

Varga, A., Hornig, R., 2008. An overview of the OMNeT++ simulation environment. In: Proceedings of the 1st International Conference on Simulation Tools and Techniques for Communications, Networks and Systems & Workshops. ICST (Institute for Computer Sciences, Social-Informatics and Telecommunications Engineering), p. 60.

Wang, K., Shao, Y., Shu, L., Han, G., Zhu, C., 2015. LDPA: a local data processing architecture in ambient assisted living communications. IEEE Commun. Mag. 53 (1), 56–63.

Wu, D., Cai, Y., Guizani, M., 2015. Asynchronous flow scheduling for green ambient assisted living communications. IEEE Commun. Mag., 64–70.

Zhou, B., Hu, C., Wang, H., Guo, R., 2007. A wireless sensor network for pervasive medical supervision. In: Proc. of the Intl. Conf. on Integration Technology, pp. 740–744.

CHAPTER 4

The Human Factor in the Design of Successful Ambient Assisted Living Technologies

**Susanna Spinsante*, Vera Stara†, Elisa Felici†, Laura Montanini*,
Laura Raffaeli*, Lorena Rossi†, Ennio Gambi***
**Dipartimento di Ingegneria dell'Informazione, Università Politecnica delle Marche,
Via Brecce Bianche 12, I-60131 Ancona, Italy*
*†INRCA, Lab. Bioinformatica, Bioingegneria e Domotica, Via Santa Margherita 5,
I-60124 Ancona, Italy*

4.1 INTRODUCTION

Ageing population is becoming a universal phenomenon. As society is getting older, all innovators and industrial players are called to develop new products, services and business models able to meet the need of the growing ageing population. The great challenges are an increasing efficiency of the global health and care systems, the improvement of the quality of life of older adults, and the creation of new opportunities for the so-called "silver market".

As a matter of fact, the market of Ambient Assisted Living (AAL) for active and healthy ageing is rapidly growing: it is clearly stated that Information and Communication Technologies (ICT) can make key contributions to an independent living of elderly people, to reduce high expenses for health and care services, to provide individual solutions, and hence to meet individual needs, to improve living standards and to open new business opportunities (Ambient Assisted Living Association, 2014). In order to address these challenges, but also to catch the opportunities described above, the European Commission has set up several initiatives to support research and innovation actions that explore and develop the ways in which ICT can meet the needs and maximize the potential of older people, thus helping Europe's growing elderly population to stay active for longer. ICT-based solutions for ageing well at home, in the community, and at work are emerging from various sectors, assembling a wide variety of products and solutions under the umbrella market of AAL (Ambient Assisted Living As-

Ambient Assisted Living and Enhanced Living Environments.
DOI: http://dx.doi.org/10.1016/B978-0-12-805195-5.00004-1

sociation, 2014). These include technologies that can be used by seniors, caregivers, health care providers and aging services providers (as first, secondary and tertiary end-users), to improve the quality of care, enhance the caregivers' experience, efficiency and cost-effectiveness. However, many barriers stand in the way leading successful innovation actions to meet market demand, especially when the target is an inhomogeneous group, as elderly people, that differs by age, sex, degree of impairment, biography, income, education, religion, culture and, especially, technology experience. The design and development of new devices and services for independent living cannot leave the human characteristics out of consideration, on the contrary they need to become aware of limitations that increase in prevalence, as one ages (Fisk et al., 2009).

This chapter is aimed at professional developers and designers, by introducing the aforementioned human centric approach as a key strategy for recognizing and removing the barriers that up to now have limited the diffusion of technological innovations among elderly people, and their extensive adoption in real-life. Moving from the definition and discussion of the human centric approach, the impact and role of user requirements in the design of AAL technologies will be discussed, referring to the practical experiences available in the literature, and the outcomes of selected relevant research projects. Such an analysis will allow to gather and identify the most relevant lessons learned, that will be delivered to the readers in the form of possible guidelines, for the true and effective implementation of technologies designed around the needs and expectations of elderly users.

4.2 THE HUMAN CENTRIC APPROACH

Population ageing is a long-term trend in Europe: the share of the people aged over 65 years is increasing in every EU Member State, with a 18.5% share in the EU-28 population estimated at 506.8 million, (EUROSTAT 2015). This trend requires immediate ways to cope with people' needs that are living more years, remaining more active and staying in their home longer than before. In the attempt to address age-related problems, the market of AAL solutions for active and healthy ageing is constantly growing and has become deeply relevant for the present and future challenges (Longhi et al., 2014) of a society with more and more consumers that are joining the ranks of "older adults".

However, a gap stands in the way to successful innovation: what is under development and what is really needed by older adults. This gap still appears related to an insufficient understanding or stereotyping of people aged 65 and over (Lee and Coughlin, 2015) and to a missed holistic view of the aging phenomena. Generally, this target of population experiences the same age-related changes in chronological biological, physiological and social dimensions but they do not represent a homogeneous group since the decay in all five senses occurs at different times and rates among persons. Therefore, gerontologists (Forman et al., 1992), approach the diversity of old age by using the chronological age as the best index to categorize sub-groups of older adults: the young old, ranging in age from 60 to 69; the middle old, comprising those individuals from 70 to 79; and the very old group that clusters people age 80 and over. The difficulty to clearly define older adults represent a great challenge in the development of ICT whereas developing successful artefacts is strictly related with an explicit deep understanding of users, their needs, expectations and limitations.

The objective of this paragraph is to introduce the human-centered perspective to professional developers and designers as a key strategy for recognizing and removing the barriers that up to now have limited the diffusion of technological innovations among older people, and their extensive adoption in real-life.

4.2.1 The User Driven Design

As people aged, it is common to experience changes in sensory modalities: loss of visual acuity and hearing, decline in working memories and attention as well as changes in touch sensitivity and movement control (Fisk et al., 2009). Many women and men are moreover affected by one or multiple chronic diseases, caring sometimes for their spouse or an older relative. At the same time such persons are involved in activities related to their daily routine and of course their own healthcare management.

An exemplifying scenario is the current life of the Smith family here described.

John Smith is a 73 year old man, married with Audrey, 70. In the last two years, John starts to suffer from high blood pressure. After the initial examinations performed in day hospital regimen, type II diabetes is diagnosed. He is very worried for his health status and he really does not know how to cope with the new circumstances of his life. Every day John must remind to take medicines, to control blood sugar levels and to take insulin. That appears very complicated and burdensome for him and very often he lost motivation to follow treatment. The main difficulty that her wife

Audrey must cope with is to maintain a constant motivation so she spends most of her time reassuring and providing emotional support to her husband. This workload has taken a psychological toll on Audrey, making her feel fatigued and overwhelmed by the stressful situation, presenting sleeping difficulty, loss of appetite, social isolation.

John and Audrey want to maintain their independence, but a huge sense of frailty and isolation affects their quality of life. On par, Thomas, their only child of 40 years, is always under pressure in caring for his family. Until now he spent a lot of time searching for suggestions on how to better assist John and Audrey but nothing seems able to calm his anxiety. His prevalent feeling is that he is never on the right track for preventing diseases and making live a healthy lifestyle to his parents.

In parallel, health care professionals (i.e. doctors, nurses, practitioners, nutritionists, pharmacist, psychologist, etc.), in charge of providing treatment and monitoring patients, are always in need to follow and track the health status of over one thousand persons like John and Audrey.

The situation described above shows the need for a more efficient and effective communication between the doctor, careers and the patient, and, at the same time, among health professionals that need to discuss together about determinants, comorbidities as well as therapy and treatment.

John and Audrey embody the so-called *primary end-user* who could use an AAL product or service to directly benefit from technology innovation, by increased quality of life. Thomas represents the *secondary end-users* group, as persons or organizations directly being in contact with a primary end-user (such as formal and informal care persons, family members, friends, neighbors, care organizations and their representatives). This group benefits from technology innovation directly when using products and services (at a primary end-user's home premise, or remotely), and indirectly when the care needs of primary end-users are reduced.

Healthcare professionals are the *tertiary end-users* as any other institutions, private or public organizations (i.e. public sector service organizers, social security systems, insurance companies), are not directly in contact with AAL innovations, but contribute in organizing, paying or enabling them.

AAL technologies offer to primary, secondary and tertiary end-users the potential to help in coping with their daily activities, by accounting for their specific needs.

The majority of elderly people prefer to stay and live independently in their own homes having control of whatever occurs within the domestic environment (safety), with the hope to remain autonomous as long as possible (independence), in a good functional status (health and wellness), being part of everyday life in their family, neighbor and community (social inclusion), free of moving to reach things or to train their body (mobility). Furthermore, caregivers, especially those involved in informal care, live in stressful contingences that negatively impacts the carer and their quality of life even with serious physical and mental health consequences (Carrettero et al., 2012). This strong impact has been conceptualized as a burden that is the first need to be addressed, followed by needs related to a recapture of leisure time, social lives, and family harmony.

It appears quite evident how these macro level needs are strictly connected to the AAL market demand, opening the way for the deployment of a needs-related technology that provides new ways for helping older citizens to live independently. Paradoxically, during developing and implementing technologies for older adults, little efforts are devoted to match these specific needs. As a matter of fact, the adoption rate of AAL products is very low, mostly due to a deep gap between what is under development and what is really needed and expected by users.

Realistically, the design of such technologies is not different from any other devices, but devices aimed at improving the health and quality of life of this target of population seem developed without a clear planning around human needs (Thielke et al., 2012).

To determine whether a design product can or cannot accommodate these human needs, designers must be aware of the target user characteristics, adopting a user driven approach. Commonly known as User Centered Design (UCD), this approach is defined in the ISO standard on Human-centered design for interactive systems (ISO 9241-210, 2010) as the iterative process of designing an item from the perspective of how it will be used and understood by users. This systematic procedure advocates strong adherence to a path during which:

1. a multidisciplinary and experienced team is engaged in the definition of the "user profile", tasks and the environment analysis,
2. an in-depth involvement of end-users is assured throughout design and iterative development process,
3. a user-centered evaluation strategy is planned to face, drive and refine the design addressing the whole user experience.

Each principle will be discussed in the following sections.

4.2.1.1 User Profile Definition

Adopting a human centric approach implies that a multidisciplinary and experienced team is engaged to go beyond the cited gap between the user profile and the designer profile. Different viewpoints can strengthen the design approach, arising awareness on constraints, realties, enablers and barriers and defining the user profile in their needs, expectations and limitations/barriers. Besides, geriatric medicine researchers are strongly recommended to be actively involved in ICT research projects to identify significant clinical outcomes, as well as to aim for cost-effectiveness of long-term care in different living settings (Lattanzio et al., 2014).

By answering some simple questions (Thielke et al., 2012), this experienced team can target technology and accomplish benefits for users (whether they are primary, secondary or tertiary users): which level of need does my device address? Will my device fully satisfy the users' needs? Are these needs addressed through my device? Not only a deep focus on needs but also on limitations and barriers contributes to characterize the user's model.

The main barriers in applying AAL technologies for seniors depend on psychological resistance of elderly users, especially related to a deep bond to their memories and their previous lifestyle, which causes an a-priori rejection of changing in behavior or habits. This condition creates the perspective that technology is an interfering, invasive and troubling complication (Lee and Coughlin, 2015; Van Den Broek et al., 2010).

In order to overcome such problems, it is strategic to spread awareness and knowledge among end-users about benefits and utility ensured by technology. Several studies (Thielke et al., 2012; Walsh and Callan, 2010; Arning and Ziefle, 2007), have effectively underlined that older adults make use of technology to reach and realize some desirable outcomes and refuse devices with an unclear evidence of real benefits. Furthermore, another limitation depends often on poor system usability and a lack of support throughout use: tools are often so complex, difficult to control, and error-inducing. Perception of friendliness, ease-of-use, clear instructional material and constant technical support are factors to face the aforementioned lack of familiarity with ICT devices, the cognitive differences, and the incoming age-related declines, thus avoiding the overwhelming features, options and information (Mitzner et al., 2010).

According to the literature review (Farage et al., 2012; Fisk et al., 2009; Morrell, 2001), in Table 4.1 some design guidelines are shown in order to accommodate the cited commonly limitations that often arise during

Table 4.1 Accommodation guidelines for age-related declines

Age-related functional changes	Accommodation guidelines
Vision decline	• High contrast (50:1) on-screen facilitates legibility.
	• Color choices in the long-wavelength end of the spectrum ("warm" colors) are preferred: if short-wavelengths ("cool" colors) are used, large contrast steps will be required to facilitate perceptibility.
	• Shades of any color used to convey information must be clearly distinct from the background.
	• Simplicity of visual presentation: no visual clutter.
	• Important information should be large, conspicuous, uncrowned, and in the central visual field.
	• Guarantee the possibility to adjust and enlarge text and graphics, as well as voice output software with clear enunciation, make computer use easier for the visually impaired.
	• Avoid distracting visual stimuli (e.g. elaborate backgrounds, flashing lights, rapid motion, or flickering).
Hearing loss	• Audio signals of at least 60 dB should reach the ear of older adult.
	• A high "signal-to-noise" ratio is necessary: the intended sound or message should be at a high enough volume with background noise kept to a minimum.
	• Volumes should be adjustable.
	• Sound frequencies in the range of 500–2000 Hz are preferable. High frequencies should be avoided in both verbal and non-verbal auditory information.
	• Alarm sounds or other auditory cues should not exceed 2000 Hz.

continued on next page

Table 4.1 (*continued*)

Age-related functional changes	Accommodation guidelines
	• An auditory signal can be reinforced by redundant cueing through another sensory channel (e.g. a telephone ring accompanied by vibration; a buzzer alarm accompanied by a flashing light).
	• If different sound cues must be used to convey information, these should be from different parts of the sound spectrum and be distinctly spaced in time. Altering the location from which sound is emitted can also help to distinguish various sound cues.
	• A slower pace of delivery aids the recognition and recall of verbal information, but the tone and pace of the delivery should be respectful and non-patronizing.
	• Use short sentences; pause slightly after each statement to facilitate comprehension by listeners who rely body language or other context cues to overcome hearing difficulties.
Motor skill diminishment	• To accommodate reduced touch perception and motor coordination, larger surfaces (e.g. for telephone dialing) are easier to actuate or depress than small ones; textured rather than smooth surfaces will increase the perception of features actuated by pressure.
	• Redundant sensory cues help signal when something has been actuated or correctly placed into position (e.g. a "click" or tone when a key is depressed; a "snap" when something locks into position).
Cognitive effects	• Visual information should be spare and uncluttered: highlight relevant information and minimize irrelevant and potentially distracting information.
	• Pictorial icons in particular should be tested with older audiences to ensure that they convey what is intended.

continued on next page

Table 4.1 (*continued*)

Age-related functional changes	Accommodation guidelines
	• The presentation format (whether textual, pictorial, or auditory) should be simple and intuitive to minimize the possibility of misinterpretation. • When presenting auditory information, simple instructions and brief messages are preferred. • A moderate pace should be employed; key information should be repeated; contextual cues should be incorporated to emphasize connections and prompt recall. • Redundant sensory cues will reinforce a correct action.

the normal ageing process. These guidelines will assist the process of user interface design, and will improve usability and acceptance of whatever AAL System.

Lastly, for a successful deploying technology among older adults, it is worth considering the possibility of stigmatization in adopting and using assistive devices. As every user, sharing information about their own health or personal data, especially if related with aging, frailty and dependence, can drive older adults away from the benefits of ICT solutions.

4.2.1.2 User Involvement and Evaluations Strategy

User involvement is a significant aspect in system development. Typically, user involvement relates to participation in activities dealing with specifying, elaborating, prioritizing, reviewing and verifying the requirements, as well as testing the developed features.

Benefits derived from this strategy are copious (Damodaran, 1996): improved quality of the system arising from more accurately identified user requirements; avoiding costly system features that the user did not want or cannot use; improved levels of acceptance of the system; greater understanding of the system by the user resulting in more effective use. Consequently, a direct contact with potential users, enables designers to learn from older people what functionalities and attributes are important to them in new products, what motivates them, and what factors would hinder the usability.

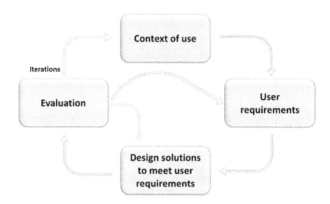

Figure 4.1 Interdependence of human-centered activities as specified in ISO 9241-210.

The UCD approach is a process consisting of four fundamental activities related to user involvement (Figure 4.1):

a) user groups are specified and the context of use is described (Activity 1: understand and specify the context of use);

b) a set of specific requirements are defined in order to create a degree of fit between device and user (Activity 2: specify the user requirements);

c) the design prototype is produced on the basis of these specifications and it is presented to the user in the form of user testing (Activity 3: Produce design solutions to meet requirement);

d) once feedbacks from the user have been received, the process begins again until all user requirements have been met (Activity 4: evaluation).

As for its iterative nature, the process requires that information are gathered from the user at each step and actions are taken based on that, in order to interpret the information correctly.

To this aim, the activities previously described, can be associated with a set of methods to be chosen and used on the basis of (Rekha Devi et al., 2012; Nedopil et al., 2013): type of users to be involved, context of use, nature of the product, constraints (such as time, effort, cost and access to users).

More than 200 methods coming from different areas of knowledge are used under the UCD framework, and it is beyond the scope of this chapter to cite all of them, but it is very useful to have in mind the common classification detailed in Table 4.2.

Among these, the common methods used to design for older adults are (Fisk et al., 2009):

Table 4.2 Classification of UCD methods

UCD phases	Output	Method
• Understand and specify the context of use	• Target definition • Context of use description	• Focus group • Survey • Questionnaires • Interview • Workshop • Brainstorming • Contextual inquiry • Delphi Method • Eye Tracking • Information Analysis • Personas • Storyboard • Thinking Aloud
• Specify the user requirements	• Context of use specification • User needs description • User requirement specification	• Scenarios • Personas • Storyboard • Task Analysis
• Produce design solutions to meet requirement	• User interaction specification • User interface specification • Implemented user interface	• Participatory design • Interview • Brainstorming • Brainwriting • Focus group • Thinking aloud • Video analysis
• Evaluate	• Evaluation results • Conformance test results • Long-term monitoring results	• Cognitive walkthrough • Contextual inquiry • Eye tracking • Focus group • Personas • Usability tests • Acceptance tests • Storyboards • Thinking aloud • Video analysis

Focus groups: are essentially discussion groups comprised of about six to twelve users. It could be very interesting and advantageous to get a group of older people together to discuss issues and requirements. These sessions can be structured with specific topics for discussion or not. A facilitator is required who can keep the discussion. This method is useful for gaining a consensus view and highlighting areas of conflict and/or disagreement.

Questionnaires: involves administering a set of writing questions to a sample population. Surveys can help determine the needs of users, and they are compose by a mix of closed and open questions. This method is useful for obtaining qualitative and quantitative data.

Interview: is a common used technique where users are questioned by an operator to gain information about their needs or requirements. Interviews are usually semi-structured based on a series of fixed questions with scope for the user to expand on their responses.

Task analysis: it is a study of what a user is required to do in terms of actions and/or cognitive processes to achieve a task. A detailed task analysis could be conducted to understand the current system and the information flows within it.

Usability testing: is a common technique where the users perform specific tasks using the product, in test environments, while one or more observers watch them. The key to interpret the results of testing is to look for general trends and behavior patterns that indicate problems with the usability.

Even though a common methodology for the evaluation of the new technologies is still lacking (Lattanzio et al., 2014), through a user driven approach it is possible to go beyond the long-standing gap between what is under development and what is really needed by target population.

In this paragraph it has been evidenced how user integration can increase market success if AAL solutions meet the target group's actual needs from the beginning. This evidence is recognized by the AAL Governance itself, that shared in 2013 a guideline (Nedopil et al., 2013) and a User Integration Toolbox to help AAL projects optimizing their user integration process.

4.3 INFORMATION AND COMMUNICATION TECHNOLOGIES IN AAL

Active and Assisted Living is an umbrella concept for a multitude of devices, services and information and communication technologies assisting citizens in their own homes, and supporting their mobility by means of innovative solutions. The presence of ICT in AAL is pervasive and manifold, ranging from the transmission of signals and data generated by sensors, to communication architectures for data delivery, to applications and services on top of them (Cubo et al., 2014; Alam and Hamida, 2014). This section briefly analyzes and discusses the different contributions provided by ICT in the field of AAL, in order to provide the reader with a handful of general reference models identifying the possible roles of ICT components and services within AAL frameworks.

AAL summarizes the use of ICT to assist people to age well at home, in the community and, if wanted, at work. On a broad level, it is possible to distinguish between:

- Assistive technologies, i.e. technologies designed to support and compensate for physical impairments, improving mobility and objects manipulation;
- Technologies for physical prevention, i.e. systems designed to evaluate the physical condition of the subject and support both physical and cognitive functionalities preservation through exercises;
- Technologies for rehabilitation, i.e. solutions to assess residual subject's capabilities and provide low-cost support to rehabilitation at home.

ICT provides the enabling building blocks for the above mentioned categories, from sensors, to reasoning components, to actuation and communication.

A sensor is composed of a fundamental transduction mechanism (sensing element), and an output interface, that includes a physical packaging, conditioning electronics, and external connections. When equipped with embedded processing capabilities and a communication interface, the sensor becomes "smart", and needs a power source to operate (Gervais-Ducouret, 2011; Guan et al., 2016). In AAL, it is possible to identify three basic types of sensors: wearable, implantable, and environmental sensors. Wearable sensors are nowadays the most common ones, based on Micro-Electro-Mechanical-Systems (MEMS) technology, and focused on location and movement (De Santis et al., 2014, 2015). Textile sensors, a specific subset of wearable ones, are getting more and more important (Trindade et al., 2015; Singh et al., 2016). Implantable sensors are basically used in the healthcare sector (e.g. pacemaker), and not yet that much spread in the broader AAL scenario, due to safety concerns they raise, but they are seen as the future evolution of sensing applied to human beings (Shasha Liu and Hung-Fat, 2013; Inman and Hodgins, 2013). Environmental sensors, including contactless ones such as depth cameras to estimate distances, are the most spread and used in the AAL scenario, being easily deployable and allowing the unobtrusive collection of a rich set of information, through approaches like sensor and data fusion (Cippitelli et al., 2015a, 2015b; Gasparrini et al., 2015).

Through reasoning, i.e. aggregation, processing and analysis operations, data are transformed into knowledge. Reasoning systems can be implemented on a dedicated device together with one or more sensors, on a mobile device the users carry in their pockets, on a home device, or on a server connected to a network. Reasoning algorithms are applied to detect and/or predict emergency situations (like falls, acute heart or blood-pressure problems in tele-health and tele-monitoring systems) (Fernandes Golino et al., 2014; Weiss et al., 2012) but also to classify, detect and predict new clusters of data, even in uncertain conditions (Schubert et al., 2015), to possibly derive information to set up a personalized health program. Sensors worn by the user or installed in the environment (e.g. in home appliances like refrigerator doors, infrared presence detectors, pressure sensors in bed and on chairs, light switches and the use of electrical devices, drawers of cupboards) (Blasco et al., 2014) generate data that need to be classified by proper algorithms, before being useful to get knowledge about the user's behavior and habits (Spinsante et al., 2015). Machine learning models and techniques allow automatic recognition of user's actions, based on signals generated by a set of different sensors, or by information-rich sensors like video and depth cameras (Rashidi and Mihailidis, 2013). Actuation includes systems and services, proactively acting for health care and monitoring, and for increasing the independent living of ageing or disable people in an assisted environment. Basically, actuation is identified with robotics. A currently very hot research field deals with assistive social robots (Broekens et al., 2009), designed for social interaction with humans, that could play an important role with respect to the health and psychological well-being of the elderly. In the next future, it is expected to develop network systems able to manage various types of robots and other agents (Samer et al., 2015). Future robots will rely on cloud computing infrastructures to access vast amount of processing power and data, thus improving their skills to better accomplish expected tasks. This will give place to so-called cloud robotics. Having more and more objects, sensors, and robots connected to the cloud and able to automatically exchange data and information (in a machine-to-machine approach), opens a number of issues related to security, i.e. ensuring that data are accessed and used only by authorized entities, that they are properly anonymized in order to avoid external intrusion into users' privacy, and that they cannot be forged or modified to make assistive services unavailable or vulnerable to security threats and violations (Memon et al., 2014). For those systems or sub-systems of an AAL platform where actions can be initiated over the network, e.g. to switch on or off certain devices in the home, or lock/unlock doors or windows, protection from unauthorized access and use is of prime importance, since malicious use can create significant damages (Porambage et al., 2015).

One of the most important aspects for design, development and market acceptance of AAL technologies and products is Human-Machine Interaction (HMI). Successful HMI requires appropriate interface technologies to satisfy specific requirements related to the abilities of users (Spinsante and Gambi, 2015). In particular, interaction with AAL systems should be as much "natural" as possible, requiring low or no effort from the user and ensuring the most effective usability (Spinsante et al., 2012; Spinsante and Gambi, 2012a, 2012b). Good examples of effective HMI devices are touch screens, successfully developed for everyone's use for gesture input on Smartphones, tablet PCs, or stationary displays. The touch screen-based interaction is very simple, it does not require an advanced mental model, and features a direct approach: the input device is also the output device. Typically, touchscreen interfaces do not require special motor skills, but on the other hand a careful and accurate design of the application is requested. Among the several devices supporting user driven interaction, the choice of the smart TV is motivated by the usually limited familiarity of elderly with personal computers or portable devices, the need for simplified and intuitive controllers, and for technology that overcomes visual and hearing impairments (Spinsante and Gambi, 2012a, 2012b). The TV is one of the most common and familiar home appliances, so the learning process can be taken up positively by the user (Spinsante et al., 2015). For smart TV equipment supporting voice recognition, it is even possible to define a set of custom commands and associate them to the corresponding browsing or control functions in the running application (Raffaeli et al., 2014).

Interfaces for AAL systems should be utilizable at any place: at home, on the move, in the car, or at public buses or trains. Advanced interfaces, so-called Natural User Interfaces (NUI), are able to recognize a sequence of words or actions performed by the user and to interpret the meaning of the situation, or the intention of the user, in the actual context.

In AAL, communications between users and systems fall within the area of interaction; when talking about communications between different devices or system components, it is intended to deal with machine-to-machine communications. Wired and wireless communication networks and protocols are the enabling elements, as well as communication standards, which play a basic role in supporting interoperability between systems, and system components (i.e. components that correctly understand and process the information exchanged over the system interfaces, and collaborate in performing services requested by users). Interoperability requires the standardization of data models, protocols, message formats, and vocabularies to enable meaningful exchange of data between services.

From a customer perspective it is of key importance that a relatively expensive assistive device is "future-proof" (Compton et al., 2012) in terms of the possibility to be used over several years and to adapt to the users's changing needs. Furthermore, systems typically need to be installed in existing homes and must be integrated into existing infrastructure. These challenges can be addressed only through modularity enabled by standardized interfaces between systems and components, and their interoperability.

Whenever an AAL system needs to transmit information to the "outside world", communications over Wide Area Networks (WAN) are involved. In the case of home AAL systems, usually the outside world is accessed through an Internet connection, be provided by means of different technologies, according to the specific practical situation, such as fixed copper lines, satellite links, or long range wireless links (Tunca et al., 2014).

Several technical challenges remain open in the field of ICT for AAL; among them, it is worth mentioning the following ones:

- Advancing integration aspects, in order to enable the dynamic configuration of services within the home environment and/or around the person, for example by simplifying and improving the procedures requested for association or leaving of devices, or their pairing;
- Implementing context-aware systems and services, enabled with predictive reasoning capabilities, in order to proactively anticipate users' behaviors and support their adjustment;
- Covering of security-related needs by means of practical solutions that can be effectively implemented in smart homes or enhanced living environments;
- Dealing with energy-related issues, in order to ensure autonomy and reliability of all the battery-powered devices and the whole platforms, also based on the design of energy-efficient communication protocols and architectures.

In order to gather a comprehensive overview of the main application areas of ICT in AAL, it is useful to resort to the knowledge base provided by the set of projects catalogues developed by the European AAL Joint Platform, and publicly available (Ambient Assisted Living Joint Programme, 2012, 2013, 2014). By analyzing the databases, it is possible to identify the following main application areas: prevention and management of users' chronic conditions at home; management and support of users' activities of daily living; support of social interactions and older adults' occupation in life; advancement of older adults' independence and mobility.

Personalized home therapies, and new care models leading to a better quality of life for people suffering from chronic diseases, aiming to prevent re-hospitalization, are the main targets of the first application area: through the use of wearable sensors, advanced signal processing algorithms and communication technologies, it is possible to unobtrusively and constantly monitor the older adult, to check his/her health status and provide early identification of possibly emerging impairments. Services supporting home therapies, memory preservation, and cognitive/physical rehabilitation, fall within this area (Gray et al., 2014).

The management of daily life activities of older adults at home is typically associated to the use of robots, especially when the user exhibits physical impairments reducing his/her autonomy. This area, however, also covers other applications, such as the provisioning of services supporting the preparation of meals or the execution of activities requiring several actions performed in a proper order/sequence. E-shopping services delivered through different platforms (from tablets and mobile devices, to Internet-connected smart TVs), also fall within this area, as they may be usefully employed to avoid the burden of going shopping for older adults, and also to help the users finding the best offers for the items needed.

The importance of social interaction and occupation in life for older adults is usually underestimated, while several studies (Nezlek et al., 2002; Courtin and Knapp, 2016) have shown how these aspects are critical in maintaining autonomy and well-being of older persons. Solutions to facilitate access and freedom to move within the working spaces, and to ensure the right environmental conditions, according to personal status and capabilities of the worker, may have a strong impact in prolonging the active presence of older adults in the workplaces. Suitably designed interfaces and tools can help them overcoming physical impairments limiting their working capabilities. Smart working environments, equipped with AAL technologies, prevent and reduce the incidence of work-related diseases

(Brach and Korn, 2012). Finally, once an older adult retires from work, AAL technologies can enable participation in society, by supporting the development of networks of online knowledge sharing and communities, and mentoring services for young people approaching work-related issues. AAL technologies may help social interaction of elderly users by supporting them in communicating with members of their family, friends and clinicians, and by providing interactive, stimulating and social-play environments to keep an acceptable level of physical activity, that may prevent the occurrence of cognitive and mental impairments.

Elderly's independence and mobility require technologies designed to enhance their residual capacities, or, in case of total absence of them, to allow the use of alternative sensory abilities (Sanchez et al., 2007). Bio-robotic solutions can overcome motor-related impairments, whereas the design of alternative interaction modalities can compensate for sensory impairments (Spenko et al., 2006). Positioning and navigation systems, together with simplified information-retrieval solutions based, for example, on proximity-technologies, such as Near Field Communication (NFC) or Radio Frequency Identification (RFID), are among the possible supporting ICT-based functionalities for individual mobility.

The innovative paradigm of the Internet of Things (IoT) (Dohr et al., 2010; Atzori et al., 2014), may have a strong impact on AAL-related applications, systems and services. Several research and market initiatives have been undertaken, affecting many different fields and business sectors. As an example, the Alliance for Internet of Things Innovation (AIOTI), that includes several industrial players across Europe, is focusing on different areas of adoption of IoT, among which also the context of "Smart Living Environment for Ageing Well" is considered. The inherent scalability featured by IoT technologies is deemed instrumental to support the increasing size of the target population addressed by AAL, i.e. older people. In a future perspective, IoT also paves the way to the exploitation of very advanced knowledge, like genomic information, to improve human health. In fact, IoT enables the massive gathering of lifestyle data that can be linked via analytics with ageing effects in cells, organs and hormonal systems, to design new preventive and personalized therapies. However, the high potential IoT has to improve health, will emerge only if IoT is unified in its services. Reliability of technology and easiness of use must be granted, to let users trust that services are available all the time, that analytics of the population's health will be used for the common good, and for clearly visible short term individual benefits, such as tangible health improvements, reduced pain and stress, better mobility.

4.4 TECHNOLOGY AND USERS IN AAL: PRACTICAL EXPERIENCES

Older people are frequently perceived and portrayed as being resistant to technology; in reality, many of them are willing to accept novel digital technologies into their lives, and to take advantage of what technology has to offer. Understanding new technologies makes older adults feel connected to others, and the world in general. Technology rejection is usually due to a missing perception of the potential benefits attainable from its usage, not necessarily because it is too difficult or time-consuming to learn. The "not worth it" impression, on the other hand, is more likely to be triggered by an unusable interface. Therefore, the objective of an effective HMI design is to generate tools, applications, and services, which would be both usable, where simplicity and ease-of-use of the interface are ensured, and useful.

This section of the chapter reviews AAL-related projects with the aim of highlighting the benefits provided by the successful involvement of users, both in the technology design and evaluation steps.

This overview mainly refers to projects that resulted from different calls published by the AAL Joint Programme, between 2009 and 2014. A huge variety of areas and topics have been covered by the calls issued in different years; the proposed overview is focused on projects in which the role of users and user-centered approach have been given a strong emphasis. Such projects address the topics of social interaction, mobility, daily life activities, and participation in the self-serve society.

As explained in Section 4.2.1.2, the user-centered design is a key factor in the development of the technological products. In fact, most of the reviewed projects include a specific work package (WP) only dedicated to the analysis of the target users' activities, and to the identification of requirements. This first step has influence on the design of the product, that has to be tested afterwards, in order to evaluate the adherence to the real needs.

For example, in the "ALIAS" Project (Adaptable Ambient LIving Assistant), users have been involved with the aim to get inputs about their needs and wishes. In fact, the objective of the project is the product development of a mobile robot system that interacts with elderly users, monitors and provides cognitive assistance in daily life, and promotes social inclusion

(ALIAS Project description, 2009). The system is designed for people living alone at home or in care facilities such as nursing or elderly care homes.

Particularly, a specific WP (ALIAS WP1 Results, 2009) has been dedicated to:

- analysis of seniors' tasks and activities;
- identification/definition of senior and relatives' requirements and preferences;
- evaluation and selection of the most adequate functions/concepts;
- evaluation of possible types of user manuals and their development (written, video, short);
- continuous testing and optimization.

Following the requirements analysis, including surveys and workshops with elderly people, professional caregivers and consultants, the field trials have been planned. These trials are aimed at the evaluation of the acceptance and functionality of the pilot, and at its continuous optimization.

"2PCS" (Personal Protection & Caring System) is a European AAL project started in 2011 and lasted 24 months (2PCS Project Home page, 2010). It includes a series of services accessible through a watch-like device. The challenge that this project seeks to answer is the significant loss of mobility of older people, caused by fear (e.g. following a recent fall event), insecurity, disorientation, lack of information and communication skills.

During requirements analysis, 2488 users from the partner countries (Austria, Germany, Italy, Switzerland and the Netherlands) were involved by using various analysis methods, such as workshops, look & feel tests, interviews, focus group discussions.

The purpose of this involvement was to understand the needs of users, both from their point of view and from the experts' one (nurses and operators). The considered users were not only elderly, but also young people, in fact the system developed under this project includes different editions:

- Private edition: the device guides the user in his spare time, offering activity parameter measurement functionalities and location-based information or can help him in emergency situations.
- Home edition: compared to the private edition, this version is more focused on the needs of the elderly, with the aim to facilitate as much as possible their active life. Through a call center service, it allows them to call a taxi, get personal assistance at home or information on activities taking place in their region.
- Business edition: it is a version of the system designed to help health care providers in nursing homes or in care institutions. Compared to

previous editions, this allows to add adaptable security features, e.g. a system for the automatic detection of falls or localization functionalities. Another example is the "SI-SCREEN" project (ELISA Home page), which essentially refers to "ELISA", the name of the project product. A specific WP deals with the analysis of the user needs, and their interaction with different input and output devices. In fact, ELISA is a software running on a tablet, which enhances social interaction, communication and information of elderly generation. The back-end is represented by a server-based social middleware enabling elderly people to access the benefits of the Social Web. The development process of the product is based on an iterative approach. Therefore, all specifications and functions have been evaluated, refined and improved after each development process.

The use of an iterative approach represents a good methodology, as it makes possible to verify step-by-step that the product under development fits with the users' expectations.

This methodology finds application also in the "ALICE" Project, whose objective is to improve the quality of life of ageing people with impaired vision, by providing a navigational assistant with cognitive abilities (ALICE Project).

End–users have been involved and given an active role in the definition of the system. The main steps of the cyclic development process are: usability testing of prototypes, analysis of outputs, modification of the prototype. For the usability tests, different metrics have been considered to improve the product, and the cycle is iterated until the final version is obtained.

In this case, the users involved in the product design have been recruited by two specialized organizations, in order to obtain as much diversity of opinions as possible.

Also in the MobileSage project (started in July 2011 and ended in January 2014), the end-users collaboration has been crucial in designing and developing the system. From the very first stage of the user requirements drafting, they were involved through focus groups work, conducted in three different countries: Romania, Spain and Norway. The participants of the focus groups were selected to represent a broad range of parameters, including age, gender, disabilities (sensory and cognitive impairments), nationality, and ICT experience and usage (Røssvoll, 2013).

The main idea of the project consists in providing older people with a personalized context sensitive tool which offers proper assistance for performing and solving everyday tasks and issues. In other words, the project provides a Help-on-Demand Service to everybody in possession of an enabled smartphone. Help can be provided both in the home environment and on travel, making the service context-based, and partly location-based.

User involvement is so important that it can represent the key objective of a project: it happens in "ExCITE", a project focused on the acceptance of a technological product by the elderly. The device under study is "Giraff", the telepresence unit developed by Giraff Technologies AB (Giraff Home page, 2009).

The study has been carried out in a number of test sites in Sweden, Italy, and Spain, in collaboration with healthcare centers, municipalities and research organizations. The users' feedback has led to the project refinement, through the development of various releases, coherently with users' requests. In fact, the project aimed to assess the reliability and usefulness of the Giraff telepresence robotic platform as a mean to support elderly and to foster their social interaction. Rather than testing the system in laboratory setting, the robotic platform has been placed in a real context of use.

In Spain in 2013 the evaluation of Giraff has been carried out in three test sites (Gonzalez-Jimenez et al., 2013). The three participants were selected by the research team, upon personal interviews with the candidates. The selection criteria took into consideration their interest and motivation, technical and practical requirements of the house, but more importantly it considered the users' social and healthcare needs.

The paper by Bevilacqua et al. (2013) describes the field test conducted in Italy and its results. The authors state that in general, the attitude of the user towards the aid remains stable being constantly positive over time. It is plausible to think that the attractiveness of Giraff contributed to the positive attitude of the user, and this attitude has affected the intention to use, and the acceptance of the robot over time. Conversely, some malfunctions may be responsible for the perceived anxiety over time related to the use of Giraff. As a conclusion, the evaluation with real end-users of the Giraff brought to light a set of user requirements and issues to be improved in the robot.

As previously mentioned, the users' contribution is not limited to the requirement analysis, but it is also crucial in the final evaluation, to verify if the output of the work reflects the initial expectations.

All the projects described above include field trials, and they have to be planned carefully.

In ALICE, for example, the navigation assistance has been tested by people from 55 to 75 years old with different levels of vision and various diseases, over two countries. In the trial, three different scenarios were designed for structured environments (urban area), semi-structured environments (parks) and indoor structured environments. For each trial scenario different situations and scenes that may occur were also predicted. The authors state that the results of the trials impact significantly the evaluation and improvement of the product.

As for ELISA, the choice of the device to employ follows the users' preferences (tablet with the bigger screen). Then, the study on usability enables to identify if ELISA is intuitive, easy to use and usable for the elderly people (ELISA WP3 Outcomes, 2009). The results showed both the positive aspects and the possible improvements.

A particular pilot is the one conducted within the "STIMULATE" project. It aims to facilitate traveling for senior citizens by choosing the best travel itinerary based on their needs and capabilities. STIMULATE enables seniors to specify their assistance needs, plan a trip, choose and optimize the transport means and itineraries, obtain secure advice, be provided with personal assistance while on the move, and obtain secure local shopping recommendations (Stimulate Project, 2010).

Smartphones are used as interaction devices. In the development, user-centered design has been exploited: senior travelers are co-designer of the system. In addition, for testing the Stimulate platform in real conditions, three pilot trips have been organized by means of this tool in three different places: Cévennes, Vienna, and Paris. The trips have been planned in detail, including for example suggestion for the most interesting local activities that the participants could perform. A demo site is also available (Stimulate Project Demo Site, 2010).

The "CONFIDENCE" project deals with a mobility safeguarding assistance service for people suffering from dementia (Schneider and Willner, 2012). The idea of the project is to offer real-time assistance via mobile phone and geo-tracking technology when people suffering from dementia suddenly lose orientation and feel insecure. After the collection phase in each country, the user requirements were brought together and classified, in order to be able to assign them to the five CONFIDENCE modules (assistance and training at home, virtual voice service, virtual video service, location tracking service and mobile community service). Thereafter the requirements were negotiated by stakeholders of the end-user organizations during a user requirements workshop. In a final step they were prioritized by the whole project consortium and ten of them selected for implementation.

End-user trials have been conducted in three countries which have different requirements due to the organizational culture of their social service systems. In this case, end-users include not only people with dementia, but also family members, home care professionals, and trusted volunteers.

4.5 LESSONS LEARNED

As it is evident from the preceding sections, the experience gained since several years through research-supporting programs, like the European AAL Joint Programme, highlighted the difficulty of bringing to the real market the huge number of prototype devices and services generated by research projects.

This awareness opens the way to some guidelines definition, to effectively account for a user driven approach during the different steps of technology design, and to avoid common mistakes leading to products that fail in meeting the users' expectation and willingness of use:

1. At first, include a multi skilled team to benefit from different perspective and ideas;
2. Define the user population as target audience;
3. Plan an evaluations strategy to guide the user driven approach;
4. Apply the human factor knowledge at every stage of the development process;
5. Choose suitable methods to face, drive and refine the user experience;
6. Involve primary, secondary and tertiary users to cover the complexity of the target;
7. Plan a robust instructional strategy within the evaluation process to reduce uncertainty and unfamiliarity with new technology;
8. Guarantee friendliness and ease-of-use of the solution;
9. Match the benefit and utility felt by users;
10. Always act for preventing users from any kind of stigmatization and discomfort.

From a technology-oriented perspective, past experiences have shown that simplification and better acceptance from the user may be obtained through as much as possible smooth integration of different systems, devices and services, aiming at a unified design that avoids fragmentation of technologies. The idea of getting a few technologies accepted as standard has to be abandoned, as it does not account for the complexity typically found in real deployments. At the same time, it is necessary to enable the coexistence of heterogeneous elements by abstracting user-friendly interfaces from the complexity of the physical layer implementation.

It is reasonable to state that the IoT paradigm will play the main role in the upcoming years, not only in the context of AAL systems and platforms, but also in many other different domains (Gubbi et al., 2013). The flexibility ensured by the opportunity to provide each single "thing" with an Internet connection, will favor the design of services and applications more and more personalized and tailored to each user's specific needs, combining both low- and high-tech devices with the additional capability of being able to learn the user's habits, choices, preferences, and adjust the offered services accordingly.

4.6 CONCLUSION

It is a common myth that seniors are "technophobes"; on the contrary, they are active users of technology if innovation is able to account for their needs, and if friendliness and tangible benefits are experienced. A user driven approach recognizes the role of human factors, user participation and iterations within the design lifecycle and comprises a variety of techniques, methods, and practices. Greater attention to this approach could help developers, researchers and providers in designing systems, products and environments for older adults, able to reach the market faster and more effectively than today, thus really addressing and consolidating the so-called silver or grey market.

ACKNOWLEDGEMENT

The Authors would like to acknowledge the IC1303 COST Action AAPELE and the Home4Dem Project (AAL JP reference: AAL 2014-1-041) in stimulating the discussion around the topic presented in the chapter.

REFERENCES

2PCS Project Home page – Call 3, 2010. Retrieved January 18, 2016, from http://www.2pcs.eu/.

Alam, M.M., Hamida, E.B., 2014. Surveying wearable human assistive technology for life and safety critical applications: standards, challenges and opportunities. Sensors 14, 9153–9209.

ALIAS Project description – Call 2, 2009. Retrieved January 13, 2016, from http://www.aal-alias.eu/content/project-overview.

ALIAS WP1 Results – Call 2, 2009. Retrieved January 13, 2016, from http://de. slideshare.net/geigeralias/alias-wp1-results.

ALICE Project – Call 2, 2009. Retrieved January 15, 2016, from www.alice-project.eu.

Ambient Assisted Living Association, 2014. Final report: a study concerning a market observatory in the ambient assisted living field. http://www.aalforum.eu/ wp-content/uploads/2014/11/AAL_Final-report_Complete_v4-00.pdf.

Ambient Assisted Living Joint Programme, 2012. Catalogue of projects. Available from: http://www.aal-europe.eu/.

Ambient Assisted Living Joint Programme, 2013. Catalogue of projects. Available from: http://www.aal-europe.eu/.

Ambient Assisted Living Joint Programme, 2014. Catalogue of projects. Available from: http://www.aal-europe.eu/.

Arning, K., Ziefle, M., 2007. Understanding age differences in PDA acceptance and performance. Comput. Hum. Behav. 23 (6), 2904–2927.

Atzori, L., Iera, A., Morabito, G., 2014. From "smart objects" to "social objects": the next evolutionary step of the Internet of Things. IEEE Commun. Mag. 52 (1), 97–105.

Bevilacqua, R., Cesta, A., Cortellessa, G., Macchione, A., Orlandini, A., Tiberio, L., 2013. Telepresence robot at home: a long-term case study. In: Proceedings of the Italian AAL Forum. Ancona, Italy.

Blasco, R., Marco, Á., Casas, R., Cirujano, D., Picking, R., 2014. A smart kitchen for ambient assisted living. Sensors 14 (1), 1629–1653.

Brach, M., Korn, O., 2012. Assistive technologies at home and in the workplace — a field of research for exercise science and human movement science. Eur. Rev. Aging Phys. Act. 9 (1).

Broekens, J., Heerink, M., Rosendal, H., 2009. Assistive social robots in elderly care: a review. Gerontechnology 8 (2), 94–103.

Carrettero, S., Stewart, J., Centeno, C., Barbabella, F., Schmidt, A., Lamontagne-Godwin, F., Lamura, G., 2012. Can technology-based services support long-term care challenges in home care? Analysis of evidence from social innovation good practices across the EU: CARICT project summary report. JRC Scientific and Policy Reports. Publications Office of the European Union, Luxembourg.

Cippitelli, E., Gasparrini, S., Gambi, E., Spinsante, S., Wahslen, J., Orhan, I., Lindh, T., 2015a. Time synchronization and data fusion for RGB-depth cameras and wearable inertial sensors in AAL applications. In: 2015 IEEE International Conference on Communication Workshop (ICCW), pp. 265–270.

Cippitelli, E., Gasparrini, S., Spinsante, S., Gambi, E., 2015b. Kinect as a tool for gait analysis: validation of a real time joints extraction algorithm working in side view. Sensors 15, 1417–1434.

Compton, M., Barnaghi, P., Bermudez, L., et al., 2012. The SSN ontology of the W3C semantic sensor network incubator group. Web Semant. Sci. Serv. Agents World Wide Web 17, 25–32.

Courtin, E., Knapp, M., 2016. Social isolation, loneliness and health in old age: a scoping review. Health Soc. Care Community 24 (1).

Cubo, J., Nieto, A., Pimentel, E., 2014. A cloud-based Internet of Things platform for ambient assisted living. Sensors 14, 14070–14105.

Damodaran, L., 1996. User involvement in the systems design process — a practical guide for users. Behav. Inf. Technol. 15 (6), 363–377.

De Santis, A., Del Campo, A., Gambi, E., Montanini, L., Pelliccioni, G., Perla, D., Spinsante, S., 2015. Unobtrusive monitoring of physical activity in AAL: a simple wearable device designed for older adults. In: ICT4AgeingWell 2015 — Proceedings of the 1st International Conference on Information and Communication Technologies for Ageing Well and e-Health, pp. 200–205.

De Santis, A., Gambi, E., Montanini, L., Raffaeli, L., Spinsante, S., Rascioni, G., 2014. A simple object for elderly vitality monitoring: the smart insole. In: 2014 IEEE/ASME 10th International Conference on Mechatronic and Embedded Systems and Applications (MESA), pp. 1–6.

Dohr, A., Modre-Opsrian, R., Drobics, M., Hayn, D., Schreier, G., 2010. The Internet of Things for Ambient Assisted Living. In: 2010 Seventh International Conference on Information Technology: New Generations (ITNG), pp. 804–809.

ELISA Home page – Call 2, 2009. Retrieved January 15, 2016, from http://www.si-screen.eu/en.html.

ELISA WP3 Outcomes: Usability Study – Call 2, 2009. Retrieved January 18, 2016, from http://www.si-screen.eu/fileadmin/si-screen.eu/data/dl_en/WP3.pdf.

Farage, M.A., Miller, K.W., Ajayi, F., Hutchins, D., 2012. Design principles to accommodate older adults. Glob. J. Health Sci. 4 (2), 2–25. 2012 Feb 29.

Fernandes Golino, H., Souza de Brito Amaral, L., Pimentel Duarte, S.F., et al., 2014. Predicting increased blood pressure using machine learning. J. Obesity 2014.

Fisk, D.A., Rogers, W.A., Charness, N., Czaja, J.S., Sharit, J., 2009. Designing for Older Adults: Principle and Creative Factor Approaches. CRC Press.

Forman, D.E., Berman, A.D., McCabe, C.H., Baim, D.S., Wei, J.Y., 1992. PTCA in the elderly: the "young-old" versus the "old-old". J. Am. Geriatr. Soc. 40 (1), 19–22. http://dx.doi.org/10.1111/j.1532-5415.1992.tb01823.x. PMID 1727842.

Gasparrini, S., Cippitelli, E., Gambi, E., Spinsante, S., Wahslen, J., Orhan, I., Lindh, T., 2015. Proposal and experimental evaluation of fall detection solution based on wearable and depth data fusion. In: ICT Innovations 2015.

Gervais-Ducouret, S., 2011. Next smart sensors generation. In: 2011 IEEE Sensors Applications Symposium (SAS), pp. 193–196.

Giraff Home page – Call 2, 2009. Retrieved January 13, 2016, from http://www.giraff.org/?lang=en.

Gonzalez-Jimenez, J., Galindo, C., Gutierrez-Castaneda, C., 2013. Evaluation of a telepresence robot for the elderly: a Spanish experience. In: Proceedings of IWINAC 2013, the 5th International Work-Conference on the Interplay Between Natural and Artificial Computation. Mallorca, Spain.

Gray, K.M., Clarke, K., Kwong, M., Alzougool, B., Hines, C., Tidhar, G., Frukhtman, F., 2014. Internet protocol television for personalized home-based health information: design-based research on a diabetes education system. JMIR Res. Protoc. 3 (1), e13.

Guan, Q., Yin, X., Guo, X., Wang, G., 2016. A novel infrared motion sensing system for compressive classification of physical activity. IEEE Sens. J. 16 (8), 2251–2259.

Gubbi, J., Buyya, R., Marusic, S., Palaniswami, M., 2013. Internet of Things (IoT): a vision, architectural elements, and future directions. Future Gener. Comput. Syst. 29 (7), 1645–1660.

Inman, A., Hodgins, D. (Eds.), 2013. Implantable Sensor Systems for Medical Applications. Woodhead Publishing Limited, Cambridge.

ISO 9241-210, 2010. Ergonomics of human-system interaction — Part 210: human-centred design for interactive systems. http://www.iso.org/iso/catalogue_detail.htm?csnumber=52075.

Lattanzio, F., Abbatecola, A.M., Bevilacqua, R., Chiatti, C., Corsonello, A., Rossi, L., Bustacchini, S., Bernabei, R., 2014. Advanced technology care innovation for older people in Italy: necessity and opportunity to promote health and wellbeing. J. Am. Med. Dir. Assoc. 15, 457–466.

Lee, C., Coughlin, F.J., 2015. PERSPECTIVE: older adults' adoption of technology: an integrated approach to identifying determinants and barriers. J. Prod. Innov. Manag. 32 (5), 747–759.

Longhi, S., Siciliano, P., Germani, M., Monteriù, A. (Eds.), 2014. Ambient Assisted Living Italian Forum 2013. Springer, New York.

Memon, M., Wagner, S.R., Pedersen, C.F., Beevi, F.H.A., Hansen, F.O., 2014. Ambient assisted living healthcare frameworks, platforms, standards, and quality attributes. Sensors 14 (3), 4312–4341.

Mitzner, T.L., Boron, J.B., Fausset, C.B., Adams, A.E., Charness, N., Czaja, S., Dijkstra, K., Fisk, A.D., Rogers, W.A., Sharit, J., 2010. Older adults talk technology: technology usage and attitudes. Comput. Hum. Behav. 26 (6), 1710–1721.

Morrell, R.W., 2001. Older Adults, Health Information, and the World Wide Web. L. Erlbaum Associates Inc., Hillsdale, NJ, USA.

Nedopil, C., Schauber, C., Glende, S., 2013. Guideline: the art and joy of user integration in AAL projects. Ambient Assisted Living Association.

Nezlek, J.B., Richardson, D.S., Green, L.R., Schatten-Jones, E.C., 2002. Psychological well-being and day-to-day social interaction among older adults. Pers. Relatsh. 9, 57–71.

Porambage, P., Braeken, A., Gurtov, A., Ylianttila, M., Spinsante, S., 2015. Secure end-to-end communication for constrained devices in IoT-enabled ambient assisted living systems. In: 2015 IEEE 2nd World Forum on Internet of Things (WF-IoT) Proceedings, pp. 711–714.

Raffaeli, L., Gambi, E., Spinsante, S., 2014. Smart TV based ecosystem for personal e-health services. In: 2014 8th International Symposium on Medical Information and Communication Technology (ISMICT), pp. 1–5.

Rashidi, P., Mihailidis, A., 2013. A survey on ambient-assisted living tools for older adults. IEEE J. Biomed. Health Inform. 17 (3), 579–590.

Rekha Devi, Kh., Sen, A.M., Hemachandran, K., 2012. A working framework for the user-centered design approach and a survey of the available methods. Int. J. Sci. Res. Publ. 2 (4).

Røssvoll, T.H., 2013. The European MobileSage project — situated adaptive guidance for the mobile elderly: overview, status, and preliminary results. In: Sixth International Conference on Advances in Computer-Human Interactions (ACHI). IARIA.

Samer, M., Moreno, J.C., Kong, K., Amirat, Y. (Eds.), 2015. Intelligent Assistive Robots. Recent Advances in Assistive Robotics for Everyday Activities. Springer Publishing Company, New York.

Sanchez, J.H., Aguayo, F.A., Hassler, T.M., 2007. Independent outdoor mobility for the blind. In: Virtual Rehabilitation, 2007, pp. 114–120.

Schneider, C., Willner, V., 2012. CONFIDENCE — a mobility safeguarding assistance service with community functionality for people with dementia. In: AAL Forum 2012, Eindhoven.

Schubert, E., Koos, A., Emrich, T., Zufle, A., Schmid, K.A., Zimek, A., 2015. A framework for clustering uncertain data. Proc. VLDB Endow. 8 (12), 1976–1979.

Shasha Liu, P., Hung-Fat, T., 2013. Implantable sensors for heart failure monitoring. J. Arrhythmia 29 (6), 314–319.

Singh, G., Chen, T.A., Robucci, R., Patel, C., Banerjee, N., 2016. Distratto: impaired driving detection using textile sensors. IEEE Sens. J. 16 (8), 2666–2673.

Spenko, M., Haoyong, Y., Dubowsky, S., 2006. Robotic personal aids for mobility and monitoring for the elderly. IEEE Trans. Neural Syst. Rehabil. Eng. 14 (3), 344–351.

Spinsante, S., Antonicelli, R., Mazzanti, I., Gambi, E., 2012. Technological approaches to remote monitoring of elderly people in cardiology: a usability perspective. Int. J. Telemed. Appl. 2012.

Spinsante, S., Cippitelli, E., De Santis, A., Gambi, E., Gasparrini, S., Montanini, L., Raffaeli, L., 2015. Multimodal interaction in a elderly-friendly smart home: a case study. In: Lecture Notes of the Institute for Computer Sciences, Social-Informatics and Telecommunications Engineering, LNICST, vol. 141, pp. 373–386.

Spinsante, S., Gambi, E., 2012a. Remote health monitoring for elderly through interactive television. Biomed. Eng. Online 11, 54.

Spinsante, S., Gambi, E., 2012b. Remote health monitoring by OSGi technology and digital TV integration. IEEE Trans. Consum. Electron. 58 (4), 1434–1441.

Spinsante, S., Gambi, E., 2015. NFC-based user interface for smart environments. Adv. Hum.-Comput. Interact. 2015.

Stimulate Project – Call 3, 2010. Retrieved January 18, 2016, from http://www.stimulate-aal.eu/stimulate/Home.html.

Stimulate Project Demo Site – Call 3, 2010. Available at: http://stimulate-demo.tripalacarte.com/frontend/index.php.

Thielke, S., Harniss, M., Thompson, H., Patel, S., Demiris, G., 2012. Maslow's hierarchy of human needs and the adoption of health-related technologies for older adults. Ageing Int. 37 (4), 470–488.

Trindade, I.G., Martins, F., Dias, R., Oliveira, C., Machado da Silva, J., 2015. Novel textile systems for the continuous monitoring of vital signals: design and characterization. In: 2015 37th Annual International Conference of the IEEE Engineering in Medicine and Biology Society (EMBC), pp. 3743–3746.

Tunca, C., Alemdar, H., Ertan, H., Incel, O.D., Ersoy, C., 2014. Multimodal wireless sensor network-based ambient assisted living in real homes with multiple residents. Sensors 14 (6), 9692–9719.

Van Den Broek, G., Cavallo, F., Wehrman, C., 2010. Ambient Assisted Living Roadmap. IOS Press.

Walsh, K., Callan, A., 2010. Perceptions, preferences, and acceptance of information and communication technologies in older-adult community care settings in Ireland: a case-study and ranked-care program analysis. Ageing Int. 36 (1), 102–122.

Weiss, J.C., Natarajan, S., Peissig, P.L., McCarty, C.A., Page, D., 2012. Machine learning for personalized medicine: predicting primary myocardial infarction from electronic health records. AI Mag. 2012, 33–45.

CHAPTER 5

Matching Requirements for Ambient Assisted Living and Enhanced Living Environments with Networking Technologies

Thomas Zinner*, Florian Wamser*, Helmut Leopold[†], Ciprian Dobre[‡], Constandinos X. Mavromoustakis[§], Nuno M. Garcia[¶]
*University of Würzburg, Germany
[†] Austrian Institute of Technology, Austria
[‡] University Politehnica of Bucharest, Romania
[§] University of Nicosia, Cyprus
[¶] University of Beira Interior, Portugal

5.1 INTRODUCTION

The increase in medical expenses caused by societal issues like demographic growth and aging puts a strong pressure on the sustainability of the health and social care system. Alternative solutions are needed to cope with a sustainable quality of life for elderly people.

The concept of Enhanced Living Environments (ELE) proposes broadening the concept of Ambient Assisted Living (AAL) to reflect more accurately the eco-system created by the combination of medical and ICT services. It aims at the prolongation of a self-conducted life of assisted persons, reducing the dependency on intensive personal care to a minimum. Thereby, it increases the quality of life for the affected group while substantially decreasing the costs for society and supporting ever-more increasing requirements coming from different stakeholders. ELE encompasses the latest developments associated with the Internet of Things, towards services designed for a better help and support for people, or as a general term, to better live their life and interact with their environment.

Both, in AAL and ELE, a multitude of heterogeneous services, involving different stakeholders, have to interact via a common network environment. Hence, infrastructures have to become pervasive, supporting an increasing number of distributed devices that will need to communicate

Ambient Assisted Living and Enhanced Living Environments.
DOI: http://dx.doi.org/10.1016/B978-0-12-805195-5.00005-3

between themselves, as well as with centralized communication endpoints. In the context of mobility, temporary co-location of devices will be exploited to build dynamic networks, without a pre-structured infrastructure, or to complement existing communication infrastructures with ad-hoc ones. The current networking infrastructures are mostly based on the Internet, and were not designed to support the varying requirements for the dynamic interaction processes between human beings, sensors and systems, e.g. in a machine-to-machine communication style. Host-based addressing is the foundation of the Internet, however, it cannot simply scale to support the dynamic connections being established, sometimes opportunistically, between various wireless devices and corresponding health services. The content itself is dynamic, thus the infrastructure has to be both flexible and content-driven, or at least elastic in nature.

Thus, AAL/ELE services need highly scalable, flexible, and dynamic networking infrastructures to cope with such requirements. This is where we witness today a raise in interest towards the use of novel technologies, like Software Defined Networks (SDN), Network Virtualization, Cloud platforms, and many others. However, the wide adoption of such technologies for the benefit of AAL/ELE has little been studied up-to-now.

The contribution of this article to understand the technological barriers hindering the widespread real-world usage of AAL/ELE systems as mentioned in Memon et al. (2014), and to identify technologies which may help to overcome them. The book chapter is structured as follows. In Section 5.2 we briefly review the application domains of AAL and ELE. Domains, applications, and stakeholders are summarized and characteristics are inferred and exemplary illustrated for the use-case *closed loop healthcare*. Section 5.3 highlights the underlying communication architecture. Based on these discussions, requirements for AAL/ELE applications are summarized in Section 5.4, while available and upcoming networking and inter networking technologies are analyzed based with respect to these requirements in Section 5.5. Key derivations are discussed in Section 5.6, and the chapter is concluded in Section 5.7.

5.2 CLASSIFICATION OF AAL/ELE DOMAINS AND APPLICATIONS

The environment in the AAL/ELE domain is very diverse and heterogeneous. This concerns the application side as well as the various involved participants. To illustrate the challenges in such an environment we highlight the following AAL/ELE examples.

The first example highlights self-monitoring of health parameters. The patient stays at his/her home, where (s)he measures parameters like weight, or metabolic age or fat percentage on a digital scale with a body analyzer. The device is connected to a personal eHealth Gateway, where a Personal Health Record (PHR) instance resides. The data are transmitted in regular time intervals to the gateway and saved in the local PHR. To achieve resilience of the monitored data the eHealth Gateway distributes copies at trusted PHR mirrors, e.g., maintained by relatives of the patient. The data is accessed to investigate time series and trends of the monitored parameters. Here, the patient and relatives are the main actors involved, and it involves a privately-owned data network.

The second example involves the integration of medical data originating from various sources. Such data sources can be divided into different types, depending on who manages the data within the system. One type of data sources are Electronic Health Record (EHR) and Electronic Medical Record (EMR) systems, where medical data is managed by the medical personnel. Other types can be the aforementioned PHR systems, where the data are managed by the patient. Another type of data sources could be the PHR systems, where it is the patient who manages the data. It should be noted that such a scenario is focused on integrating the medical data, which is located "somewhere" in the network and can be stored in multiple copies. In this case, the patient shares his/her data with other users in the network through setting adequate content access rights. After that, such data can be downloaded by a doctor, analysis services or medical systems with which the patient (the owner of the information) wishes to share his health data.

These examples illustrate, that monitoring data of patients gathered by heterogeneous sensors play an essential role in AAL/ELE. Various AAL/ELE stakeholders own or have access to different sensors and monitored data. This data has to be shared between the different stakeholders to generate additional value. These interactions between the stakeholders may occur in regular intervals, event-based, or on demand resulting in a dynamic interaction process. Further, the examples highlight different involved domains, the medical domain and the home domain.

In the following, we want to broaden this view on AAL/ELE involving the main application scenarios and stakeholders from our point of view.

For that, we distinguish between four application domains: (i) medical domain which deals essentially with medical health data and thus imposes dedicated requirements on systems and processes; (ii) care domain which focuses on specific services to support the daily life for people with special needs and requirements; (iii) lifestyle which describes further non-critical services to support daily life; and finally (iv) safety & security which includes dedicated monitoring and control functions at home to improve the safety of people with special needs. The different domains are highlighted in Table 5.1.

In relation to these domains, the following potential stakeholders are considered: (1) AAL/ELE customers (e.g. elderly persons), (2) physicians, (3) caregivers, (4) family members and friends, (5) public authorities, (6) service providers and (7) sensors. We extend the definition of stakeholders presented in Rashidi and Mihailidis (2013) by family members and friends, public authorities and sensors. The latter one is referred to as a passive stakeholder. Other possible stakeholders in the AAL/ELE context such as manufacturers and advocacy groups are omitted. These stakeholders represent a heterogeneous group in the AAL/ELE environment, ranging from customers in AAL/ELE systems to experts who monitor, guide or help people, and sensors that record or pass on information to other people. All participants have different system requirements and usage purposes. A comprehensive analysis of actors and stakeholders can be found in Nedopil et al. (2013).

Based on the specific scenario and the corresponding AAL/ELE function we identify the stakeholders that play an essential role. The resulting classification illustrated in Table 5.1 is discussed in the following. AAL/ELE customers affect all domains. Experts like physicians or caregivers are primarily relevant to their expert area. Family members are active primarily in the care domain and lifestyle. Public authorities are involved in the regulation of health care systems. Service providers and sensors can be found in all application domains.

The next sections provide a description for each domain and discuss the resulting challenges for today's ICT systems. Further, we characterize the domain by defining different key attributes: (1) the actors involved, (2) the required services for the operation, (3) the required architecture and (4) general system requirements. Based on this analysis, we detail a specific AAL/ELE scenario, namely patient monitoring.

Table 5.1 Domains and involved stakeholders. Stakeholders which play an essential role for an application domain are marked with an "x"

	Customer	Physicians	Caregivers	Family and friends	Public authorities	Service providers	Sensors
Medical domain	x	x				x	x
Care domain	x		x	x	x	x	x
Lifestyle	x			x		x	x
Safety & security	x			x	x	x	x

5.2.1 Involved Domains and Key Attributes

In the following we discuss the presented application domains in detail.

Medical Domain

One of the main applications in the eHealth and AAL context is telemedicine. The main driver which differentiates this kind of service from any other service classes is the inherent treatment of personal medical health data. The main focus is to support health treatment processes through telecommunication services so that physicians can deal with patients separated in space and time. Collected vital parameters of patients can be monitored online, anytime, anywhere, on any device being used by physicians. Further on, electronic systems can analyze trends and correlation of collected health data such as blood pressure, glucose level, weight, mental activity, or physical fitness and thus offer additional information for physicians to improve health care processes for both the patient and the physician. Additionally, a physician has the possibility to get in contact with the patient by different means like, email, text messages (pre-defined messages, free text), chat, phone service, or even multimedia based communication services.

For the collection of health data, different sensors are used. Sensor networks such as body area networks and home networks are important building blocks of a telemedicine platform. User identification, privacy and security and finally usability are essential service features to be considered. At the physician side data representation, user–interface and interoperability with other health applications and even legacy systems are important issues to be considered.

The treatment of medical health data imposes stringent requirements on the system design and implementation processes driven by privacy and security requirements and could even go to medical product validation processes.

Finally, it is important to note, that next generation health services based on digital platforms have to support interoperability and close interworking with further applications and users from other domains like care givers, insurances, etc.

Care Domain

This domain summarizes applications for the care and welfare of people. It includes the daily care of elderly people and the care of people with specific requirements, conditions and even diseases. Thus, services and data from the medical domain are of basic importance for the care domain as well to implement health monitoring and health treatment processes.

In addition to that, additional services to support individual care for social, mental and physical fitness, even physical security, are important functions in the care domain. Hence, beside the typical vital parameters, more general parameters like the personal feeling and the movement patterns of the assisted people are of interest. Monitoring the actual situation of a person at home (fall monitoring), prevention of dangerous situations (fall prevention, alarming of different events), nutrition motivation and support, and even support for the management of the daily living are functions which can be supported by next generation communication services.

Further on, easy to use communication services with other stakeholders such as family members, friends, care organizations, and public authorities extend the requirements for the ICT platform by a broader usage of interactive multimedia services.

The following applications may be seen as a relevant subset of applications in the care domain:

- care services of daily life without a physician involved
- care for mental and physical fitness
- care for diseases (dementia, alzheimer, etc.)
- monitoring and stimuli for nutrition, drugs, etc.
- services for motivation of elder people and social care such as social interaction via interactive multimedia services (chatting with other people, singing, playing games, etc.)
- monitoring dangerous situations at home and habits by sensors and remote management of different functions at home (smart home)

Lifestyle

Another field of application are services for living and lifestyle. In addition to the domains mentioned above, entertainment services like TV and interactive multimedia are vital services that represent a high share in the daily use. This category contains also all the helping functions in coping with daily life, such as shopping and performing of administrative procedures and the handling of everyday activities.

Since a non-discriminatory and inclusive usability of communication services is one of the essential requirements of any public service, also in the AAL/ELE context, dedicated services from public authorities have to be offered. Especially ill or elderly people have specific needs for information and services from the public. This imposes additional system requirements on usability, data provisioning and interoperability on existing services in the eGovernment and smart city context. Some examples in this application domain are:

- communication and social networking
- interactive multimedia and TV services
- electronic tools for the needs of everyday life such as electronic shopping aids, assistance with administrative formalities and use of e-Government services, guidance in cities, parks, and other public places.

Safety & Security

This category includes services that contribute to the safety and protection of elder people. The category includes also services that may need to work outside the home such as activity monitoring for family members and dedicated monitoring services to prevent dangerous situations, such as the monitoring of oven and gas in the kitchen or the front door.

The following services may be considered as a relevant subset of services in this domain:

- status information for family & friends (location of people, social activity, etc.)
- warning/alarm services (gas, electricity status, etc.)
- identification of abnormal situations

Applications in this domain may provide critical data for other domains, e.g., in the event of an alarm or warning.

Summary

The different domains involve various actors, require diverse network services and network architectures and have to consider different criticality of the data for transmission and storing. The following system requirements can be summarized:

- Actors in the communication process: Different actors (physician, patients, family members, care giver, etc.) will use the system at the same time, imposing different requirements on the technical platform with respect to QoS parameters, real-time behavior, or sensitiveness of data.
- Required services: (real time) Monitoring of the user by sensors and wearable devices, bidirectional communication services, monitoring of specific events, data collection functions, and alarm service infrastructure.

- Required architecture: body area network, home network, data collection function, and access to a network service.
- System requirements: flexible communication services for patient-physician interaction, customizable user interfaces (treatment specific defined by a physician), reliability and accuracy of sensor data, increased privacy.

5.2.2 Closed Loop Healthcare as Typical AAL/ELE Application

The goals of any closed loop healthcare system (Modre-Osprian et al., 2014) including patient monitoring are: (a) a high recall in detecting emergencies immediately; and (b) high precision, to prevent invalid emergency detections and alerts as a consequence of misinterpretations. Requirement (a) is mandatory to provide a trustworthy service quality to the affected persons in case of emergency situations. Requirement (b) is essential for economic reasons, since invalid emergency alerts may unacceptably increase care costs and decrease trustworthiness. Further, it is desirable to extend a pure emergency detection service by an emergency prediction service, which attempts to recognize a critical health condition before it escalates into an emergency. As a reaction to the detection of such critical situations, the service may assist the person in preventing the emergency, e.g., by suggesting appropriate medication.

In general, closed loop healthcare involves periodically transmitting routine vital signs and, in some cases, alerting signals when vital signs cross a specific threshold. Depending on the type of usage and the specific environment, the accuracy of the monitored data may vary. Additional functions like the data recording and analysis may allow to trace anomalies, and to infer specific illnesses. The current work done in closed loop healthcare includes, among others, home monitoring (Lee et al., 2000), wireless systems for digitized EKGs (Khoor et al., 2001), hospital-wide mobile monitoring systems (Pollard et al., 2001), mobile telemedicine (Hung and Zhang, 2003; Pattichis et al., 2002), and real time home monitoring of patients (Mendoza and Tran, 2002).

A variety of approaches previously made attempts to address the issues of reliable and efficient message delivery from deployed sensors to central processing units (an analysis is shown in Braem et al. (2008)). The problem is finding a trade-off between reliability and energy efficiency, because any closed loop healthcare system will need to maximize the amount of delivered messages, with minimum energy consumption. In an ad-hoc environment, the success of message delivery is not only related to the consumed power, but also depends on the cooperation of neighboring devices. As specified in Varshney (2007), it is impossible to use a single method to coordinate multiple entities in a dynamic and complex environment. Apparently, closed loop healthcare has become an interdisciplinary topic and needs more intelligent technologies than other subjects (e.g., artificial intelligence). For example, the Ambient Cardiac Expert (ACE) monitoring system (Gondal et al., 2007) is a cardiac closed loop healthcare system which collects physiological data observed by sensor networks (together with gene expression data) to predict the heart failure rate. Clinical data monitored by attached sensors on patients' bodies is used to generate training data to predict the odds of heart failure.

Hospital Domain

For the use-case closed loop healthcare, physicians are typically interested in the monitoring of time series of sensor information in high accuracy. Thus, they are able to track problems, and even identify specific illness. Further, detailed logging information of the environment and the data monitoring have to be recorded to fortify the confidence of the data.

Although the usability of applications and corresponding sensors is an important feature, physicians put the accent on accuracy and traceability of the available data set.

Care Domain

Care givers are more interested in periodical information, to be able to react if something is happening. Sensor accuracy has not be so high since it does not aim at identifying specific illnesses, but to be able to react if problems occur. Further, emergency predictions may be of interest.

Sensors and actors should not be perceived by the end-user in the care domain, since they typically aim at long-time monitoring of health parameters. In case of alarms, care givers are able to check health parameters of the elderly fast and uncomplicated. Hence, the involved stakeholders in the care domain may prefer usability over accuracy and reliability.

Lifestyle

Patients use monitoring apps for actual self-check. In most cases, they are interested in their current condition, e.g., after running, or a short time series showing for instance the cardiogram in the last 30 minutes.

Accuracy in such cases is not so important, as the user/patient cares only about a rough trend of the result. Typically, he is not interested in the exact value, but in a specific interpretation like the current condition on a qualitative scale, the raise of a problem, or he wants to know if regularly training enhances his pulse frequency. For that, the end-user places a large accent on usability.

Safety & Security

Taking safety and security into account, the end-users are mostly interested in an identification of emergency situations. This includes suddenly occurring emergencies like falls or accidents, and impending dangers like deteriorating health parameters, which might be prevented with an appropriate emergency prediction. Further, spouse and relatives may profit from telemonitoring services, since they are able to check the health condition of their loved ones.

Sensors and applications have to be integrated in the daily life of the elderly resulting in high usability demands. Accuracy and reliability of the system are also of high importance, since relatives or emergency services are typically not on site and can check the state of the end-user.

5.3 COMMUNICATION SERVICES TO SUPPORT AAL/ELE INFRASTRUCTURE

Within the above-mentioned domains, communication services between the different actors are an integral and crucial part of the entire AAL/ELE system. Involved networks includes body area networks, sensor networks, middleware in the home network, the wide area network as well as local networks at the involved actors. The various stakeholders, their individual actors and the networks are shown in Figure 5.1. Communication between the devices in body area networks and sensor networks is done based on Machine-to-Machine (M2M) techniques (Chen et al., 2014). Available data like EKG, blood pressure and temperature is typically transmitted to data collection functions using local wireless technologies (e.g., WLAN, RFID, ZigBee) or via cellular networks. Such technologies are at the heart of projects such as eMotion ECG or TruVue, and are used to monitor, for example, an elderly's well-being and detect critical situations that prompt care givers take actions (Park and Jayaraman, 2007). The data is usually sent from the patient, using a combination of point-to-point and point-to-multipoint networking technology, to a central repository. The communication layer

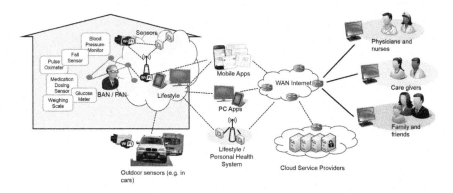

Figure 5.1 The different actors in AAL/ELE require an adequate network infrastructure.

needs to combine cross-layer management with underlying transfer mechanisms, over transmission protocols.

In addition, multimedia equipment like smart TVs and other user equipment communicate using fixed or wireless networking technologies. Specific middleware and set-top boxes are additionally integrated to provide AAL/ELE services as well as typical network functions like connectivity, Network Address Translation (NAT) or firewalling. Users and AAL/ELE services are further connected to actors in the medical and care domain, as well as to the typical consumer services like entertainment.

We understand an ICT-based AAL/ELE system as an ICT platform which supports independence, increases safety and supports health care for people with special care needs. This includes any person which can be served by smart networking techniques that can ensure an easier and safer living at home. Thus, a next generation AAL technology platform has to support the needs of a broad range of applications and their context and requirements, offering services and supporting a wider set of potential stakeholders. In particular, the AAL/ELE networking infrastructure has to be adaptable to incorporate AAL/ELE application demands and extensible to integrate a wide range of requirements coming from stakeholders.

To better illustrate Figure 5.1, we refer to the closed loop healthcare scenario. A patient wears trendy and non-intrusive sensors, usually in the form of smart bracelet (that features built-in electrodes or biosensors for reading and recording single-channel electrocardiogram (ECG) and temperature measurements, and time and location). The wearable device sends monitoring readings over short- and medium-range communication (i.e., using WiFi, Bluetooth, or ZigBee). If the users also carries a smartphone, the bracelet would connect with it, and in turn the smartphone becomes a communication hub, sending the data out towards a processing unit (via broadband communication, i.e., 3G or 4G). Otherwise, the bracelet still is able to send the data out whenever an opportunity occurs (e.g., whenever a WiFi Access Point is encountered).

On the Cloud/processing side, each patient is described by a patient/medical/psychological profile (e.g., an electronic health record). When the patient uses the medical service, the data from the smart bracelet is used to learn/construct this record (e.g., his daily walking routine is linked to his typical ECG rhythm — a medical profile of the patient, what is considered for this patient to be "normal" medical condition).

This personalized medical model of the patient can be further used to detect unusual situations — e.g., when his heart rate gets off the charts (i.e., by comparison with the medical profile) an alarm can be raised to the care giver. His usual medical record can be used by his physician to identify a possible health problem and to establish a personalized medical pro-active treatment to prevent possible health conditions in the future).

For this example scenario, the medical service just described has to put the patient entirely in control of his personal health data. The patient should be able to control what data/alarms can be seen by whom, or what data is to be sent to which end user. This separation between end users leads to a better control the medical data. For example, the patient might feel more secure if he does not show his current location to everyone in the family (for privacy reasons), but only shares it with his care giver.

Figure 5.2 depicts the standards and organizations working on enabling the introduced scenario as introduced by (Drobics et al., 2012). Different standards specify communication protocols for data sources and sensors, application host devices, WAN devices and health reporting network (HRN) devices. The multitude of available standards and frameworks results in a complex environment making it difficult for end users to pick the right products for his needs. Organizations like the Continua Health Alliance aim at providing a survey of interoperable products and supporting the end users.

5.4 REQUIREMENTS OF AAL/ELE APPLICATIONS

In the following, we derive general requirements that are imposed on the technical infrastructure by AAL/ELE applications. It is based on the classification and characterization conducted in the previous sections.

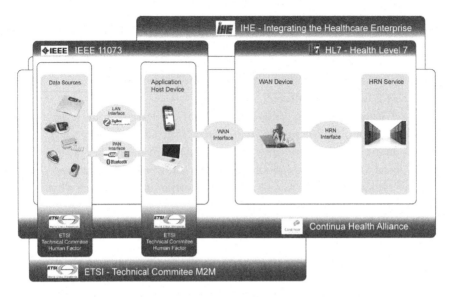

Figure 5.2 Technical overview of relevant standards and involved organizations.

Requirement 1 (SLA): AAL/ELE Need Dedicated Service Level Agreements (SLA) Between Actors and Network Service Provider

With respect to the many different and diverse AAL/ELE applications, we have to differentiate between the following communication service classes, which impose different requirements on the network:

1. Exchange of data, e.g., medical health data;
2. Real time communication, either as peer-2-peer communication or multi-peer communication based on voice, video and data;
3. Sensor data exchange — M2M communication;

The requirements of these applications have to be mapped to the underlying technical parameters of the system infrastructure, e.g., network Quality of Service (QoS) parameters like maximum bandwidth, minimum jitter, maximum packet loss or the average packet delay. In many cases, however, it may be sufficient that an upper limit for the provided QoS is ensured. For the example of health monitoring it is desirable that the transmission of data is done in a timely manner so that the transmitted information can be incorporated within the medical, simultaneously-held consultation. This provides at the network level the requirement for an upper limit on the packet delay. More general, in the next generation communication world, as exemplified by the AAL/ELE services, additional QoS parameters are becoming increasingly important. These are for instance the maximum end-to-end delay, and the reliability/availability of the service:

- Maximum end-to-end delay: this means that all parts of the system infrastructure have to be considered. A notification of a sensor may traverse different networks until it is processed and an appropriate action is triggered. Hence, this includes the transmissions and processing delays of the involved components.
- Reliability/availability: This includes the awareness of the status of the system and its components, e.g., the application has to know, whether a sensor is active or not, or whether the generated data is valid or not.

The heterogeneous environment and the huge variety of AAL/ELE applications exacerbate the fulfillment of these requirements. This includes varying channel conditions of wireless technologies resulting in bandwidth bottlenecks or energy-saving mechanisms of mobile devices increasing the delay.

In addition to the QoS requirements as summarized above, AAL/ELE applications stimulate additional system requirements, which are essential for achieving real added value for the end-user and the different stakeholders in the communication scenarios. These requirements are presented in the following.

Requirement 2 (Costs): Low Upfront Infrastructure Investments for the User's Premises Equipment

This requirement highlights the need for re-usability of existing network elements like Set-Top-Boxes, home gateways, smartphones, tablets, or special purpose devices for the different application scenarios. The different services should be independent of the final HW/SW platform, and the system should adapt the user interface to the device capabilities. This allows a step-wise introduction of different AAL/ELE services and is the basis for a positive business case per service. Thus, remote maintenance and service update mechanisms of the network elements in the home network will be important functions. Initial attempts are TR069 or the OSGi Alliance supporting the "dynamic download of Apps", meaning the support for integration of dynamic software components (often called bundles) into AAL platforms. Adaptation and extensibility are important properties of AAL/ELE platforms: Consider a platform being installed on the house premises of an elderly to support him with various activities. The platform includes the programs to support specific functions (remind the patient about medication, supervise some activity, etc.). However, in the future, we would want the platform to add additional functionality (i.e., the patient develops new symptoms for which additional support is required, or the family want to add additional monitoring sensors, etc.). The "dynamic download of Apps" is all about creating the infrastructure with minimum functionality (i.e., low upfront investment), and have the possibility to support dynamic download and execution of additional software components in the future, when and if needed.

Requirement 3 (Usability): Intuitive User Interfaces, Enhanced Usability Due to Self-* Capabilities and Easy Operation/Configuration of the Service

Usability is essentially important for two reasons: (i) on the one hand to support the requirements of handicapped end-users (ill or very old); but even more important, (ii) to ensure a very high data accuracy. We refer here to two aspects: First, it is the usability of the user interface, which is an important feature of AAL applications, since most end users (including the medical personnel) may not be familiar with the use of technology. Second, there is the data accuracy, which is highly relevant for usability. If a data item from a sensor is not valid according to the process definition, the corresponding information is potentially rendered useless. This also includes usability issues related to the configuration and setup of the AAL/ELE application and easy maintenance.

For the medical area, the usability requirement further addresses the appropriate representation of the data. This is required to get the relevant information with added value for the user. Further, the demands for data set accuracy are more stringent, i.e., to missing or changed data.

Requirement 4 (Security): Privacy and Data Security to Implement Different Security Levels for AAL/ELE Services

If specific security levels have to be defined and even validated, dedicated network architectures have to be defined beforehand to do so. For instance, the patient's vital parameters have to be transported via the network without any related to the actual identity of the patient.

Requirement 5 (Sensor Interoperability): Sensors — Interoperability for Data Collection

Many different types of sensors or even special purpose equipment from different markets and industries like health, care, smart home, smart grid, entertainment, games, or business, request an interoperability of the different sensor devices within the home network. This includes also wearable devices connected in a personal area network or sensors interconnected by a home LAN. Despite the interoperability of sensors with each other this foremost includes the interoperability between sensors and a data collection function in the home network. The data sources must have different interfaces ranging from the analogue signal to wired and wireless interfaces. Accordingly, appropriate adapters have to be defined. It is important to note, that it is not clear if sensor data has to be immediately converted into a common format at the sensor or only in a "transferable" signal; e.g. an analog signal into a digital signal. The processing of the data is then performed at a later stage at a server somewhere in the network.

Requirement 6 (Data Characteristics): Sensors — Data Transmission Characteristics

Based on application requirements data will be sent a) continuously in a well defined order, or b) only when specific levels are passed or events happened, or c) only if requested by a user. By this, the amount of data to be transported will be limited, saving networking resources but contributes also to the privacy requirement. The network, however, has to be able to support specific requirements of the sensors like guaranteed delivery or a maximum latency.

Requirement 7 (Application Interoperability): Interoperability at Application Level Between Sensor Devices and Back End Systems

Data generated by sensors in the home network has to be exchanged with back-end systems. Dedicated standards have been developed for ensuring this interoperability above network level like DICOM, IHE, HL7, and/or Continua (Rogers et al., 2010). For a more detailed analysis of such standards, their roles and challenges for interoperability, we further refer to Moorman (2010).

DICOM (or, Digital Imaging and Communications in Medicine) is a standard for handling imaging data. The standard assists communication between various image based modalities and accessories to each other. It provides reliable protocols for integration of image data between imaging, nonimaging modalities, devices and systems. The functional elements broadly comprise of Protocols, Objects, Services, Service Class and Conformance (National Electrical Manufacturers Association and others, 1993).

For managing non-imaging data, HL7 (Health Level Seven) provides protocols for exchange, management and integration of clinical and administrative electronic health data. Health Level Seven is considered by many as the accepted global standard for exchange, integration, sharing and retrieval of electronic health information in Hospitals.

IHE (or, "Integrating the Healthcare Enterprise") is more of a strategy to integrate various health-related workflows, using standards such as DI-COM and HL7 (or, as Henderson et al. (2001) defines, it is a "multi-year initiative that creates the framework for integrating applications, systems and settings across the entire healthcare enterprise"). IHE accomplishes the integration by a four stage process: a) interoperability problem identification; b) integration profile specification; c) implementation and testing and d) integration profile conformance statements (Kuperman, 2011).

5.5 NETWORKING TECHNOLOGIES AND THEIR IMPACT ON THE AAL/ELE REQUIREMENTS

This section highlights the drawbacks of current networking infrastructures, and discusses why they do not meet the requirements of AAL/ELE services. Afterward, different networking technologies and paradigms, seen as building blocks for a future Internet, are presented. The focus of this section is on the question whether network technologies today can meet requirements of future ALA/ELE infrastructures. The results are summarized in Table 5.2.

Drawbacks of Current Networking Infrastructures

AAL/ELE services rely on the use of current telecommunication infrastructures, such as mobile and fixed telephone operator networks and the Internet. The Internet itself was built around a "best-effort" philosophy, meaning that no guarantees are provided concerning the data transmission. For many services, this is acceptable when no specific QoS requirements have to be fulfilled, enough network bandwidth is available or the applications don't generate high volumes of data.

Different strategies are available to deal with the "traffic management problem" beyond best effort in networks. This includes over-provisioning, reserved bandwidth — either on physical links based on transmission technologies like WDM or dedicated IP protocols like RSVP —, priorities — e.g. Ethernet priority bit or priority fields of the IP —, or flow-control mechanisms between sender and receiver. For traffic management reasons, even dedicated network architectures are implemented by network operators like the well dimensioned IPTV network infrastructure. The IP-based

Table 5.2 Impact of specifc technologies and functionalities on the defined requirements. ** denotes that the requirements can be fulfilled completely, * that the provided functionalities may provide a significant improvement towards the fulfillment of this requirement.

	SLAs	Costs	Usability	Privacy/security	Data inter-operability	Sensors	Application interoperability
DSFs	*				**		**
Network virtualization	*	*	*	*			
SDN	*	*	*	*		*	
Application-awareness	*	*		*		*	
Cloud/NFV		*					

IPTV service is based on a network infrastructure separated from the "public" Internet network elements to ensure proper system performance. This differs to the progressive download of video packets used to transport a video stream over a best effort network service like Apple-TV. Here the application level implements additional mechanisms to cope with best-effort network behavior.

Although these techniques may address at least some of the introduced requirements, they have several limitations resulting in a limited deployment and market acceptance. Among these are the missing support of content-centric networking, security, end-to-end QoS support via different network domains, Interdomain Name-Based Routing, or the IP hourglass bottleneck.

Dynamic Software Frameworks (DSFs) to Support AAL/ELE

Such software frameworks provide a dynamic component model, where application and components (called bundles) can be dynamically integrated and removed without a reboot. Communication between the components is typically realized using an abstraction layer. This allows the flexible and easy interconnection of appliances and devices in a home network. Examples of such systems are the specification by the Open Service Gateway initiative (OSGi, OSGi Alliance (2015)) and the flexible Smart Home service delivery platform provided by the Home Gateway Initiative (HGI, Rogers et al. (2010)).

The OSGi specifications standardize secure and reliable service delivery and provisioning, for remote life cycle management of services, and for bridging different networking standards. Applications and components, coming in the form of bundles for deployment (or plug-ins), which are tightly coupled, dynamically loadable collection of classes, jars, and configuration files. OSGi was originally conceived to be a gateway for managing smart appliances and other Internet-enabled devices in the home. From there, several efforts continued towards adopting OSGi as an open and standard platform for telematics services, with applications ranging from mobile phones to the open-source Eclipse IDE. Various OSGi-based middlewares, like Sensor Node Plugin System (SNPS) (Di Modica et al., 2013), the Alcatel Lucent's M2M E2E solution, and others, look at sensors and services able to be used and composed over the Internet, providing support for composition in complex applications.

OSGi, however, provides limited interoperability support. For example, in SNPS, Base Stations (BSs) implement the logic for locally managing several attached sensors. Many such BSs may be attached to different physical networks, but in this case the communication between the two bundles are implemented as Remote Services (R-OSGi). This means, that independent services have to be developed by a BS manufacturer to provide the bridge between proprietary protocols and a SNPS allowed data format. R-OSGi provides for this a specific OSGi bundle, offering the support for remote communication with other bundles living in different runtime contexts. The resulting communication between come at the cost of additional complexity in the application development.

Hence, other initiatives addressing the interoperability between different technologies emerged. One such initiative is the Home Gateway Initiative (HGI), who aims at providing ways to deliver services in the digital home. HGI is an open forum launched by a number of telephone and manufacturing companies in 2004, with the aim to release specifications of the home gateway. The initiative takes as a basis the work undertaken within existing bodies such as ITU-T, Broadband forum, DLNA, or OSGi Alliance. It aims at producing requirements for a residential gateway enabling end-to-end delivery of services. To ensure interoperability it closely works together with manufacturers.

At the basis, a universal template facilitates interworking between home devices and smart home applications. This universal template is a component of a logical abstraction layer used to provide smart home services to broadband consumers. The aim of the abstraction layer is to allow smart home applications authored by different companies to easily connect to devices using one of several smart home interface technologies. The applications do not need to know which interface technology is used, but only the device capabilities that are described in the template. The home gateway (HG) plays a central role in the digital home, interconnecting computers, devices on the home network, and the Internet, all while supporting Quality of Service and remote management. Service providers are increasingly looking to deliver HG-based consumer services such as energy management, media server, and home network diagnostics. Pairing the dynamic and modular OSGi technology with HG-specific Application Programming Interfaces (APIs) and protocols will greatly extend the service capabilities of the home gateway.

In 2009, the OSGi Alliance and the Home Gateway Initiative (HGI) made a partnership to enable broadband service providers to offer more flexible applications to residential customers. Under the agreement, OSGi

Service Platform will eventually be integrated into the home gateway, creating a software execution environment that will facilitate the deployment of new service capabilities into the digital home.

Different other initiatives and projects rely on OSGi like Qivicon, also targeting at home networks, or UNIVERSAAL and OPENAAL (Wolf et al., 2010), both aiming at sensor nodes and middleware in the AAL/ELE context. The EU-funded UNIVERSAAL project aims at producing an open platform along with a standardized approach for making it technically feasible and economically viable to develop AAL solutions. For that, it defines and provisions a reference implementation of a platform that facilitates the realization of AAL systems.

Similarly, the joint open-source initiative openAAL develops a middleware for ambient-assisted living scenarios based on research results of several German and international projects. The goal is to have a platform that enabling the easy implementation, configuration and situation-dependent provisioning of flexible, context-aware and personalized IT services.

Impact on the Requirements: These technologies have an impact especially on interoperability of sensors and applications and may fulfill the requirements 5 and 7. Due to reusability of data functions and sensors in other contexts, the amount of necessary equipment and therewith the costs can be reduced, and the usability may be increased (requirement 2 and 3).

Network Virtualization

Network Virtualization (NV) enables the operation of multiple logical networks upon a shared physical infrastructure (Chowdhury and Boutaba, 2010). It permits distributed participants to create almost instantly their own network with application-specific naming, topology, routing, and resource management mechanisms. The role model includes physical infrastructure providers, virtual network providers and operators, and also application service providers and enables an automatic interaction between the different roles including brokering of virtual networks with certain SLAs (Meier et al., 2011). VN is thus seen as an enabler for application tailored networks with specific resource guarantees across multiple separated administrative domains.

Impact on the Requirements: NV may provide resource guarantees for a virtual network and enables the logical separation of different virtual networks and thus primarily addresses the requirements 1 and 4. It fulfills requirement 1 as long as the involved applications are well-known and can

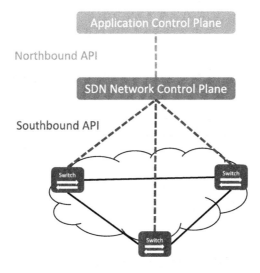

Figure 5.3 SDN Interfaces.

be controlled. Otherwise additional mechanisms are required to enable a control of specific applications within a virtual network.

Software-Defined Networking

The goal of Software-Defined Networking (SDN) is to increase flexibility and innovation in the network and, thus, to improve the efficiency of network operation and the service quality as well as lead to reduce of CAPEX and OPEX. This is facilitated by the removal of the network control plane from the distributed network devices to a logically-centralized control entity, which enables the introduction of new open interfaces between the application, the data-plane, and the control plane (Jarschel et al., 2014). With these interfaces, the network control plane can be realized as a freely programmable software, which can essentially be described as an operating system for the network. The network operating system, often called "controller", is responsible for all forwarding decisions within the network it controls. The network devices forward the traffic according to the rules set by the controller.

Figure 5.3 illustrates the relationship of involved control planes and interfaces. The "Southbound-API" represents the interface between data- and control-plane. Current SDN implementations often use the Open-Flow protocol (McKeown et al., 2008) as a realization of this interface. The OpenFlow protocol handles the communication between the individual network devices and the controller. Each of the network devices maintains a set of "flow rules" matching individual network flows in so called "flow tables". The term "flow" refers in this context to packets matching a general set of header fields either out of layers 2 to 4 of the ISO/OSI stack or headers defined by the operator of the network. Additionally, a flow rule contains a set of one or more actions that define how a packet matching the rule should be handled as well as flow statistics.

When a packet reaches an OpenFlow-enabled SDN switch, it is buffered and the packet header is matched against the rules in the flow table. In case of a successful match, the action(s) specified in the rule are executed. If there is no matching rule in the flow tables, the packet is either dropped or an OpenFlow "packet-in" message containing the packet header is sent to the controller for processing. The controller calculates the action the network element should take with regard to the packet and communicates it. Furthermore, the controller can specify a flow rule and send it to the network element(s). This way all following packets of the flow are treated the same way by the network and the controller does not need to be involved any longer.

The controller can also introduce new flow rules or modify existing ones without being triggered by an incoming packet. For example, the controller may adhere to a pre-programmed schedule or implement a network policy. This is where the flexibility of SDN comes into play. Where traditional network devices would have to be reconfigured by an administrator, SDN enables the automatic and seamless implementation of changes in the forwarding behavior of the network. These changes can be triggered by external entities via the other key SDN interface — the "Northbound-API". This interface makes application-awareness in the network feasible as it opens up a communication channel between the applications using the network and the controller, which can then utilize information provided by the applications to adapt its policy and the network traffic on different levels of granularity.

Research related to AAL/ELE aims at enabling a less complex management of home networks (Kim and Feamster, 2013) or a dedicated resource control for specific applications (Zinner et al., 2014; Jarschel et al., 2013).

Impact on the Requirements: SDN allows a dynamic, more centralized control of the network and its data flows and thus addresses requirement 1. The externalization of the network control plane reduces network equipment costs and impacts requirement 2. Due to the vendor-independent access to networking hardware usability is simplified (requierement 3). Additionally, SDN may also be used to separate data flows and to control network resources. Thus, it also addresses requirement 4.

Application-Aware Networking

Application-Aware Networking (AAN) is an approach to improve the service quality in networking scenarios with limited resources (Staehle et al., 2010; Wamser et al., 2014; Jarschel et al., 2013; Qazi et al., 2013; Ferguson et al., 2013; Huang et al., 2013; Georgopoulos et al., 2013; Thakolsri et al., 2009). Application needs are incorporated in the network management decision in a dynamic way. Thus, in the case of limited resources, a quality-related decision can be made that enhances the service quality.

For that, services are divided into groups with similar quality demands according to their requirements. The packet forwarding in the network is carried out in relation to these groups with regard to the available resources. The requirements are determined on the basis of application information. Application information are, for example, status of the application (e.g. idle, active, downloading, content synchronizing), type of application (e.g. chat, browsing, interactive application) or inherent application parameters (e.g. video buffer level, precaching ability). By monitoring the application information, decisions and actions are taken that affect packet processing, packet forwarding or the network settings. The aim is to better address critical applications and to distribute the traffic according to the actual requirements of the applications, i.e., by allocating more network resources to them.

A key component to realize AAN is application-specific monitoring. This application monitoring can be done on network entities inspecting the data packets (Wamser et al., 2014), or by passing relevant monitoring from the application to the network via specific interfaces (Zinner et al., 2015).

Impact on the Requirements: Based on application information, a better and more accurate quality-of-service agreement can be made (requirement 1). Furthermore, with respect to requirement 2, the targeted use of application information may help to distinguish between critical and non-critical applications, in order to reduce the utilization of resources and, ultimately, the costs. Furthermore, safety-critical applications might be detected on the network and can be specifically treated with the help of other networking technologies (requirement 4). Further on, by utilizing application information, the network might have the ability to better support specific requirements of sensors like guaranteed delivery or a maximum latency (requirement 6).

Cloudification/Network Function Virtualization (NFV)

Cloud Computing (Vogels, 2008; Fox et al., 2009) describes the idea of outsourcing computing power and storage in Internet data centers. Services or small devices are virtualized and pushed into the Internet, to work in large-scale datacenters. This allows to access an almost unlimited number of computing resources. For service providers, the cloud paradigm brings a good maintainability and the ability to scale the utilized resources on-demand with respect to their current requirements. For the AAL/ELE environment, the cloud paradigm can provide both scalability and energy savings as well as computing power to support of a large number of devices.

NFV in this context allows dynamic and programmable network functionality. Dedicated functions are virtualized, implemented in software, and executed on standard server hardware. Just as in the cloud, not only services on the Internet can be instantiated but also parts of the network architecture can be provided in a flexible way. Features that are present and written in software, can be used interchangeably on different hardware, which reduces costs.

The FP7 project Fusion (Griffin et al., 2014) particularly works towards cloud-like systems for interactive services and execution resources. To achieve this goal, design issue, necessary interfaces, and algorithms for dealing with service provisioning and scaling are discussed. In the area of AAL/ELE a dynamic provisioning of interactive services in the cloud might allow to transfer specific services from the user's premises to the cloud. The trend to utilize the cloud to centralize specific functions recently jumped over from the application domain to telecommunication networks, where it is known as NFV (Chiosi et al., 2012). Here, network providers want to leverage standard IT virtualization technology to consolidate data plane packet processing and control plane functions in WAN networks instead of using proprietary middleboxes. In the context of Cloudification/NFV, SDN is seen as an enabler for a flexible forwarding of data flows to specific virtualized functions and to allow a proper service chaining.

In the EU H2020 INPUT (INPUT Consortium, 2015) project, this idea is further contemplated. The aim of this project is to develop personalized cloud services. It means that for each user, an individual cloud application can be made available. Personalized services can better meet the exact requirements and can be specifically modified to meet the conditions of the individual person. An additional aspect that comes here into play is the better security in personalized, encrypted cloud instances.

Impact on the Requirements: The scalability provided by using and paying for resources on demand is seen as enabler to reduce capital expenditures for equipment and also the operational expenditures. Hence it primarily addresses requirement 2.

5.6 KEY DERIVATIONS

The above definitions and requirement analysis aims to create an understanding of the applicability of upcoming networking technologies for AAL/ELE. Although much research has been conducted, AAL/ELE systems have not made it yet to a widespread real-world usage. We argue that the corresponding technical requirements of operating such systems have not been fully understood yet. Hence, there is a lack of a clear identification of the requirements that are needed for assessing technical concepts for their suitability of AAL/ELE.

In order to provide an assessment methodology, we (1) identify four domains that essentially need to be taken into account in an AAL/ELE system. Furthermore, we (2) determine the corresponding users and actors that interact with each other. We (3) discuss possible application cases and define, based on the cases, (4) seven technical requirements which have to be fulfilled to ensure a proper market acceptance.

Following this, we investigated promising networking technologies, challenging whether they fulfill each AAL/ELE identified individual requirements. Our analysis shows that none of the existing technologies is able to fulfill all the presented requirements, i.e., that several technologies have to be combined to enable a proper acceptance of AAL/ELE.

Further, there may also be requirements which may not be fulfilled by one specific technology, but where several technologies might be combined, e.g., network virtualization, SDN and application awareness. The communication between the involved technologies, however, needs to be realized using open interfaces to enable fast innovation and adaptation of new technologies. We believe all these will need to be addressed in the near future, to advance properly the AAL/ELE domain towards its true market potential.

5.7 CONCLUSION

The increase in medical expenses due to societal issues like demographic ageing, puts strong pressure on the sustainability of health and social care systems. Different AAL/ELE technologies are today being developed, but systems do not yet take place at a relevant scale.

The chapter is a step towards a better understanding of the requirements of AAL/ELE, its domains and stakeholders. Based on an inductive approach we derived seven requirements highlighting the need for specific SLAs, a high degree of flexibility in the involved networks, and a good usability. In a second step we evaluated several technologies against the requirements and identified which requirements can be fulfilled by them. This approach can be adapted to help to classify other technologies and gauge the potential benefits of using them in the context of AAL/ELE. Their main features can be identified and weighted, and the implementation of the system can be planned accordingly.

ACKNOWLEDGEMENTS

This work was partly funded by COST Action IC1303 — AAPELE, Architectures, Algorithms and Platforms for Enhanced Living Environments. The authors alone are responsible for the content.

REFERENCES

Braem, B., Latré, B., Blondia, C., Moerman, I., Demeester, P., 2008. Improving reliability in multi-hop body sensor networks. In: SENSORCOMM'08. Second International Conference on Sensor Technologies and Applications, 2008. IEEE, pp. 342–347.

Chen, M., Wan, J., González, S., Liao, X., Leung, V., 2014. A survey of recent developments in home M2M networks. IEEE Commun. Surv. Tutor. 16 (1), 98–114.

Chiosi, M., et al., 2012. Network functions virtualisation introductory white paper. In: SDN and OpenFlow World Congress.

Chowdhury, N.M.K., Boutaba, R., 2010. A survey of network virtualization. Comput. Netw. 54 (5), 862–876.

Di Modica, G., Pantano, F., Tomarchio, O., 2013. SNPS: an OSGi-based middleware for wireless sensor networks. In: Advances in Service-Oriented and Cloud Computing. Springer, pp. 1–12.

Drobics, M., Dohr, A., Leopold, H., Orlamünder, H., 2012. Standardisierte Kommunikation in der IKT bei AAL und eHealth. In: Technik für ein selbstbestimmtes Leben.

Ferguson, A.D., Guha, A., Liang, C., Fonseca, R., Krishnamurthi, S., 2013. Participatory networking. In: ACM SIGCOMM 2013 Conference — SIGCOMM '13. ACM Press, New York, New York, USA, p. 327. http://cs.brown.edu/~rfonseca/pubs/sigcomm13.pdf. http://dl.acm.org/citation.cfm?doid=2486001.2486003.

Fox, A., et al., 2009. Above the clouds: a Berkeley view of cloud computing. Rep. UCB/EECS, 28, 13. Dept. Electrical Eng. and Comput. Sciences, University of California, Berkeley.

Georgopoulos, P., Elkhatib, Y., Broadbent, M., Mu, M., Race, N., 2013. Towards network-wide QoE fairness using OpenFlow-assisted adaptive video streaming. In: ACM SIG-COMM 2013 Workshop on Future Human-Centric Multimedia Networking — FhMN '13. ACM Press, New York, New York, USA, p. 15. http://dl.acm.org/citation.cfm?doid=2491172.2491181.

Gondal, I., Sehgal, S., Iqbal, M., Kamruzzaman, J., 2007. Ambient cardiac expert: a cardiac patient monitoring system using genetic and clinical knowledge fusion. In: 6th IEEE/ACIS International Conference on Computer and Information Science, 2007. ICIS 2007. IEEE, pp. 496–501.

Griffin, D., et al., 2014. Deliverable d3.1 — initial specification of algorithms and protocols for service-oriented network management. http://www.fusion-project.eu/deliverables/deliverable_3.1_ver-1.0-public.pdf.

Henderson, M., Behlen, F.M., Parisot, C., Siegel, E.L., Channin, D.S., 2001. Integrating the healthcare enterprise: a primer: Part 4. The role of existing standards in IHE. Radiographics 21 (6), 1597–1603.

Huang, T., Johari, R., McKeown, N., 2013. Downton abbey without the hiccups. In: ACM SIGCOMM 2013 Workshop on Future Human-Centric Multimedia Networking — FhMN '13. ACM Press, New York, New York, USA, p. 9. http://dl.acm.org/citation.cfm?doid=2491172.2491179.

Hung, K., Zhang, Y.-T., 2003. Implementation of a WAP-based telemedicine system for patient monitoring. IEEE Trans. Inf. Technol. Biomed. 7 (2), 101–107.

INPUT Consortium, 2015. Eu h2020 input: in-network programmability for next-generation personal cloud service support. www.input-project.eu.

Jarschel, M., Wamser, F., Höhn, T., Zinner, T., Tran-Gia, P., 2013. SDN-based application-aware networking on the example of YouTube video streaming. In: 2nd European Workshop on Software Defined Networks (EWSDN 2013). Berlin, Germany.

Jarschel, M., Zinner, T., Hoßfeld, T., Tran-Gia, P., Kellerer, W., 2014. Interfaces, attributes, and use cases: a compass for SDN. IEEE Commun. Mag. 52 (6), 210–217.

Khoor, S., Nieberl, J., Fügedi, K., Kail, E., 2001. Telemedicine ECG-telemetry with Bluetooth technology. In: Computers in Cardiology 2001. IEEE, pp. 585–588.

Kim, H., Feamster, N., 2013. Improving network management with software defined networking. IEEE Commun. Mag. 51 (2), 114–119.

Kuperman, G.J., 2011. Health-information exchange: why are we doing it, and what are we doing? J. Am. Med. Inform. Assoc. 18 (5), 678–682.

Lee, R.-G., Chen, H.-S., Lin, C.-C., Chang, K.-C., Chen, J.-H., 2000. Home telecare system using cable television plants — an experimental field trial. IEEE Trans. Inf. Technol. Biomed. 4 (1), 37–44.

McKeown, N., Anderson, T., Balakrishnan, H., Parulkar, G., Peterson, L., Rexford, J., Shenker, S., Turner, J., 2008. OpenFlow: enabling innovation in campus networks. Comput. Commun. Rev. 38 (2), 69.

Meier, S., et al., 2011. Provisioning and operation of virtual networks. Electron. Commun. EASST 37.

Memon, M., Wagner, S.R., Pedersen, C.F., Beevi, F.H.A., Hansen, F.O., 2014. Ambient assisted living healthcare frameworks, platforms, standards, and quality attributes. Sensors 14 (3), 4312–4341.

Mendoza, G., Tran, B., 2002. In-home wireless monitoring of physiological data for heart failure patients. In: Engineering in Medicine and Biology. Proceedings of the Second Joint 24th Annual Conference and the Annual Fall Meeting of the Biomedical Engineering Society EMBS/BMES Conference, 2002, vol. 3. IEEE, pp. 1849–1850.

Modre-Osprian, R., Pölzl, G., von der Heidt, A., Kastner, P., 2014. Closed-loop healthcare monitoring in a collaborative heart failure network. eHealth 2014, 17–24.

Moorman, B., 2010. Medical device interoperability: standards overview. Biomed. Instrum. Technol., 132–138.

National Electrical Manufacturers Association and others, 1993. Digital Imaging and Communications in Medicine (DICOM). Parts 1–10. The Association.

Nedopil, C., Schauber, C., Glende, S., 2013. Knowledge Base: AAL Stakeholders and Their Requirements. Ambient Assisted Living Association.

OSGi Alliance, 2015. Open service gateway initiative. https://www.osgi.org/.

Park, S., Jayaraman, S., 2007. Wearable biomedical systems: research to reality. In: IEEE International Conference on Portable Information Devices, 2007. PORTABLE07. IEEE, pp. 1–7.

Pattichis, C., Kyriacou, E., Voskaride, S., Pattichis, M., Istepanian, R., Schizas, C.N., 2002. Wireless telemedicine systems: an overview. IEEE Antennas Propag. Mag. 44 (2), 143–153.

Pollard, J., Rohman, S., Fry, M., 2001. A Web-based mobile medical monitoring system. In: International Workshop on Intelligent Data Acquisition and Advanced Computing Systems: Technology and Applications, 2001. IEEE, pp. 32–35.

Qazi, Z.A., Lee, J., Jin, T., Bellala, G., Arndt, M., Noubir, G., 2013. Application-awareness in SDN. In: ACM SIGCOMM 2013 Conference — SIGCOMM '13. ACM, pp. 487–488.

Rashidi, P., Mihailidis, A., 2013. A survey on ambient-assisted living tools for older adults. IEEE J. Biomed. Health Inform. 17 (3), 579–590.

Rogers, R., Peres, Y., Müller, W., 2010. Living longer independently — a healthcare interoperability perspective. E&I, Elektrotech. Inf.tech. 127 (7–8), 206–211.

Staehle, B., Hirth, M., Pries, R., Wamser, F., Staehle, D., 2010. YoMo: a YouTube application comfort monitoring tool. In: New Dimensions in the Assessment and Support of Quality of Experience for Multimedia Applications. Tampere, Finland, pp. 1–3.

Thakolsri, S., Khan, S., Steinbach, E., Kellerer, W., 2009. QoE-driven cross-layer optimization for high speed downlink packet access. In: Special Issue on Multimedia Communications, Networking and Applications. J. Commun. 4 (9), 669–680.

Varshney, U., 2007. Pervasive healthcare and wireless health monitoring. Mob. Netw. Appl. 12 (2–3), 113–127.

Vogels, W., 2008. Head in the clouds — the power of infrastructure as a service. In: First Workshop on Cloud Computing and in Applications (CCA'08) (October 2008), vol. 5.

Wamser, F., Zinner, T., Iffländer, L., Tran-Gia, P., 2014. Demonstrating the prospects of dynamic application-aware networking in a home environment. In: Proceedings of the 2014 ACM Conference on SIGCOMM. ACM, pp. 149–150.

Wolf, P., Schmidt, A., Otte, J.P., Klein, M., Rollwage, S., König-Ries, B., Dettborn, T., Gabdulkhakova, A., 2010. OpenAAL — the open source middleware for ambient-assisted living (AAL). In: AALIANCE Conference. Malaga, Spain, pp. 1–5.

Zinner, T., Hoßfeld, T., Fiedler, M., Liers, F., Volkert, T., Khondoker, R., Schatz, R., 2015. Requirement driven prospects for realizing user-centric network orchestration. Multimed. Tools Appl. 74, 413–437. http://dx.doi.org/10.1007/s11042-014-2072-5.

Zinner, T., Jarschel, M., Blenk, A., Wamser, F., Kellerer, W., 2014. Dynamic application-aware resource management using software-defined networking: implementation prospects and challenges. In: IFIP/IEEE International Workshop on Quality of Experience Centric Management (QCMan). Krakow, Poland.

CHAPTER 6

Recent Advances in Remote Assisted Medical Operations

Loizos Toumbas*, Constandinos X. Mavromoustakis[†],
George Mastorakis[‡], Ciprian Dobre[§]
*University of Nicosia Medical School, Cyprus
[†] University of Nicosia, Cyprus
[‡] Technological Educational Institute of Crete, Greece
[§] University Politehnica of Bucharest, Romania

6.1 INTRODUCTION

The concept of robotics originates from the stuff of science fiction. Robots were first envisioned by the Czech playwright, Karel Capek, in 1921 as machines designed to replace monotonous work. In fact, the word robot originates from the Czeck word robota, which translates to 'forced labor'. Since then popular science fiction has anticipated robots as becoming ultra intelligent and humanoid, like R2D2 and C3PO from the star wars universe, with multiple capabilities, which have included medical ones. The reality so far has been slightly underwhelming. Nevertheless, robots have made their way into society and play important roles in industry. They are used to perform dangerous, repetitive tasks, which also require accuracy (e.g. automobile assembly and hazardous waste handling in nuclear industries), and in addition, tasks that demand great precision and dexterity (e.g. computer chip manufacturing). It was only a matter of time before robotics would find a use in medicine.

The first pioneers in the field of surgical robotics were Scot Fisher and Joseph Rosen. In the late 1980s they combined developments in virtual reality and the ability to interact with 3-dimensional virtual scenes with the concept of surgical robotics (Fisher et al., 1984). Together they hypothesized the ability to perform telepresence surgery with the use of robotic arms from a remote position. The term telepresence can be defined, in essence, as the ability to give surgeons the sense that they are operating on something directly in front of their eyes, which could in fact be on the other side of the room, or even thousands of miles away. Thus, surgeons could operate on a patient who would otherwise be completely

Ambient Assisted Living and Enhanced Living Environments.
DOI: http://dx.doi.org/10.1016/B978-0-12-805195-5.00006-5

inaccessible. The research by Scot Fisher and Joseph Rosen was conducted in cooperation with the National Aeronautics and Space administration (NASA) and funded by the Defence Advanced Research Project Administration (DARPA) with the main objective of providing remote surgical assistance and healthcare to astronauts in space, or soldiers on the battlefield.

In parallel to these developments, in 1988 to 1989, the development of laparoscopic surgery burst onto the medical scene. Laparoscopic or keyhole surgery is now widely used and provides significant benefits to patients, being less invasive than conventional open surgery. Nevertheless it does have its negatives and recently laparoscopy has reached a plateau in terms of its development (Ballantyne, 2002). Laparoscopy creates a number of difficulties for the surgeon, in that, the surgeon loses a sense of touch so that it becomes harder to judge how much force needs to be applied, the surgeon is limited to a 2-dimensional view, and has reduced dexterity due to a limited range of motion and the fulcrum effect (the tool endpoints move in the opposite direction to the surgeons hands, making laparoscopic surgery less intuitive and harder to learn). This means that many complex procedures are difficult to perform laparoscopically unless the surgeon is well experienced. It has since become clear that surgical robotics and telepresence could be used to address these difficulties.

In essence, therefore, the aim of surgical robotics is not to try and replace the surgeon with a fully automatic machine that can carry out surgeries autonomously, but rather augment the surgeon's capabilities to make more complex surgeries simpler for the surgeon and safer for the patient. Indeed, it is believed that not only will robotic and telepresence surgery address the limitations of laparoscopic surgery, but expand the frontiers of surgery and continue to create surgical innovations. It is not at all implausible that robotics could be used in nearly all surgeries in the future. This would in turn require major changes to surgical training, as robotics will no longer be a field for the specialist, but a key component to surgical training. Furthermore, the recent development of robotic surgical simulators and telementoring also offer new solutions to even help lower the learning curve so that less-experienced surgeons are more prepared for surgery on patients.

6.2 THE DEVELOPMENT AND HISTORICAL BACKGROUND OF CURRENT SYSTEMS

A robot is a device, which can be programmed to carry out specific tasks or sequences, either automatically, or via manual control from a computer-based and or mechanical interface. It is a feat of mechanical, electrical and informatics engineering. In surgery, robots can be programmed to aid in the positioning and manipulation of surgical instruments, in order to aid the surgeon in more difficult surgeries and tasks. Again, currently surgical robots have not been programmed to act independently, but rather under the control of the surgeon.

The first system to be developed for clinical use was a modified PUMA 200 industrial robot (Programmable Universal Manipulation Arm; Unimation, Stanford, California, USA) in 1985 (Kwoh et al., 1988). It was developed to perform CT-guided brain biopsies with increased accuracy. Other early systems include the PROBOT (Harris et al., 1997) (Imperial College, London, UK), which was developed to perform prostatic resections with ultrasound guidance, and ROBODOC (Paul et al., 1992) (Integrated Surgical Systems, Sacramento, California, USA) was approved in 1992 for hip replacement surgery. These early systems, however, never became commercially viable because the computer interfaces used were too limited, and the length of time needed to plan and setup operations was too long.

Currently there are two main systems that have been developed and approved by the US Food and Drug Administration (Satava, 2002): the Zeus® system (Computer motion, Goleta, California, USA) and the da Vinci® surgical system (Intuitive Surgical, Sunnyvalle, California, USA). These machines act as extensions of the surgeon, and both are described as *master-slave manipulators*. The two systems are both composed of two components connected by data cable and a computer.

The first component is the *surgeon's master control*, which is the robot's user interface. The user interface has a number of functions. It allows the surgeon to view the surgical field from a 3-dimensional image relayed by an endoscopic camera inside the patient's body, so that the surgeon feels they are directly at the surgical site. Additionally, the surgeon can control the movements of the robotic arms using handles or joysticks, called surgical manipulators. These are designed to filter out the tremor of the surgeon and use motion scaling, whereby the surgeons large natural movements are converted to ultraprecise micro movements, increasing the surgeon's accuracy and precision. Finally, there is the control panel for focusing the camera, adjustment of motion scaling and any accessory units.

The second components of the master-slave systems are the robotic arms, situated in close proximity to the patient and the operating table, known as the *Patient-side slave robotic manipulators*. It is through these robotic arms that the surgeon can manipulate the surgical instruments and the camera, which are inserted into the patient through laparoscopic ports.

Presently there is a monopoly in the surgical robotic market, as the company Intuitive Surgical now owns both systems. In addition, after the merging of the companies in 2003, the Zeus platform was discontinued. The da Vinci robot had a number of advantages over the Zeus platform that led to this. One of the most important reasons being that the da Vinci system used a 3D immersive camera, whereas the Zeus robot displayed the surgical field on a 2D screen, and therefore, did not solve the problem of depth perception that arises in laparoscopic surgery. In addition, the da Vinci's manipulators offered superior control and dexterity. The da Vinci system controls the surgical instruments using microarticulations (*EndoWrist®*), these mimic the actions of the wrist, allowing for more natural movements with 7 degrees of freedom (the largest amount of movement possible at a joint). In contrast, the camera of the Zeus robot was voice controlled, and the other arms controlled by a haptic interface. This did not allow as stable and precise navigation as the da Vinci robot.

The latest generation of da Vinci systems is the da Vinci Xi™ (Intuitive Surgical, 2014), which was released in 2014. It is designed in a more ergonomic way in order to save space and reduce the size. The robotic arms are thinner and arranged in a specific way to allow multiquadrant procedures. In addition, it has been upgraded over the years to include other features, such as fluorescence imaging and near infrared technology, which allows for better visualization of vessels, blood flow and related tissue perfusion, and bile ducts. One of the major drawbacks of the system, from which there has yet to be a breakthrough, is the lack of haptic feedback. In that it has not been possible yet to recreate the sensations of touch to the user.

6.3 TELEPRESENCE AND TELESURGERY

The development of these surgical robotic systems brings with it the possibility of controlling these robots remotely, also described as telepresence surgery. In essence, the surgeon would be able to be transported to anywhere where they are needed, without having to travel anywhere. The surgeon could control the robotic arms in the comfort of their own office, from a console thousands of miles away from where the actual surgery is taking place.

Indeed, this now a reality. In September 2001, the first transatlantic procedure, known as operation Lindbergh, took place (Marescaux et al., 2002). A surgeon in New York (USA) operated on a patient in Strasbourg (France). The patient was a 68-year-old woman suffering from abdominal pain with ultrasound evidence of gall bladder stones. The gallbladder was removed by remote robot-assisted laparoscopic cholecystectomy. It was performed using the Zeus robot. Two robotic arms for instrument manipulation and the one robotic arm for the endoscopy camera were based in Strasbourg (at the patient's side), and the surgeon's console in New York. In addition, the procedure required both the rapid and accurate transmission of information, in order to convert video images into electronic signals. In order to achieve this, two computers using high-speed communication channels connected the two locations. These were established through asynchronous transfer mode (ATM) technology (France Telecom/ Equant's, Paris, France) and a high-speed fibreoptic connection with an average latency delay of 155 ms, despite a round-trip distance of more than 14,000 km. The surgery was completed in 54 minutes without difficulty or complications, and the surgeons found the procedure to be safe and reliable. The patient recovered well and was discharged two days after the operation.

The surgery was a great success, and proved that telepresence surgery was now a reality. Indeed, since then a telesurgery center has been set up in Canada with a distance of 400 km between the two stations (Anvari, 2007). The service has successfully treated 21 patients in a number of different procedures. Thus, as well as improving a surgeon's ability to perform a surgery, robot systems can be used to carryout surgeries from remote locations. Notwithstanding, there are a number of issues with telesurgery, namely the delays between operator input and instrument reaction over very long distances (refer to Figure 6.1). This is described as the latency factor. The estimated total delay compatible with a safe operation is roughly 330 ms. This was demonstrated in Paris to Strasbourg animal studies (Marescaux et al., 2001). Other studies, in dry laboratory conditions, have show that the time taken to complete tasks was significantly increased with delays greater that 500 ms (Rayman et al., 2006). Nevertheless, more recently pig nephrectomies (kidney removal) have been performed using the da Vinci system with latencies between 450–900 ms (Sterbis et al., 2008). The surgeons in the study found that even with these latencies telesurgery was possible with prior experience.

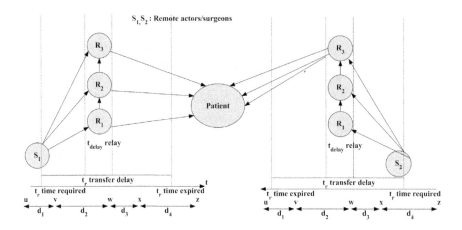

Figure 6.1 Current relay considerations for telepresence in deep space missions.

The surgery between Strasbourg and New York overcame the latency factor through the use of dedicated telecommunication lines. The mean delay was 150 ms, proving that the surgery can be carried out over transoceanic distances when using a high bandwidth and dedicated telecommunication lines. Difficulties arise, however, when dedicated fibers are not available, for example in space, remote locations like a battlefield or a ship, or places without sufficient resources or infrastructure. These are ironically the places where telepresence would be most useful.

The use of setting up ground based Internet protocol systems is also expensive. In operation Lindbergh the cost for 1-year availability of the ATM lines ranged between $100,000 and $200,000, which alone could make such an operation too costly. Not to mention the other expenses for setting up the telecommunication lines, human resources, involving a varied group of professionals, from surgeons and computer scientists to engineers, and of course the price of the robotic system. The costs could, however, decrease as technology improves. In addition, while the cost may be excessive for the sole benefit of growing current surgical practice, it may be more beneficial in the long term. The availability of telepresence could increase access to healthcare, speed up and improve surgical training and efficiency, and as a consequence improve the patient's overall outcome. Hence, in the future the initial investment and maintenance costs may prove to be worthwhile and less costly than currently believed.

An alternative to land–based systems could be the use of geosynchronous satellite systems. Satellites could provide access to remote locations, or to where ground based Internet protocol systems are unavailable due to limited resources. Furthermore, satellites could provide a flexible and mobile back-up if there is a communications failure. The main problem, however, with satellites is the latency issue. Signal delays can be approximately 10 times greater than land based systems, and exceed 500 ms. One study, however, comparing the ground to satellite performance of left internal mammary artery dissection in pigs, found no significant differences in the quality of surgery (based on a global rating scale) (Rayman et al., 2007). The study used the Zeus bot, and a 10 Mb/s bandwidth for both the ground and satellite network telecommunications. The round-trip distance of the study was 4000 km between London, Ontario, to Halifax, Novo Scotia. However, while the ground network latency was 200 ms, the satellite latency was roughly 600 ms.

The ground-based portion of the experiment had high audiovisual communication quality and the endoscope picture was clear with no pixilation. Upon switching to satellite imaging, the imaging was still clear, but with minor pixilation every 5–10 minutes, requiring the surgeon to wait until the next clear frame and then continue. There was also a noticeable difference in the latency, but the operator quickly adjusted to this. These included, movement-to-view delays, with a move and wait strategy employed to counteract this. Similarly, double-handed manipulations had to be done in a slow, deliberate manner to minimize mistakes. Trying to rush or hurry the surgery also resulted in more errors. Despite these differences, however, the study found that there was no significant difference to quality of the surgery (including respect for tissue, time and motion, instrument handling, slow of operation and control of bleeding) or the length of left internal mammary artery dissected.

This suggests that satellite-based telesurgery as shown in Figure 6.2, could be a viable backup to ground-based systems, and maybe a viable alternative in the future. However, surgeons need to be aware of the latency issues and how to overcome them, especially when the latency is greater than 500 ms. It was suggested in the article that predictive or virtual reality methods could provide a possible way of preparing surgeons for these situations. The effect of bandwidth degradation on the surgery was also investigated, and it was found that the quality of the surgery positively correlated with the bandwidth. In this respect alternative paths may be explored so that different and more flexible pathways may be used (i.e. Mobile Peer-to-Peer nodes or simply Peer-to-Peer nodes) as shown in Figure 6.2. There was a significant decrease in quality at bandwidth 4 Mb/s where

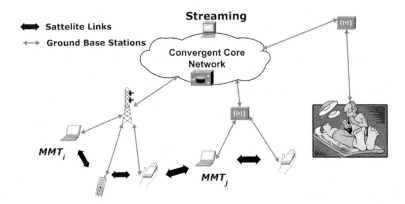

Figure 6.2 Different infrastructures used during the communication of the components of the surgical system.

the surgery could no longer be performed. It is important to note that the cost of satellite bandwidth is high, and purchasing it privately could cost as much as $100,000 per month. This would make satellite use even more expensive than ground-based systems. Moreover, there have yet to be any human trials involving satellite surgery, only animal studies. Overall, therefore, it seems that satellite surgery is theoretically possible, but not realistic at this time.

6.4 TELESURGERY IN EXTREME ENVIRONMENTS

The development of remote assisted operations opens up the opportunity to treat patients that need urgent attention no matter where they are, as long as a robotic system is nearby. This could be lifesaving in situations where it is not possible to get to a standard hospital or time is limited. This could include saving the life of a soldier wounded on the battlefield, to providing surgical attention to an astronaut on a deep space mission. There are number of barriers, however, in the way of making this type of surgery a reality. These include, but by no means limited to, adjusting the robotic systems for use in extreme environments, and constructing robust connections so that teleoperation runs smoothly and with as little latency as possible. There has been a multitude of research into these areas, with some exciting developments.

6.4.1 Remote Surgery in Space and Beyond

Figuratively speaking, astronauts on a deep space mission would not have access to standard medical care, but could possibly be operated on from earth by telesurgery. Surgical emergencies must be accounted for if humans are to safely explore space. Many potentially fatal injuries and illnesses (e.g. appendicitis, intracranial haematoma or kidney stones) can occur suddenly in healthy individuals. Studies conducted by NASA have estimated that the probability of a serious illness or injury during a 30-months-long Mars mission with six astronauts is 90% (Allen et al., 2003). The loss of bone mass in space is also well documented, and could lead to serious fractures, which would need surgical attention. These astronauts will not be able to be brought back to earth, but could be operated on board a spaceship or space station remotely from earth via telesurgery. Surgical support needs to be considered if any such mission is ever undertaken.

It has been shown that telesurgery on earth, using a satellite orbiting the earth, is theoretically possible. Satellite-based transmissions travel at the speed of light (300,000 km/s), meaning that transmissions to earth or anything orbiting earth, like say a space station, would happen with almost no delay. However, when this signal is broadcasted further the delay increases. For example, operating on a patient on the moon would theoretically have a 1 s delay, and operating on someone on Mars, which is approximately 72 million kilometers away, a signal delay of 6 minutes, meaning that no real time procedure could be carried out effectively using telesurgery (Haidegger et al., 2011). A possible solution could be to have a crewmember trained in robotic surgery present on any mission.

Another barrier to telesurgery in space is the viability of zero-gravity surgery. Experiments have been conducted in such conditions. In 2006 aboard the European Space agency Airbus A-300 Zero-G aircraft, a lipoma was removed from a patient (New Scientist, 2006). The plane climbed and dived sharply 25 times to create brief periods of weightlessness, simulating zero gravity in space, the operation took place only during these periods. In order to carry out the operation the doctors received weightlessness training, were strapped to the walls of the aircraft during the operation, and used equipment fitted with magnets to fix the objects to the metal operating table. It was the first time that it was demonstrated that a human being might be able to be operated on in space. It opened the door for the possibility of conducting more complex operations using surgical robots, remotely controlled from earth.

This was, however, only a simple operation and more complex surgeries require more equipment and more time. As of yet more invasive procedures have yet to be reproduced in humans under zero gravity conditions, let alone using robotic systems. The majority of surgeries would have to be conducted laparoscopically, rather than with open surgery. Open surgery would be complicated because the weightlessness makes it very difficult to deal with body fluids. In open surgery the surgical incisions are larger and cause more trauma to the patient, whereas the laparoscopic approach avoids this by only requiring small ports of entry to pass the equipment through, meaning less blood loss and release of bodily fluids. While there have been no such human trials, research has shown that laparoscopic operations in weightlessness can be carried out using pigs (Kirkpatrick et al., 2009). These experiments, however, did not incorporate telesurgery.

Another question is whether robotic surgery could work in extreme conditions. NASA has carried out a number of operations using robotics in extreme conditions, but again only using animals or human simulations. The commercial robotic systems currently being used (e.g. Da Vinci) were deemed too large and heavy, taking up too much space and requiring too much manpower for use in extreme environments, where such commodities would be few and far between. A spaceships weight for example is crucial in its ability to take off, and a large robot would take up important space and weight. A surgical robot in such missions would have to be small, lightweight and easy to use and set up. Alternative robotic systems have been developed (Kirkpatrick et al., 2009) including the Raven robot by the BioRobotics Lab (University of Washington, USA), and the M7 by Stanford research International (Menlo Park, California, USA). The M7 robot consists of two robotic arms with 7 degrees of freedom, and weighs only 15 kg. Its development started in 1998, and it is equipped with motion scaling (1:10), tremor filtering and haptic feedback. Similarly, the Raven has two arms, each of which holds a stainless steel shaft for different surgical instruments. Both have been designed to be easily assembled and with the intention of use in extreme conditions, either for space experiments or for military use. The raven has also undergone much recent iterations, that includes, a 3D ultrasound imaging system, and a warning system for real-time navigation and recognition of important anatomical structures (to avoid any errors due to communication delay).

These robotic systems have been tested in the NASA's undersea laboratory, called Aquarius. Aquarius is situated 19 meters below the sea and has been used to simulate the environmental conditions that would be found in space and other extreme environments. A series of missions have been carried out since 2001, called the *NASA Extreme Environment Mission Operations* (NEEMO). The most significant of these missions, regarding the simulation of medical procedures with robotic systems, using teleoperation and telementoring, include the 7th, 9th and 12th NEEMO missions (Haidegger et al., 2011).

The 7th NEEMO mission occurred in 2004, and included a number of simulated procedures using the Zeus bot (Thirsk et al., 2007). This included ultrasound examination of abdominal organs and structures, ultrasound guided abscess drainage, repair of vascular injury, cystoscopy (gall bladder visualization), kidney stone removal and laparoscopic cholecystectomy. The Zeus bot was controlled 2500 km away with a signal delay of 100 ms to 2 s. Despite the time delay, the non-trained personnel in the Aquarius were able to complete these tasks with a skilled telementor. In the 9th NEEMO mission the M7 robot was used (Haidegger et al., 2011). It had to be set up and used to complete an abdominal surgery on a simulation. This time, however, a microwave satellite connection to mimic a moon to earth connection was used. The time delay, however, increased to 3 s throughout the procedure. The crew also investigated how effective telemedicine could be used to diagnose extreme injuries and the surgical management of fractures. During these experiments the effects of fatigue and the different stressors of being and working in an extreme environment were also measured.

In 2006 the 12th NEEMO mission investigated suturing, the insertion of needles into phantom vessels and other basic principles of laparoscopic surgery, using both the M7 and the Raven robots (Broderick and Doarn, 2007). The research was ground breaking in that the needle insertion was the first action to be performed semi-autonomously by a robotic system. The needle was positioned with ultrasound, and then programmed to insert by itself upon the surgeons command using peddles to insert and retract. The ability to program robotic systems into carrying out semi-autonomous functions could be crucial in overcoming and aiding operations where there would be significant latency or communication issues. A wireless connection was used in the mission without any significant difficulties with a lag time of up to 1 s. In addition, before the mission took place a competition was run for middle school students, and successful students were selected to actually operate the Raven robot during the mission. They were given the task of picking up small foam blocks from a table, which they achieved with great success, demonstrating the ease at which these robotic systems can be used.

Overall the NEEMO missions have been very successful in beginning to investigate how remote robotic surgical systems could be used in extreme situations like space. Despite many questions arising from the research, many significant steps have been made regarding the experience gained in teleoperation, the control and use of surgical robots, as well as other issues involved in space surgery, such as the behavior of organs and body liquids in weightlessness.

6.4.2 Bringing the Operating Table to the Battlefield

Surgical intervention directly on the battlefield or in natural disasters has the potential to save many lives that would otherwise be lost. In these situations patients can often need urgent medical care, especially surgical intervention, at the scene of the injury. The problem is that getting the patient, in such extreme and remote locations, to the appropriate medical specialist in time can be a logistical nightmare. It requires the right infrastructure and human resources to be in place and in close proximity to the patient. This is often not the case in extreme situations like say an earthquake or on the battlefield. The development of telesurgery offers a solution to this, creating the possibility of being able to treat these patients directly at the site of injury.

One surgical system, which has been developed for such situations, is the trauma pod. The trauma pod is a semi-automated telerobotical system that has been designed to be rapidly deployable, to perform lifesaving surgery and to stabilize wounded soldiers on the battlefield (Garcia et al., 2009). It is intended to be deployed when treatment in a combat hospital cannot be provided in time, or where it is too dangerous for medical personnel to approach. Examples of life saving treatments include the ability to treat the loss of airway, haemorrhage and other acute injuries, such as pneumothorax. The robot would hopefully be able to go beyond currently available first aid treatments by providing an airway for the patient, inserting an intravenous line, performing haemostasis, or handling damaged tissues and placing monitoring devices.

In phase I trials the trauma pod was used to perform specific portions of surgical procedures, using a custom made robotic system with 13 subsystems. This included placing a shunt in a major abdominal (e.g. iliac) vessel and performing bowel anastomosis. In addition, the second objective was to perform a full CT scan in the field during an operation. All these procedures were carried out using phantoms in a controlled environment. While in this experiment the surgical system and the control site were in adjacent rooms; it is envisioned that the surgical cell could be deployed in the battlefield while the surgeon operates from a safer location. The experiment was successful and the surgeon was able to demonstrate bowel closure and shunt placement, interestingly, with no other human assistance. The procedure took 30 minutes from start to finish and was conducted by a surgeon with prior experience of the system.

Not only was the surgeon successful in controlling the robotic arms and tools for the actual surgery, but also for other tasks required. The trauma pod system was developed to be able to automate some support functions in the operating room, like the ability to change tools, dispense of equipment and track supplies. In the future this could mean that operations with tasks, requiring the coordination of a surgical staff, might be performed automatically by a unified control interface controlled by the surgeon. In essence, there might not just be a surgical robot, but a scrub–nurse robot too with different functions. This would be essential to performing surgical procedures in remote locations where there might not be other medical personnel at hand or the location is too dangerous.

Despite this initial success there still remain a number of questions that need to be overcome to make trauma surgery with a robotic system a reality. Firstly there is the problem of sterilization between operations. The parts that come in contact with patient need to be removable and easily sterilized. This could be especially problematic when there are multiple patients requiring medical attention. In addition, so far only parts of operations have been completed. Future systems need to be able to complete full procedures. They must be capable of exploring wounds, dissection, suction and irrigation, being able to manipulate larger tissues or supplies (e.g. laparotomy pads). These actions require varying levels of force, which may be difficult to recreate using robotics. Anaesthesia is another big issue, which is almost always required in any major surgical situation. The trauma pod system needs to be able to provide anaesthesia in a safe way that can be easily controlled by the operator of the robotic system. This will also require the aid of an anaesthetist to direct the operator, or to control part of the operation, or the operator to be specialized in both surgery and anaesthesia. The robot system itself is also quite complex with many manipulators. Their operation needs to be smooth and occur without collision. Additionally, the more parts there are the more error that can occur. The failure rate and overall performance of such a system needs to be carefully evaluated to ensure the safety of the patients it is handling and robustness of the system.

Finally, another major consideration is how the system will be controlled remotely. On the battlefield, and in other extreme environments, dedicated landlines would not be in place. The system needs to be operated wirelessly. Satellite based connections and their disadvantages have already been discussed in detail, but there are also other wireless options. Telesurgery could be conducted using an unmanned aerial vehicle in combination with a field deployable surgical robot like the trauma pod. This was demonstrated in the High Altitude Platforms Mobile Robotic Telesurgery project (Lum et al., 2007).

The project was conducted in the desert of Simi Valley, California using the university of Washington surgical robot system. A desert was chosen to recreate the environmental conditions that could be realistically faced when using this system and how such conditions (e.g. dusty winds and high temperatures) would affect the electronics and computer hardware. How the operation would be powered in such remote conditions was also taken into account. In the experiment two surgeons used a robotic system to carry out gross manipulation tasks on an inanimate model 100 m away. This was carried out using a wireless communication link through the PUMA UAV designed by Aerovironments. The experiment could have occurred up to a distance of 2 km (the UAVs full range), but a 100 m distance was chosen for the convenience of the experiment.

The surgeons were able to operate successfully via the wireless communication, but with a limited bandwidth and variable time delays. The maximum delay for control of the robot was 20 ms, and for the video stream 200 ms, well below the latency requirement suggested for safe telesurgery (500 ms). The main problem in the initial testing, however, came when trying to maintain a full bandwidth. The bandwidth had to be scaled back, but the surgeons did not feel this significantly affected their performance, noting only some increased pixilation in the video stream. Nevertheless the project was significant in showing that mobile robotic telesurgery is possible in extreme environments, and has the potential in the future to save human lives in isolated or extreme environments. In fact, the pentagon has already invested $12 million in a project to help develop the trauma pod, and their future hopes of the system have been captured in a concept video produced by the Defense Advanced Projects Agency (DARPA) (YouTube, 2006).

6.5 THE ETHICAL AND LEGAL CONSIDERATIONS OF TELEPRESENCE

The relationship between a patient and a telesurgeon creates a new precedent in medicine, where for the first time a patient could be operated from by someone from the other side of the globe. This requires careful ethical and legal considerations (van Wynsberghe and Gastmans, 2008). Problems arise when a surgery is not successful and errors occur. In such a situation it may be difficult to assign the blame. For example, in the scenario where the telesurgeon has been recommended by the patient's physician, who has the main responsibility when something goes wrong? If the on-site surgeon (overseeing the telesurgery) does not carry out any of the surgery themselves do they still have accountability if something goes wrong? On the other hand maybe it is the on-site surgeons job to oversee the operation and safeguard their patient. Therefore, if the on-site surgeon deemed the surgical procedure unsafe or dangerous they could stop the operation. The on-site surgeon, however, may not be experienced in telesurgery and not be in a position to make an educated decision as to when the conditions are not safe.

The telesurgeon carrying out the operation also faces a dilemma. If the telesurgeon is also held partially responsible for the patient's clinical outcome, it may be unclear to them where their responsibility to the patient ends. Post surgery would the surgeon be able to contact their patient to assess their wellbeing? Or would this be beyond their responsibility, and interfere with the patient's physicians care? In theory it may be straightforward to blame the person who made the mistake, however, in the operating theater mistakes are often not clear-cut. The complication may have risen from a number of factors. Furthermore, when there is a technical difficulty, with the robotic system or the network system, is it fair to blame the surgeon for the complications? It may be that legal responsibility falls with the robotic manufacturers or the network provider.

The other worry is the affect that telesurgery will have on the operating surgeon, and how this will affect the patient-doctor relationship. When there is such a geographical barrier to communication it may be hard for the telesurgeon to keep up face-to-face interaction, and this may detract from the patient's care. The fear is that the patient will no longer be seen as a patient, but as an object or a physical task to be completed. The lack of physical contact may, therefore, dehumanize the patient because the telesurgeon has not had enough contact to form an emotional connection on some level with their patient. This may result in the telesurgeon feeling that they have a reduced responsibility to the emotional and physical needs of the patient. Furthermore, it may lead to the telesurgeon making decisions that may not be in the best interests of the patient. This highlights the necessity for the telesurgeon to try and understand the needs of their patient and to build a relationship with them. This could be achieved by teleconferencing with the patient, but by also having a good working relationship with the patient's on-site doctor.

All of these ethical dilemmas bring about the main issue of standardization, and just how these matters will be resolved legally. There needs to be some form of legislation to govern how reimbursement and malpractice is handled in telesurgery. Of course because telesurgery by nature can transcend regional and even national boundaries, differences in local legislation may add confusion to proceedings and conflicts of jurisdiction may arise. It therefore, seems sensible to form some sort of international governing body of telesurgery and telemedicine, to create international legislation and address the issues pertaining to responsibility, the relationship between the telesurgeon and the patient, as well as liability of the network and surgical system manufacturer. Moreover, before telesurgery can be widely embraced by the international medical community, and become routine, the nonmaleficence (potential for not harming the patient) of the patient needs to be assured. Statistically significant data needs to be gathered and assessed as to the safety of these procedures, and to address the ideal conditions or minimum requirements for performing a telesurgical procedure.

Finally, for the true potential of telepresence, and all it entails, to be realized this technology needs to be made available to the people that need it most. Quintessentially people from under-developed countries and rural areas, where specialized surgical intervention is not available. Unfortunately, these are the people who are least likely to have access because telesurgery requires specialized technology and skills. Underdeveloped countries, therefore, are unlikely to see telesurgery become a reality for many decades to come. Commitment to health, therefore, on an international level needs to be achieved to make the full use of this technology possible. After all the chief objective of telesurgery is to provide specialized surgical intervention to those that would otherwise not be able to receive it. The cruel irony is that this may not be possible, unless the international community bands together and tries to make this a reality.

6.6 IS ROBOTIC SURGERY VIABLE? — ADVANTAGES, DISADVANTAGES AND FUTURE DIRECTIONS

Since the US Food and Drug Administration approved the first robotic surgical device in 1994 there has been a surge in interest (Barbash and Glied, 2010). Between 2007 and 2009 the number of da Vinci systems rose from 800 to around 1400, an increase in 75%, and in other countries doubled, from 200 to nearly 400. The number of robot-assisted procedures performed worldwide has almost tripled, from 80,000 in 2007 to 205,000 by 2010. Since 1994 there have been 461 clinical trials involving surgical robotics (PUBMED clinical trials search for "Robotic surgery" from 1994 through to 2015). The American College of Surgeons at its 90th Annual Clinical Congress even devoted one of its opening scientific sessions to the application of robotic surgery in general surgery and surgical specialties (Lowenfels, 2005). At the talks it was highlighted that due to advanced research and development of new, sophisticated machinery robotic surgery was undergoing a rapid period of growth, but that this was being held back by the costs of the system and a number of other problems. Nevertheless, it seems that many agree that surgical robotic systems will continue to comprise a growing part of surgery.

Already, robotic surgery has been successfully used in a wide range of procedures. This includes robotic gastrointestinal surgery, robotic urologic surgery, robotic gynaecologic surgery, robotic cardiothoracic surgery, robotic oncologic surgery and robotic paediatric surgery (Garrison, 2005; Diana and Marescaux, 2015). In turn within each field different operations have been performed, with varying success. To go into depth about each individual surgery is beyond the scope of this article. The general consensus is, however, that robotic surgery has allowed for the completion of more complex procedures, which normally would not be able to be carried out laparoscopically, with increased precision and with a minimally invasive approach. The more complex the procedure the more appropriate becomes the use of robotic surgery. Clinical trials have shown that mortality, morbidity, and hospital stay all compare favorably with laparoscopic results. However, for many procedures more randomized control trials and prospective trials are needed in order to compare whether robotic surgery is actually superior to conventional surgery.

From the procedures carried out using robotic surgery so far, a number of advantages have emerged. Firstly, robotic surgery offers a 3-dimensional view, whereas laparoscopic surgery is limited to a 2-dimensional view, which can impair depth perception. In laparoscopic surgery movement of the instruments occurs opposite to the screen due to a mirror image effect, but movements using robotic surgery are more intuitive. A movement to the right on the surgical robot monitor actually translates to a movement to the right. Another benefit is better camera control. In laparoscopic surgery the camera is controlled by an assistant, which can make it unstable. This is not the case with the robotic arm, which can hold the camera in position and allow solo surgery. The instruments used in laparoscopic surgery are also very straight, allowing for less degrees of freedom to move the instruments. By contrast, the tip of the robotic arm holding the instruments has microwrists, which allow for more flexible movements. These movements mimic the actions of the human wrist, making more complex movements possible. Robotic surgery is also more comfortable, in general, for the surgeon. The surgeon is seated comfortably at the console station, while the laparoscopic surgeon often has to adopt uncomfortable positions during an operation. Over the course of a day with many operations this can be very tiring, but robotic surgery reduces the physical demand of the surgeon, and this may make the surgeon less prone to errors. Finally, the summative affect of all these advantages is that it makes learning how to carry out robotic surgery easier in comparison to laparoscopic surgery, which has a steeper learning curve.

There are, however, a number of negatives that had impeded the spread of robotic surgical technology and prevent it from becoming a routine part of surgical care. Namely, that the cost is too high. The average cost of the da Vinci Surgical System is between $0.6 million to $2.5 million, the price of the instruments and accessories is $700–$3,200 per procedure, and maintenance charges are between $100,000 to $170,000 per year (Barbash and Glied, 2010). It is believed that cost-effectiveness will be achieved in the long term by reducing postoperative recovery and hospital stays. However, further evidence is needed from well-designed, large randomized trials to really determine whether patients actually benefit from robot assisted procedures compared to more open surgical approaches. Consequently, at this moment in time it is really not possible to say that robotic surgery is any better than conventional techniques. This is not to say that robotic systems are no good, but that more doctors and patients may need to embrace robotic surgical treatments before its virtues and flaws can be fully comprehended. As time goes on the cost of technology often decreases, and this may provide the push for surgical robotics to expand. The main problem impeding this is that the company that produces the da Vince surgical robot, Intuitive Surgical, has a monopoly in the market. Other systems are currently in different stages of development, but until these arrive and the da Vinci system has real competition, prices are most likely to remain high, and whilst these prices remain high it may prove difficult for hospitals and governments to invest in a still, very much unproven technology.

Other disadvantages include the bulkiness of the robotic equipment currently in use, and the amount of space they take up in the operating theater. Operation times are also often increased when compared to other laparoscopic procedures. In addition, the development of haptic technology has yet to truly recreate the feel of tissues through force feedback. All these problems, however, may be solved with time. As technology improves so will the size of the surgical robots most probably decrease and haptic technology improve. Moreover, more experience using the robotic system will decrease operating times as set up becomes quicker.

The Current focus in medical robotic research is to try and make these systems more lightweight, flexible, and as dexterous as possible. This is being done to try and facilitate a reduction in trauma by making the entry incisions as small as possible and limited to a single entry point, or even limited to just natural orifices (e.g. transorally, transvaginally). These new surgical interventions are referred to as SILS (single-port laparoscopic surgery) and NOTES (natural orifice transluminalendoscopic surgery). The use of conventional laparoscopy in these procedures would be too difficult because of such a small entry point making control of the instruments and the camera almost impossible. A surgical robot, however, could provide easy swapping of surgical instruments, improved stability, better visualization and better control of the instruments. Numerous robotic systems are presently being developed for this type of new ultra-minimally invasive surgery, including the ARAKNES and other snake-like robots (Bergeles and Yang, 2014).

6.7 ROBOTIC SURGERY AS A TRAINING TOOL

Current surgical training involves getting surgical experience by a supervised trial and error method. This system has been in place and remained unchanged for the past century. This can be both frustrating for the surgical trainee and even dangerous for the patient. Surgical training depends on caseload, which depending on each person's circumstances can prolong surgical training and compromise the safety of the patient. However, this could be about to change. All operations involving surgical robotic systems can be simulated. Surgical trainees will now be able to use surgical robots to hone their skills and learn operating procedures by practicing on 3-dimensional, virtual reality visual simulations created from CT or MRI images, and soft tissue models that recreate the texture and consistency of the human body through haptic technology. In addition, using image-guided simulations, surgeons can practice a surgery, before it takes place, on 3-dimensional reconstructions of the anatomy of their actual patient. This could also allow surgeons to refine or revisit skills needed for uncommon procedures. One pitfall of this type of training, however, is that these models do not recreate the natural movements of a patient (e.g. breathing) or what happens with the manipulation of soft tissues during a surgical procedure. While, therefore, these type of simulations can be an adjunct to learning, they cannot yet replace real surgical experience.

Robotic systems are, therefore, expected to significantly speed up the learning curve for surgeons, and there is research indicating that this may be the case (Hanly et al., 2004). Training can become shorter, and at the same time patient safety will increase, as surgical errors are reduced. Whereas surgical robotic training is considered a specialty now, in the future it may be part of formal training. In fact, training curriculums are being put together in various teaching hospitals. One such example is the university of Texas Southwestern Medical Center at Dallas. The center has created a Surgical Skills Training Laboratory, set aside specifically for surgical robotic training (Arain et al., 2012). They have developed a comprehensive, standardized, proficiency-based robotic training curriculum, which seems to be quite successful in helping to improve the skills of beginners in robotic surgery and their comfort with such systems. In addition, displaying how such a curriculum could be easily developed and implemented elsewhere. It, therefore, doesn't seem to unrealistic to foresee that robotic surgery could make its way into standard training for surgeons and help improve the learning curve. As discussed earlier, this would benefit both the trainee surgeons and patients, as less experienced surgeons will have the opportunity to practice and hone their skills outside of the operating room. Thus trainee surgeons would be free to practice more and not be constrained by the operating room timetable, increasing their experience and improving patient safety.

6.8 CONCLUSION

In conclusion, surgical robotics holds the promise of revolutionizing surgery by making surgeries less invasive and safer for the patient, and by augmenting the abilities of the surgeon to give them more dexterity, vision, comfort and precision. Moreover, the use of robotics also introduces the concept of remote control. Telesurgery can allow for the teleportation of a surgeon and their unique skills to anywhere where those skills are sorely needed. Indeed, this has already proven to be a reality with dedicated landlines. Barriers, however, to the expansion and widespread adoption of this technology exist. There are issues with latency when using wireless connections, with studies only being done in animal or simulated models, preventing the use of this technology in remote locations. In addition, as of yet there is no truly realistic haptic system. Probably the most significant obstacle, however, is the cost of the system and its maintenance. Nevertheless, just with any new technology, costs are initially high, but always become more reasonable as time goes by and competitors join the market. Likewise, as time progresses so will the technology improve and robotic systems become sleeker, more effective and issues with telesurgery addressed.

The field of robotics has already transformed the industrial world, and it seems only a matter of time before medicine and surgery join this new age of high speed and precision technology. In the future, the challenges for robotic surgery and remote operations may, therefore, may not be making the technology work, but making it work for humanity. The potential of this technology is enormous. Its expansion, therefore, needs to be handled with care. Telesurgery could one day be a key component in allowing humans to explore deep space safely, save the lives of injured soldiers on the battlefield, and allow rapid access to life saving surgical interventions for victims of natural disasters, like earthquakes. Nevertheless, this technology will be wasted if research is not directed towards making this technology available to the people that would benefit from it most, namely people from less economically developed countries and more isolated areas where under normal circumstances these surgical interventions would not be possible. To conclude, therefore, further research and advancements in this technology should be made with this ethical consideration in mind.

ACKNOWLEDGEMENTS

I would like to thank the faculty at Saint George's University London Medical School delivered by the University of Nicosia. In particular, special mention must go to my main tutor Dr. Mavromoustakis who guided me throughout this project, as well as, all the doctors and surgeons who have inspired my interest in medicine, surgery and its future directions.

REFERENCES

Allen, C.S., Burnett, R., Charles, J., Cucinotta, F., Fullerton, R., Goodman, J.R., et al., 2003. Guidelines and Capabilities for Designing Human Missions. Designing Human Missions. NASA/Johnson Space Center TM. 2003-210785.

Anvari, M., 2007. Remote telepresence surgery: the Canadian experience. Surg. Endosc. 21 (4), 537–541.

Arain, N.A., Dulan, G., Hogg, D.C., Rege, R.V., Powers, C.E., Tesfay, S.T., et al., 2012. Comprehensive proficiency-based inanimate training for robotic surgery: reliability, feasibility, and educational benefit. Surg. Endosc. 26 (10), 2740–2745.

Ballantyne, G.H., 2002. The pitfalls of laparoscopic surgery: challenges for robotics and telerobotic surgery. Surg. Laparosc. Endosc. Percutan. Tech. 12 (1), 1–5.

Barbash, G.I., Glied, S.A., 2010. New technology and health care costs — the case of robot-assisted surgery. N. Engl. J. Med. 363 (8), 701–704.

Bergeles, C., Yang, G.Z., 2014. From passive tool holders to microsurgeons: safer, smaller, smarter surgical robots. IEEE Trans. Biomed. Eng. 61 (5), 1565–1576.

Broderick, T.J., Doarn, C.R., 2007. Final report to USAMRMC - TATRC – Grant W81XWH-06-1-0084. Science Support – NASA Extreme Environment Mission Operation 9: Evaluation of Robotic and Sensor Technology for Surgery in Extreme Environments.

Diana, M., Marescaux, J., 2015. Surgical robotics. Br. J. Surg. 102 (2), e15–e28. [Cited 2015 July 4th]. Available at: http://onlinelibrary.wiley.com/doi/10.1002/bjs.9711/full#bjs9711-fig-0003.

Fisher, S.S., McGreevy, M., Humphries, J., Robinett, W., 1984. Virtual environment display system. In: Crow, F., Pizer, S. (Eds.), Proceedings of the Workshop on Interactive 3-Dimensional Graphics. ACM, New York, pp. 1–12.

Garcia, P., Rosen, J., Kapoor, C., Noakes, M., Elbert, G., Treat, M., et al., 2009. Trauma Pod: a semi-automated telerobotic surgical system. Int. J. Med. Robot. 5 (2), 136–146.

Garrison, A.W., 2005. Robotic surgery: applications, limitations, and impact on surgical education. MedGenMed 7 (3), 73.

Haidegger, T., Sandor, J., Benyo, Z., 2011. Surgery in space: the future of robotic telesurgery. Surg. Endosc. 25 (3), 681–690.

Hanly, E.J., Marohn, M.R., Bachman, S.L., Talamini, M.A., Hacker, S.O., Howard, R.S., et al., 2004. Multiservice laparoscopic surgical training using the daVinci surgical system. Am. J. Surg. 187 (2), 309–315.

Harris, S.J., Arambula-Cosio, F., Mei, Q., Hibberd, R.D., Davies, B.L., Wickham, J.E., et al., 1997. The Probot — an active robot for prostate resection. Proc. Inst. Mech. Eng. H 211 (4), 317–325.

Intuitive Surgical, 2014. da Vinci Xi Surgical System. [Internet]. Sunnyvale California (United States): [cited 2015 July 3rd]. Available from: http://www.intuitivesurgical.com/products/da-vinci-xi/.

Kirkpatrick, A.W., Keaney, M., Kmet, L., Ball, C.G., Campbell, M.R., Kindratsky, C., et al., 2009. Intraperitoneal gas insufflation will be required for laparoscopic visualization in space: a comparison of laparoscopic techniques in weightlessness. J. Am. Coll. Surg. 209 (2), 233–241.

Kwoh, Y.S., Hou, J., Jonckheere, E.A., Hayati, S., 1988. A robot with improved absolute positioning accuracy for CT guided stereotactic brain surgery. IEEE Trans. Biomed. Eng. 35 (2), 153–160.

Lowenfels, A.B., 2005. Robotics. Highlights of the American College of Surgeons 90th Annual Clinical Congress. [Internet]. Medscape CME Conference Coverage: [cited 2015 July 4th]. Available from: http://www.medscape.org/viewarticle/498575.

Lum, M.J., Rosen, J., King, H., Friedman, D.C., Donlin, G., Sankaranarayanan, G., et al., 2007. Telesurgery via Unmanned Aerial Vehicle (UAV) with a field deployable surgical robot. Stud. Health Technol. Inform. 125, 313–315.

Marescaux, J., Leroy, J., Gagner, M., Rubino, F., Mutter, D., Vix, M., et al., 2001. Transatlantic robot-assisted telesurgery. Nature 413 (6854), 379–380.

Marescaux, J., Leroy, J., Rubino, F., Smith, M., Vix, M., Simone, M., et al., 2002. Transcontinental robot-assisted remote telesurgery: feasibility and potential applications. Ann. Surg. 235 (4), 487–492.

New Scientist, 2006. Doctors remove tumour in first zero-g surgery. [Internet]. United Kingdom: 27th September [cited 2015 July 3rd]. Available from: http://www.newscientist.com/article/dn10169-doctors-remove-tumour-in-first-zerog-surgery.html#.VZaxMGD6RSV.

Paul, H.A., Bargar, W.L., Mittlestadt, B., Musits, B., Taylor, R.H., Kazanzides, P., et al., 1992. Development of a surgical robot for cementless total hip arthroplasty. Clin. Orthop. Relat. Res. 285, 57–66.

Rayman, R., Croome, K., Galbraith, N., McClure, R., Morady, R., Peterson, S., et al., 2006. Long-distance robotic telesurgery: a feasibility study for care in remote environments. Int. J. Med. Robot. 2 (3), 216–224.

Rayman, R., Croome, K., Galbraith, N., McClure, R., Morady, R., Peterson, S., et al., 2007. Robotic telesurgery: a real-world comparison of ground- and satellite-based Internet performance. Int. J. Med. Robot. 3 (2), 111–116.

Satava, R.M., 2002. Surgical robotics: the early chronicles: a personal historical perspective. Surg. Laparosc. Endosc. Percutan. Tech. 12 (1), 6–16.

Sterbis, J.R., Hanly, E.J., Herman, B.C., Marohn, M.R., Broderick, T.J., Shih, S.P., et al., 2008. Transcontinental telesurgical nephrectomy using the da Vinci robot in a porcine model. Urology 71 (5), 971–973.

Thirsk, R., Williams, D., Anvari, M., 2007. NEEMO 7 undersea mission. Acta Astronaut. 60 (4–7), 512–517.

van Wynsberghe, A., Gastmans, C., 2008. Telesurgery: an ethical appraisal. J. Med. Ethics 34 (10), e22.

YouTube, 2006. Battlefield Surgery 2025 part 1/3. [Internet]. San Bruno California (United states): December 7th [cited 2015 July 3rd]. Available from: https://www.youtube.com/watch?v=C4wjAlprgBc.

CHAPTER 7

Cloud Based Smart Living System Prototype

Ivan Chorbev*, Vladimir Trajkovik*, Rossitza I. Goleva†, Nuno M. Garcia‡

*Faculty of Computer Science and Engineering, University Ss Cyril and Methodius, ul. Rugjer Boskovic 16, Karpos 2, 1000 Skopje, The former Yugoslav Republic of Macedonia
† Technical University of Sofia, Kl. Ohridski blvd 8, Faculty of Telecommunications, Department of Communication Networks, 1756 Sofia, Bulgaria
‡ Instituto de Telecomunicações, Universidade da Beira Interior, R Marquês d'Ávila e Bolama, 6200-001 Covilhã, Portugal

7.1 INTRODUCTION

Ambient Assisted Living (AAL) as well as Enhanced Living Environments (ELE) aim in improving the quality of life and the independence of elderly and people with disabilities using appropriate technology integration (Cardinaux et al., 2011). The goal is achieved by smart procedures and algorithms applied for the data intended to end-user, caretakers, personal nursing services, medical doctors, and family members that will enable better communication and information provision in an efficient way. The transfer of the people to the medical centers and the visits of the caretakers could be optimized using recent technologies.

Many research groups have already put al lot of efforts in developing new pervasive devices for AAL and use of Ambient Intelligence to integrate these devices into platform that forms the environment for the end-user (Memon et al., 2014). Because the technologies are changing the devices should have standard interfaces, support standards protocols whereas the environment should be open to the extensions and be scalable. Using the new technology the human being could express the necessity of help and support by improving the information delivery and communication capabilities. Social networking is a good example for the use of advanced information and communication technologies in connecting people and organizing community activities.

The aim to increase the quality of life is going together with the Quality of Service (QoS) and Quality of Experience (QoE) for AAL/ELE. In

Ambient Assisted Living and Enhanced Living Environments.
DOI: http://dx.doi.org/10.1016/B978-0-12-805195-5.00007-7

spite the requirements for interoperability and conformance to the standards, the AAL platform should be usable, reliable, providing correct data, be cost-affordable, secure, and with different levels of privacy support. All requirements (Sun et al., 2009) should be evolved in the development cycle, system analyses, design, implementation, testing, validation, enhancement, and deployment to citizens, caregivers, healthcare, IT industry, researchers, and governmental organizations of AAL systems for the seek of the end-users benefit.

Because the system is aiming to support information services, the first task is to incorporate already existing data for medical care like electronic health record (EHR). It is a collection of electronic health information about an individual patient and/or groups of people. Medical doctors and nursing services used to this information flows and rely on it in their everyday workflow. Medical centers support this records and depending on the country, local legislation, differences in religions and customs the information could vary. The AAL/ELE platform integrates the currently collected health care information in paper and electronic medical records (EMR) throughout special customized interfaces. Mostly the patient maintains a personal health record (PHR). It contains a complete and accurate summary of an individual's medical history online or offline and the access to it is secured. By aiming to integrate all the existing data records and allowing its further enhancement by new technological data like videos, special images, data from continuous monitoring the data flow to the AAL/ELE platform is becoming significant (Kryftis et al., 2015; Pop et al., 2015). Big data analyses uses different algorithms for data processing and management and allows better and more secure data modeling of the platforms. Furthermore, data from the person environment could became also part of the platform helping to the health cares and patients to perform in a customized way. In some countries, this process is still in an early stage whereas in other countries local legislation and community have already implemented it and enabled a comprehensive view on health data (Knaup and Schöpe, 2014). Decisions made by physicians, nurses, patients, and informal caregivers could be based on these health data in the favor of patients.

AAL/ELE platform seems to become a promising alternative to the current care models. Due to the slow processes in the social environment, the integration and management process is heterogeneous and slower in comparison to the technology development. There are sources (Pop et al., 2015) where AAL interoperability services are classified like: 1) notification and alarming; 2) health services; 3) voice and video communication services. Further classification is based on type of media transmitted (video, image, voice, raw data, data streams, messages, movements) and the way the data is transmitted (real-time, near real-time, non-real-time, question/answer, command/response, broadcasting, multicasting, reliable multicasting). So, the support of diversified services is intended (Stainov and Goleva, 2015; Autexier and SHIP, 2015; Autexier and Hutter, 2015) not only to elderly and people with disabilities, but also to the people who track an everyday healthy lifestyle. The services could be also classified in a special way and level of abstraction. A convenient way to integrate EHRs, PHRs, AAL, home care and self-care systems is the approach of integrating them in a highly connected, robust and reliable cloud platform, offering uninterrupted availability. Furthermore, the platform could be client/server or cloud/foggy (Stainov and Goleva, 2015).

The system for assisted living we propose is based on the cloud technology and uses stationary, mobile, web-based and broadband access. The capability of the broadband mobile access almost worldwide provides dislocation of the electronic care environment. Internet-based storage of data changes the data modeling of the system allowing free data mobility in space and time. Also, the proposed system allows 24 hours monitoring, increased prevention of medical emergencies, timely response at emergency calls, patient notification in different scenarios, transmissions of the collected bio-signals (blood pressure, heart rate) automatically to medical personnel. In other words the proposed system model creates an opportunity for increasing patient's health care and medical capacity of the health care institutions reducing the costs and improving the QoS and QoE at the same time.

When approaching stakeholder requirements it is important to distinguish different groups of stakeholders. There is a categorization of AAL Forum in three groups:

- Primary users or end-users that are senior citizens or disabled people having installed AAL solutions at home
- Secondary users or informal caregivers like family members and friends, nursing companies and organizations that are using AAL services to help end-users
- Tertiary users or institutions and organizations from public and private sectors like health and insurance organizations, health agencies, governmental bodies that need special collection and statistical services on the AAL end-user data that will allow decision taking

7.2 STATE OF THE ART

New generations aim and are capable to increase dramatically the use and functionality of consumer electronic devices trying simultaneously to reduce the cost. The already defined interoperability among IT-based public services and vast use of wireless smart devices in Internet of Things are indispensable for not only enhanced but also for assisted living. Many IT and nursing companies are contributing in creation and use of new assistive enhanced frameworks, smart cloud-based solutions. E-medical services supports medical doctors in their every day work throughout wireless sensor devices while smart home systems improve user convenience and better living.

7.2.1 Assistive Technology Architectures

Assistive technologies lead to the development of many different architectures and frameworks worldwide last decade. All of them are more or less specific and had been developed based on the local demands and data. The aim of the following state of the art analyses is to demonstrate the common characteristics and actors among AAL/ELE systems that allow creation of a new solution in next sections.

The term Tele-health is used in describing integrated information system for delivery of remote health care communication services. There are many aspects in Tele-health that could be considered as remote sensing, remote investigation, remote health care services and applications for home care, long-term care services, health prevention, self-care and social services integration. The innovation and functionality within Tele-health is increased, due to the development of new technologies such as smart devices, video IP streaming and broadly available wireless Internet access. Tele-health can be delivered synchronously or asynchronously, i.e. patients and doctors interact in real-time or non-real-time. Some of the services require reaction. Other services cloud be considered informative and do not require any reaction from the end-user. The level of interactivity is also different and depends on the circumstances. EIP-AHA and the 3 Million Lives project in the UK (3 million lives, 2012) are examples of Tele-health systems in the European Union that include the commitment to provide services for people with chronic diseases in about 30 European regions in 2014.

There are many different solutions for wheel chairs (Autexier and SHIP, 2015; Autexier and Hutter, 2015) and some of them are incorporated into

AAL laboratories worldwide. The NavChair Assistive Wheelchair Navigation System (Koren and Borenstein, 1999) presents assisted mobility throughout home. It implements navigation assistance algorithms for calculation of the environment, collecting inputs through various data sensors. Mobile robots implementation in healthcare for people with motor disabilities is proposed in Pires et al. (2001).

Visually impaired people need different type of navigation. The interaction between sensors, such as accelerometers, gyroscopes, actuators and ICT technologies including GPS or voice/image recognition allows different indoor and outdoor services (Yelamarthi et al., 2010; Kim and Cho, 2012). Blind people need navigation that could be based on a microcontroller and sensors (Bousbia-Salah and Fezari, 2007), but some of the data processing could be performed in the cloud (Cooklev et al., 2015; Cooklev, 2015). In many cases like cerebral palsy or Alzheimer there is a need of intelligent self-training and mobility navigation through robots to increase rapidly the quality of the patients life (Kirchner et al., 2015; Metzen et al., 2014).

AsTeRICS (Nussbaum et al., 2012) is an Assistive Technology Rapid Integration and Construction Set system. It is an open source platform for people with disabilities that support flexible solutions. The AsTeRICS platform is based on sensing components with different type of signals.

Dynamic software frameworks provide a dynamic component model. Application and components, so called bundles, can be dynamically integrated and removed without a reboot. Communication between the components is typically implemented at abstract layer (Autexier and Hutter, 2015). This allows flexible and easy interconnection of appliances and devices in a home network. Examples of such systems are the specification by the Open Service Gateway initiative (OSGi) and the flexible smart home service delivery platform provided by the Home Gateway Initiative (HGI). Different projects rely on such technologies like Qivicon targeting at home networks (Spinsante and Gambi, 2012), or UNIVERSAAL and OPENAAL (Schmidt et al., 2010) which specifically aim at sensor nodes and middleware in the AAL/ELE context.

7.2.2 Assistive E-medical Services

The medical services are also changing towards AAL/ELE solutions. Medical doctors, health care providers, patients and citizens use gathered and disseminated data to analyze conditions and situations and as an evidence in taking decisions. In recent years, there has been a clear change in demand patterns that is pushing for a broad use of these technologies leading to intensive research and implementation of "e-health".

Telemedicine, as defined, should promote the sharing of (patients') information between healthcare professionals, allowing for remote analysis of diagnostic images, audio signals and good resolution video in real-time that help the "tele-health care professional" to reach a diagnosis. In order to establish the communication, telemedicine uses transmitted images, voice and other data to enable consultation, education and integration of medical services over a distance (Tavares and Lapão, 2006). These systems can be used with success to deliver radiology, dermatology and eco-cardiology services to patients who are in distant places (Levy et al., 2003), (Mikalsen et al., 2009). Benefits of telemedicine include immediate access to medical expertise no matter where the patients are, timely diagnosis and treatment, and the elimination of the need for patients and clinicians to travel long distances between rural areas and central medical centers. Both benefits can and should be measured. At the same time, however, open issues still remain, related to the evaluation of telemedicine usability, according to the different users' perspectives (Spinsante et al., 2012).

The necessity of the interdisciplinary research and engineering is proven by the importance of the health care to individuals and society. It has high financial impact on the life and many public and private organizations are trying to deal with this problem. Business solutions are necessary but also should be economically applicable. Many different approaches have been developed as patient-centric pervasive environments that add the perspective of the end-user into the system design. It allows not only patient but also medical and nursing personnel to take decisions based on the access of data in spite of place, time and mobility. Therefore, the virtual AAL/ELE environment is a place where end–users and health cares meet and negotiate can make a distinction between every day life and medical treatment and investigation allowing all participants to follow their normal daily activities.

7.2.3 Wireless Sensor Networks Related Systems

Wireless and wired sensor networks are very popular research area nowadays with vast implementation in AAL and ELE. Standardization of technologies like ZigBee, 802.15.x including body and personal communication are driving the technology to new levels of development and implementation. There are many open issues in the area like security and reliability that need special attention but the price of the sensors is decreasing and its economical applicability is becoming obvious. Furthermore, there is no need to consider all health case data critical and to try to transmit everything in real-time. There are many cases in the every day monitoring that generated data in specific traffic patterns in non-real-time (Goleva et al., 2016, 2015a, 2015b).

A complex personalized AAL solution with tele monitoring, health management, mental monitoring, mood assessment and physical exercises is shown in Takács and Hanák (2007). The system AlarmNet (Wood et al., 2008a) is presenting AAL residential monitoring network for adaptive healthcare taking into consideration the diverse needs of the patients. Project EMERGE (Kleinberger et al., 2009) is demonstrating emergency monitoring applications using sensors for elderly people needing special attention.

AAL platform CareTwitter (López-de-Ipiña et al., 2010) presents adoption of passive RFID tags as tiny databases where a log of a person can be stored and other users can access and manipulate. The data is encoded and transferred to the public micro-blogging service Twitter. The tweets published by CareTwitter are never made publicly available. They can be followed only by users authorized by the residents or their family.

An AAL system for blood glucose management and insulin therapy is proposed in Jara et al. (2011). It is based on Internet of things and defines set of services for end-user monitoring and connection to the diabetes information system. Special personal device is developed together with environment gateway, web portal, and the management desktop application. A monitoring system called SINDI (Mileo et al., 2010) is equipped by pervasive sensors and intelligent algorithms for data combination and processing. The system is capable to predict risky states and to generate alarms based on the computational model, medical knowledge and clinical profile of the end-user. Home monitoring and care giving commercially available technology BeClose (2015) is developed for the elderly. It indicates normal condition to the end-users and personnel using sensors.

A promising area of AAL/ELE applications is smartwatches where the wrist-worn devices report fitness data. Devices or specialized activity tracking wearables are increasingly equipped with sensors and algorithms for activity recognition. Active aging and active living are considered a modern way of behavior (Tonchev et al., 2015; Hidoussi et al., 2015). It is a matter of time to map disease prevention and active living. This will lead to the evolvement of home/business/car/environment automation data with body/personal data as well as health insurance bodies. There are many separate systems already on the market and possible integration is coming in near future (Goleva et al., 2016, 2015a).

In this chapter, a general architecture of the AAL system is proposed aiming to support not only the elderly and people with disabilities but also people that like to track their daily activities and look at short and long-term perspectives.

7.2.4 Smart Home and Health Monitoring Systems

Smart home and health monitoring systems integrate different sensors in home appliances and health monitoring devices in order to gather data, and create knowledge for the end-users. The knowledge can be used for different recommendation or additional information in case of needs.

The authors had already published some views on the topic in Trajkovik et al. (2015), Stainov and Goleva (2015), Goleva et al. (2016) presenting there ideas and different experiments in sensor and peer-to-peer computing. The architecture uses fixed, mobile, web-based and broadband access. Mobile devices collect raw data from environment and body area network (BAN) as well as some special condition-related data as traveling, weather, holidays. Raw data is processed well in advance, the noise is filtered and fusion healthcare analyzes algorithms are performed. The aim is to obtain correct decisions based on the raw data analyses and historical data stored and processed previously. Links to the social networks are also considered. As a result, the system generates set of recommendations to the end-users and health care personnel.

Some of the systems are more advanced like "Smart Monitor" (Frejlichowski et al., 2014). It uses video capturing and face recognition that is essential in some of the health centers and special cases of home automation. Some authors call such system intelligent AAL platform because it is capable to detect the move of the objects, track, recognize and react without the support from the personnel. The video content and sensor data are analyzed and triggered actions are performed. In this case, the services are not only monitoring but also managing, i.e. proactive. The service could be implemented in remote monitoring and surgery especially in remote locations like liners in the sea, arctic expeditions, and mountain trekking.

Some of the platforms could be considered strange like "Smart Kitchen" (Blasco et al., 2014; Krieg-Brückner et al., 2015). Being part of complete

AAL system such functionality lead to the overall quality of living to the end-user. The platform is based on the sensors and switches that allow control of the devices and things in the kitchen using different technologies like WiFi or ZigBee. A serious approach to the problem requires abstract layer definition (Krieg-Brückner et al., 2015).

Many of the new systems are context-aware (Oh, 2015; Kryftis et al., 2015; Pop et al., 2015). Most of them are cloud-based and are not bounded with the specific area. These systems are used for crime or insurance event prevention. The platforms consist of sensor network and a data server that could export data to the cloud. Fusion algorithms for data processing could be applied in the cloud (Wood et al., 2008b). Context-awareness of the services could be applied in social networking.

7.3 GENERAL ARCHITECTURE OF A SYSTEM FOR ASSISTED LIVING

7.3.1 Data Retrieval in AAL Systems

The general architecture of the AAL system has its access, edge and core parts. The access part is based on the sensor area networks mostly in or around the human body. Smart phones or fixed controllers could be used as gateways to the edge and core parts of the network. Whereas the access part of the network is collecting information and partially could preprocess it the gateway has the responsibility to store for a while the information and transmit reliably to the edge network. Sensors measure the physical and physiological signals of the body. The solutions depend very much on the technologies and some of the technologies are interoperable only through the gateway, i.e. controller (Goleva et al., 2016; Liolios et al., 2010). The sensors could measure in discrete-time and continuous-time. The reporting time and protocol are a matter of configuration.

Depending on the customer requirements, sensors could measure and transmit different data. Whereas real-time video is very demanding by means of communication channel bandwidth sensors for temperature are simple and transmit few bytes. Very often, the data collected needs to be compressed locally in order to avoid congestion in communication chan-nel. On the contrary, big data processing could be transferred to the cloud or to the processing servers at the edge of the platform (Kryftis et al., 2015; Pop et al., 2015; Cooklev et al., 2015; Cooklev, 2015). Sensors for location management could be also considered. The edge part of the platform consists of one or many virtualized servers capable to use the data for analyses throughout special algorithms that will allow further data presentation in a readable format to different platform stakeholders (Nugent et al., 2011).

7.3.2 Data Processing in AAL Systems

The data-preprocessing phase includes checking whether the received data is from the right sensor of a Wireless Sensor Network (WSN). Part of the data could be compressed/decompressed as well. The results of this phase are still considered raw sensor data and the data is sent to the next stages. The next step is data filtering, including noise reduction out of the received signals. Removing the noise could be performed by different algorithms depending on the signal and noise nature. Since there is a possibility of noise within the proposed system model caused by movement of the user it could be reduced using adaptive filters. Additionally, fusion algorithms for data filtering and validation like Kalman filter, Bayesian statistics, and Dempster-Shafer methods (Mohammad and Mohebbi, 2009) are also applicable at edge network aiming data acquisition.

The next phase is data verification including generalization and additional data calculations. This phase of the data processing workflow aims to formalize and give an overview about the real situation of the user. In case the results are critical, a message is sent to the responsible medical staff and the user. There is a tendency to move this data processing to the cloud or fog (Stainov and Goleva, 2015). It is expected that 5G technology will improve vastly the capacity of the communication channels and this will allow fast and reliable data collection in data centers and in the fog.

Important aspects in this kind of systems are timely transmission of data and the type of data that should be transferred, which has a very important role in determining the strategy to be used in case of delay. Based on the nature of the signals, whether it is a continuous transmission or not, they can be divided into:

- Wavy-dependent data that requires non-continuous data transfer. As a result, the transmission of data can commence only when data is requested from the system or periodic monitoring is scheduled (Goleva et al., 2016)
- Wavy-independent data that requires continuous transmission of data for long periods of time, especially in applications like remote surgery and emergency events (Boytsov, 2011) in real-time or near-real-time

One possible and nice cooperation of the AAL platform is its interconnection with home area network (HAN) and social networks. Both technologies tend to transmit multimedia contents like video, music, images, data, and voice including home automation data (Stainov and Goleva, 2015). As a matter of fact, possibly time-critical data transmissions (such as those related to dangerous events or alarms) have to coexist with delay-tolerant ones. This requires setup of the priorities in the data transmission throughout reconfiguration. The process of HAN integration depends very much on the interoperability between technologies. Some standardized solutions like ZigBee can coexists in a heterogeneous environment. At this point QoS and QoE issues are very important. There is also a possibility for future integration with existing 112 service. The development of the platforms is still in early stage in order to consider this issue.

Very clear distinction has to be done between real-time and non-real-time services and necessity to broadcast, multicast, acknowledge or retransmit the data. It is essential not only because the capacity of the communication channel is limited but also because of the emergency of given events. In this case, peer-to-peer applications within the fog might be very helpful. Data model needs to define precisely what, when and for how long is stored and where.

Data received from the sensors may have been damaged, so the following approaches can be executed: (1) to accept the data, (2) remove and wait for the next load, (3) seek to reload. The selection of the proper strategy depends on the nature of data. If the data is irrelevant or it is a type of data that loads often, then the second option is acceptable. If measurements are for more sensitive data, the third option should be applied.

Energy consumption also needs to be carefully analyzed to make an informed decision. One approach to save energy is the application of an efficient compression scheme. This method achieves a reduction of the amount of information that will be transmitted (Di Francesco et al., 2011). The following algorithms can be used in WSN: Lempel-Ziv 77 (LZ77) and Lempel-Ziv-Welch (LZW) (based on the statistical methods for prediction by partial matching), while Burrows-Wheeler Transform (BWT) reduces the amount of information and optimizes data compression. Benchmark tests done on these algorithms show that the algorithm LZ77 has maximum speed and minimum memory used (Barr and Asanović, 2006). Energy conservation could be performed in sensor network by letting the sensors to sleep when possible.

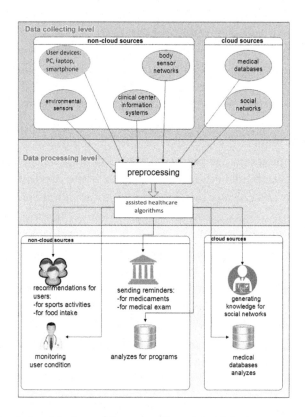

Figure 7.1 Logical architecture of the system for assisted living.

7.3.3 Logical Architecture of a System for Assisted Living

The logical structure of the proposed architecture is shown on Figure 7.1. It is tightly related to the data model and processing. Three parts could be distinguished in the system. The first is for data gathering health parameters via sensor networks. Part of the data could come from end-user smart devices like smart phones, tablets, holters. After collection of the data to the gateway it is intended to be preprocessed by assisted healthcare algorithms. The data is incorporated with the existing data in medical data centers as well. The results are sent back to the end-users, stored in the data center and used later for additional statistical analyses.

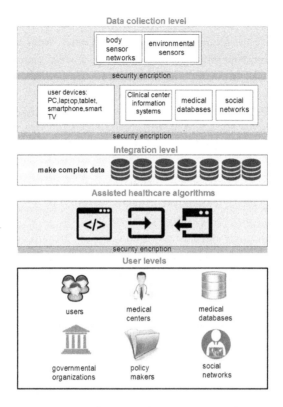

Figure 7.2 Physical architecture of a System for Assisted Living.

7.3.4 Physical Architecture of AAL System

An idea for physical architecture of AAL System is presented on Figure 7.2. There is a level of actors and application interfaces, sensors for data collection and management, data preprocessing level at the edge and core parts of the platform. Special place is defined for security and authorization regardless the fact that security should be applied at all levels. The data collected and processed is related to the different institutions in different ways:

- Clinical centers that need to monitor the data and patient condition and provide recommendation if necessary
- Medical databases that collect data from all stakeholders and analyze data for medical management, insurance companies, governmental organizations and non governmental organizations that act in the field

- Government organizations that analyze and work on local legislation, tried to harmonize the local to the international rules, work at standard conformity and interoperability
- Policy makers that can analyze data and create strategies and recommendation to the government and international bodies
- Social networks that even not considered seriously and reliable support the process of data analyses and end-user social activity
- Services for environmental data that help the health cares in their daily work

All other data collection and processing devices and networks using different technologies need to be interconnected through proper interfaces and be interoperable with the existing platform. On the other hand, the platform is open to the new devices and technologies through appropriate gateways at application, data gathering, pre- and post-data processing.

7.3.5 Security Issues

System security plays a significant role at any layer of the platform. Data access authorization should be managed by end-user too because the access to the medical information of non-medical and nursing related organizations has to be prevented and carefully allowed. Security requirements could be identified at:

- Data storage or foggy computing using peer-to-peer communication (Stainov and Goleva, 2015) where the presentation of the data arrays to the servers bounds the data access from the non-registered users. For peer-to-peer computing data could be stored for the limited amount of time. Usually data centers are reliable and even not being accessed all the time due to the lost connectivity the data is stored there reliably
- Data confidentiality should be defined at different layers starting from the application and end-user because people are generally sensitive to the openness of the personal information. Recent cryptographic algorithms allows reliable transfer everywhere
- Authentication is developing vastly using biomedical markers and this allows better system design and high end-user confidence
- Access control is combined by authentication and authorization. There is a need for managed services provided to the end-users especially to the disabled people to gather with this issue
- Privacy concerns needs a development of the configurations in patterns that will allow end-users easy navigation and access management

7.3.6 Validity of Information

Vitality and importance of the information gathered as well as its correctness are important aspects of the platform. Some of the historical raw data from sensors like temperature could not be a matter of interest after few months or years. Some of the historical data concerning pulse and blood pressure could be an early sign for preventive actions and medication. Some of the data collected could be wrong due to the errors or noises. The role of the data fusion algorithms at this stage is essential. Data fusion should avoid false alarms keeping the track on the important data at the same time. A simple data classification identifies:

* The most reliable information that is collected in medical data centers and coming from vital sensors
* The less reliable information collected through social network that is more important for the enhanced environment
* The least important information coming from the end-user and other sources that could not be considered while taking important vital decisions. It could be used for data analyses in different social bodies

7.3.7 Cloud Implementation

AAL/ELE platform proposed is cloud-based and the implementation in the virtual environment has many advantages (Maurer et al., 2013; Calheiros et al., 2011; Belgacem and Chopard, 2015) like integration, heterogeneous characteristics, scalability, extensibility, reliability and disaster recovery. Cooperation to other cloud-based technologies could be done in the cloud without changing the infrastructure at the access network. Big data analyses and high processing algorithms could be performed at high performance servers. The virtual data management by virtual machines allows energy management at data center level.

On Figure 7.3 there is a proposal for three cores system architecture using REST-full HTTP endpoints and JSON data format. There are three type of components as sensors, processors and triggers. The exchange of data and management information goes through messages. The extensibility of the cloud allows almost unlimited processing time and capacity. This makes lightweight data collection and processing in the access and edge parts of the platform and high performance data management in the cores. Furthermore, the access parts could be based on different technologies.

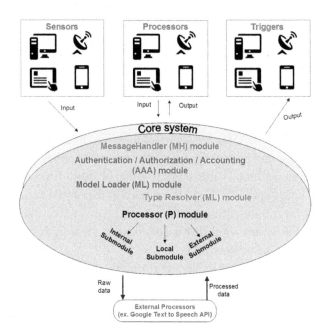

Figure 7.3 Core system architecture.

The proposed system has a Graphical Design Tool (GDT) in the form of web-based application, used for service design, and modeling. It is a drag&drop modular tool with intuitively described rules and component interaction that has incorporated the most generic usage patterns. It supports multiple end-devices as traffic sources behaving on the standardized interfaces and using standardized protocols.

The core system consists of message handler for HTTP request processing including encapsulation of JSON data and used for synchronization within the system. The AAA module supports security protocols using OAuth 2.0. The model loader module starts the deployed subsystem as well as rules and components for the type resolver. The type resolver decides where to send data for processing (Figure 7.4) depending on the type of the component.

The processor's module consists of internal, external and local processing submodules. The role of the modules is to process external components like Google Text to Speech API or Google Translate API, to support core system locally or to process components internally. On Figure 7.4, message processing is shown. All messages are de-serialized from the JSON object, request authentication from AAA module, load the profile, deploy model

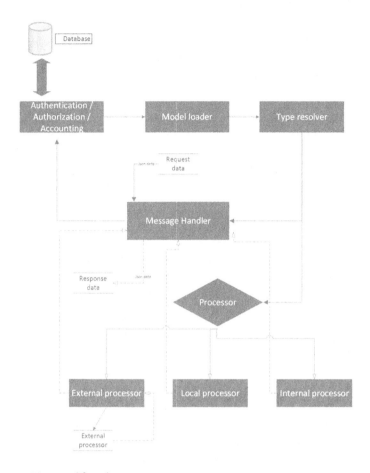

Figure 7.4 Message lifecycle.

and parse the components and rules for interaction. The type resolver composes the response of requests processing depending on the component type.

7.3.8 Information Integration Aspects

Information integration is essential and is performed within the infrastructure of the system. There is a need to set rules for integration very carefully because the final results could lead to proactive behavior of the system or data misleading. The integration could be done:

- To the layer above maintaining the consistency with the formal logic, i.e. deductive approach
- To the layer below maintaining the consistency with the non-logical reasoning, i.e. inductive approach
- By the use of holistic approach that cares for the development of the system as a whole. The hardware, the software, and the technology should be developed and enhances as a whole system

For the purposes of proper information integration, some degree of autonomy should be provided on the lower levels of the system (actuators, sensors) in order to solve some part of the tasks on the same level.

It is important to utilize the cognitive capabilities of the lower levels of the system such as:

- patterns identification and recognition within the data flow generated by the sensors and actuators
- scene interpretation: classification, conceptualization
- reasoning
- planning
- smart control
- complex goal-directed control
- learning on all levels

These cognitive capabilities should be integrated within the general system's framework and directed towards the achievement of the overall goal of the system.

7.4 USE-CASE SCENARIOS

The general use-case scenario of the proposed AAL/ELE platform is presented on Figure 7.5.

One interesting use-case showing multiple functionalities of the described system prototype includes a student suffering from diabetes (or various heart conditions, cardiomyopathy, Angina pectoris, Brady arrhythmia, atrial fibrillation). Whilst being active at the campus doing university activities, physiological parameters are continually monitored by wearables. Simultaneously, the smart-phone is constantly monitoring student position on the map. While moving outside, GPS positioning is utilized, but indoors, the indoor navigation algorithm using the smartphone sensors and WiFi positioning supplements the lack of GPS signal. Via the smartphone and its Internet link, all gathered data is regularly synchronized with the

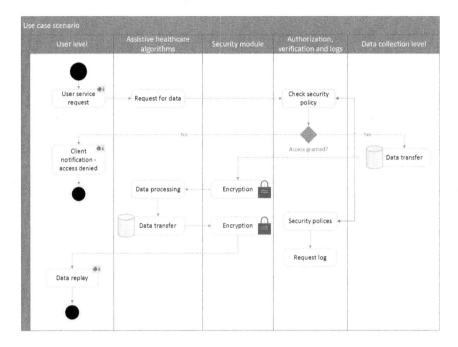

Figure 7.5 General use-case scenario.

cloud system. The system reminds the student to take his medications at the scheduled time, takes care of medication counter indications etc.

In case of a sudden abnormal shift in the physiological parameters, or a fall event detection due to sudden drop in the blood sugar level, the algorithm implemented in the cloud signals an alarm. The cloud AAL system tries to contact the student and check on him, directing him to the nearest infirmary based on his location. In case of impossibility of contacting the student due to loss of conciseness, an emergency team is dispatched, again guided by his exact location transmitted by the student's smartphone. The emergency team itself is directed using the same indoor navigation tool used by the student. (See Figure 7.6.)

On the contrary, medical or para-medical personnel can remotely access the data being triggered after predefined thresholds or events and could recommend quickly appropriate activity and therapy. Furthermore, all events will be recorded for post-processing and analyzes for the seek of the end-user. The proposed solution is very useful in preventive medicine when the silent signs before crisis could be monitored carefully. Applied at profes-

Figure 7.6 Student healthcare monitoring.

sional activities the system could alarm the management of the companies for health problems of the personnel due to the working conditions.

7.5 CONCLUSIONS

This chapter opens a door for pre causing and preventive medicine as well as strict monitoring of the patients with health problems based on the proposed AAL/ELE platform. The use of sensors at the end–user environment, data collection and processing at the edge and core parts of the systems allows modular structure, easy technological enhancements and heterogeneity at the access part. The system is more reliable in comparison to the existing architectures on the market. It implements fusion algorithms for precise statistical analyses and generation of decision-making information for the health cares and end–users. Configuration is based on different profiles and scenarios. People data tracking is made silently and did not disturb the normal life of the people. Preliminary analyses show low data traffic patterns in all cases when the data does not contain video and high resolution images. The system is open and uses standardized interfaces and protocols. The economical feasibility of the system is essential for its further deployment.

5G technology together with Internet of Things are promising vast end–device development performing vital signal tracking. Standardization and interoperability of the devices will allows their integration to the cloud or fog and post processing of the data collected. Processing in the cloud could be easily advanced due to the high performance computing technology and power servers. The capability to implement such a power signal processing at the end-device level is very limited. Furthermore, agencies, governmental organizations, health care bodies will obtain vital information that will allows better use and planning of the personnel occupancy and availability. Integration at cloud level is virtualized and could not influence the performance of the end-user applications.

The proposed system has high applicability in health monitoring, helping people with disabilities, preventive medicine, and people aiming active aging. It can support better living to the people with difficulties, alarm the health cares and health institutions for the possible health problems. Clinical centers could be also virtualized letting the personnel and the patients to consult remotely. The system is essential in all cases when the presence of the medical doctor is not possible on liners, in space, on arctic expeditions. Statistical analyses implemented in the platform could lead to the better epidemic prevention providing maps of the ill patients and patients with symptoms. Part of this information could be made publically available especially in case of fly. Health monitoring virtualization leads to the new health care management procedures that will need further profiling and analyses.

ACKNOWLEDGEMENTS

This work was partially financed by the Faculty of Computer Science and Engineering at the Ss.Cyril and Methodius University, Skopje, the former Yugoslav Republic of Macedonia and is supported by the networking activities provided by the ICT COST Action IC1303 AAPELE.

REFERENCES

3 million lives, 2012. Recommendations from industry on key requirements for building scalable managed services involving telehealth, telecare & telecoaching. http://3millionlives.co.uk/wp-content/uploads/2013/03/3ML-Indsutry-Recommendations-for-Scalable-Services.pdf.

Autexier, S., Hutter, D., 2015. SHIP — a logic-based language and tool to program smart environments. In: LOPSTR, pp. 313–328.

Autexier, S., Hutter, D., 2015. Structure formation in large theories. In: CICM, pp. 155–170.

Barr, K.C., Asanović, K., 2006. Energy-aware lossless data compression. ACM Trans. Comput. Syst. 24 (3), 250–291.

BeClose, 2015. http://beclose.com/. Accessed 17 January 2015.

Belgacem, M., Chopard, B., 2015. A hybrid HPC/cloud distributed infrastructure: coupling EC2 cloud resources with HPC clusters to run large tightly coupled multiscale applications. Future Gener. Comput. Syst. (ISSN 0167-739X) 42, 11–21.

Blasco, R., Marco, A., Casas, R., Cirujano, D., Picking, R., 2014. A smart kitchen for ambient assisted living. Sensors 14, 1629–1653. http://dx.doi.org/10.3390/s140101629.

Bousbia-Salah, M., Fezari, M., 2007. A navigation tool for blind people. In: Innovations and Advanced Techniques in Computer and Information Sciences and Engineering, pp. 333–337.

Boytsov, A., 2011. Context reasoning, context prediction and proactive adaption in pervasive computing systems. PhD. Department of Computer Science, Electrical and Space Engineering, Luleå University of Technology, Sweden.

Calheiros, R.N., Ranjan, R., Buyya, R., 2011. Virtual machine provisioning based on analytical performance and QoS in cloud computing environments. In: Parallel Processing (ICPP).

Cardinaux, F., Bhowmik, D., Abhayaratne, C., Hawley, M.S., 2011. Video based technology for ambient assisted living: a review of the literature. J. Ambient Intell. Smart Environ. 3 (3), 253–269.

Cooklev, T., 2015. Making software-defined networks semantic. In: WINSYS, pp. 48–52.

Cooklev, T., Darabi, J., McIntosh, C., Mosaheb, M., 2015. A cloud-based approach to spectrum monitoring. IEEE Instrum. Meas. Mag. 18 (2), 33–37.

Di Francesco, M., Das, S.K., Anastasi, G., 2011. Data collection in wireless sensor networks with mobile elements: a survey. ACM Trans. Sens. Netw. 8 (1), 7.

Frejlichowski, D., Gościewska, K., Forczmański, P., Hofman, R., 2014. "SmartMonitor" — an intelligent security system for the protection of individuals and small properties with the possibility of home automation. Sensors 14, 9922–9948. http://dx.doi.org/10.3390/s140609922.

Goleva, R., Stainov, R., Savov, A., Draganov, P., 2015a. Reliable platform for enhanced living environment. In: First COST Action IC1303 AAPELE Workshop Element 2014, in Conjunction with MONAMI 2014 Conference. Wurzburg, 24 Sept. ISBN 978-3-319-16291-1. Springer International Publishing, pp. 315–328.

Goleva, R., Stainov, R., Savov, A., Draganov, P., Nikolov, N., Dimitrova, D., Chorbev, I., 2016. Automated ambient open platform for enhanced living environment. In: Loshkovska, S., Koceski, S. (Eds.), ICT Innovations 2015. ELEMENT 2015, Ohrid, FyROM, 1 Oct. In: Advances in Intelligent Systems and Computing. Springer International Publishing, Switzerland, pp. 255–264.

Goleva, R., Atamian, D., Mirtchev, S., Dimitrova, D., Grigorova, L., Rangelov, R., Ivanova, A., 2015b. Traffic analyses and measurements: technological dependability. In: Mastorakis, G., Mavromoustakis, C., Pallis, E. (Eds.), Resource Management of Mobile Cloud Computing Networks and Environments: Information Science Reference, Hershey, PA, pp. 122–174.

Hidoussi, F., Toral-Cruz, H., Boubiche, D., Lakhtaria, K., Mihovska, A., Voznak, M., 2015. Centralized IDS based on misuse detection for cluster-based wireless sensors networks. Wirel. Pers. Commun. 85 (1), 207–224.

Jara, A.J., Zamora, M.A., Skarmeta, A.F.G., 2011. An Internet of things-based personal device for diabetes therapy management in ambient assisted living (AAL). Pers. Ubiquitous Comput. 15, 431–440.

Kim, Sung Yeon, Cho, Kwangsu, 2012. Electronic cane usability for visually impaired people. In: Information Technology Convergence, Secure and Trust Computing, and Data Management. In: Lecture Notes in Electrical Engineering, vol. 180, pp. 71–78.

Kirchner, E.A., Fernandez, J., Kampmann, P., Schröer, M., Metzen, J.H., Kirchner, F., 2015. Intuitive interaction with robots — technical approaches and challenges. In: SyDe Summer School, pp. 224–248.

Kleinberger, T., Jedlitschka, A., Storf, H., Steinbach-Nordmann, S., Prueckner, S., 2009. An approach to and evaluations of assisted living systems. In: Using Ambient Intelligence for Emergency Monitoring and Prevention. Universal Access in Human-Computer Interaction. In: Intelligent and Ubiquitous Interaction Environments, vol. 5615. Springer, Berlin/Heidelberg, pp. 199–208.

Knaup, P., Schöpe, L., 2014. Using data from ambient assisted living and smart homes in electronic health records. Methods Inf. Med. 53, 149–151.

Koren, Y., Borenstein, J., 1999. The NavChair assistive wheelchair navigation system. In: Rehabilitation Engineering. IEEE.

Krieg-Brückner, B., Autexier, S., Rink, M., Nokam, S.G., 2015. Formal modelling for cooking assistance. In: Software, Services, and Systems, pp. 355–376.

Kryftis, Y., Mavromoustakis, C.X., Mastorakis, G., Pallis, E., Batalla, J.M., Rodrigues, J.J.P.C., Dobre, C., Kormentzas, C., 2015. Resource usage prediction algorithms for optimal selection of multimedia content delivery methods. In: ICC, pp. 5903–5909.

Levy, S., Jack, N., Bradley, D., Morison, M., Swanston, M., 2003. Perspectives on telecare: the client view. J. Telemed. Telecare 9 (3), 156–160.

Liolios, C., Doukas, C., Fourlas, G., Maglogiannis, I., 2010. An overview of body sensor networks in enabling pervasive healthcare and assistive environments. In: Proceedings of the 3rd International Conference on PErvasive Technologies Related to Assistive Environments. Samos, Greece, 23–25 June.

López-de-Ipiña, D., Díaz-de-Sarralde, I., García-Zubia, J., 2010. An ambient assisted living platform integrating RFID data-on-tag care annotations and twitter. J. Univers. Comput. Sci. 16 (12), 1521–1538.

Maurer, M., Brandic, I., Sakellariou, R., 2013. Adaptive resource configuration for cloud infrastructure management. Future Gener. Comput. Syst. (ISSN 0167-739X) 29 (2), 472–487.

Memon, M., Wagner, S.R., Pedersen, C.F., Beevi, F.H.A., Hansen, F.O., 2014. Ambient assisted living healthcare frameworks, platforms, standards, and quality attributes. Sensors 14, 4312–4341.

Metzen, J.H., Fabisch, A., Senger, L., Fernandez, J., Kirchner, E.A., 2014. Towards learning of generic skills for robotic manipulation. Künstl. Intell. 28 (1), 15–20.

Mikalsen, M., Hanke, S., Fuxreiter, T., Walderhaug, S., Wienhofen, L., 2009. Interoperability services in the MPOWER ambient assisted living platform. In: Medical Informatics Europe (MIE) Conference. Sarajevo: August 30–September 2.

Mileo, A., Merico, D., Bisiani, R., 2010. Support for context-aware monitoring in home healthcare. J. Ambient Intell. Smart Environ. 2 (1), 49–66.

Mohammad, S., Mohebbi, N., 2009. A comparative study of different Kalman filtering methods in multi sensor data fusion. In: International Multi-Conference of Engineers and Computer Scientists IMECS. 18–20 March, Hong Kong.

Nugent, C.D., Galway, L., Chen, L., Donnelly, M.P., McClean, S.I., Zhang, S., Scotney, B.W., Parr, G., 2011. Managing sensor data in ambient assisted living. J. Comput. Sci. Eng. 5 (3), 237–245.

Nussbaum, G., Veigl, C., Acedo, J., Barton, Z., Diaz, U., Drajsajtl, T., Garcia, A., Kakousis, K., Miesenberger, K., Papadopoulos, G.A., Paspallis, N., Pecyna, K., Soria-Frisch, A., Weiss, C., 2012. AsTeRICS — towards a rapid integration construction set for assistive technologies. In: Everyday Technology for Independence and Care. In: Assistive Technology Research Series, vol. 29. IOS Press, pp. 766–773.

Oh, Y., 2015. Context awareness of smart space for life safety. Int. J. Smart Home 9 (1), 135–140.

Pires, G., Honório, N., Lopes, C., Nunes, U., Almeida, A.T., 2001. Autonomous wheelchair for disabled People. In: Robotics & Automation Magazine. IEEE.

Pop, F., Dobre, C., Cristea, V., Bessis, N., Xhafa, F., Barolli, L., 2015. Reputation-guided evolutionary scheduling algorithm for independent tasks in inter-clouds environments. Int. J. Web Grid Serv. 11 (1), 4–20.

Schmidt, W., Otte, A., Parada, J., Klein, M., Rollwage, S., König-Ries, B., Dettborn, T., Gabdulkhakova, A., 2010. OpenAAL — the open source middleware for ambient-assisted living (AAL). In: AALIANCE Conference. Malaga, Spain, March 11–12.

Spinsante, S., Gambi, E., 2012. Remote health monitoring by OSGi technology and digital TV integration. IEEE Trans. Consum. Electron. 58 (4), 1434–1441.

Spinsante, S., Antonicelli, R., Mazzanti, I., Gambi, E., 2012. Technological approaches to remote monitoring of elderly people in cardiology: a usability perspective. Int. J. Telemed. Appl. 2012. http://dx.doi.org/10.1155/2012/104561. ID 104561.

Stainov, R., Goleva, R., 2015. Intelligent backhauls for P2P communication in 5G mobile networks. In: The 11th Annual International Conference on Computer Science and Education in Computer Science. June 04–07: Boston, MA, USA.

Sun, H., De Florio, V., Gui, N., Blondia, C., 2009. Promises and challenges of ambient assisted living systems. In: International Conference on Information Technology: New Generations (ITNG), pp. 1201–1207.

Takács, B., Hanák, D., 2007. A mobile system for assisted living with ambient facial interfaces. Int. J. Comput. Sci. Inf. Syst. 2 (2), 33–50.

Tavares, L., Lapão, L., 2006. HPC telemedicine's service improves access to pediatric cardiology in central Portugal: leadership, organization and training as critical success factors — people really matter! In: MIE.

Tonchev, K., Sokolov, S., Velchev, Y., Balabanov, G., Poulkov, V., 2015. Recognition of human daily activities. In: IEEE International Conference on Communication Workshop (ICCW). June 8–12.

Trajkovik, V., Vlahu-Gjorgievska, E., Koceski, S., Kulev, I., 2015. General assisted living system architecture model. In: Mobile Networks and Management. In: Lecture Notes of the Institute for Computer Sciences, Social Informatics and Telecommunications Engineering, vol. 141, pp. 329–343.

Wood, A., Stankovic, J., Virone, G., Selavo, L., He, Z., Cao, Q., Doan, T., Wu, Y., Fang, L., Stoleru, R., 2008a. BContext-aware wireless sensor networks for assisted living and residential monitoring. IEEE Netw. 22 (4), 26–33.

Wood, A., Stankovic, J., Virone, G., Selavo, L., He, Z., Cao, Q., Doan, T., Wu, Y., Fang, L., Stoleru, R., 2008b. Context-aware wireless sensor networks for assisted living and residential monitoring. IEEE Netw. 22 (4), 26–33.

Yelamarthi, K., Haas, D., Nielsen, D., Mothersell, S., 2010. RFID and GPS integrated navigation system for the visually impaired. Circuits and Systems (MWSCAS). In: 53rd IEEE International Midwest Symposium.

CHAPTER 8

AAL and ELE Platform Architecture

Rossitza I. Goleva*, Nuno M. Garcia†, Constandinos X. Mavromoustakis‡, Ciprian Dobre§, George Mastorakis¶, Rumen Stainov‖, Ivan Chorbev**, Vladimir Trajkovik**

**Technical University of Sofia, Kl. Ohridski blvd 8, Faculty of Telecommunications, Department of Communication Networks, 1756 Sofia, Bulgaria*
†*Instituto de Telecomunicações, Universidade da Beira Interior, R Marquês d'Ávila e Bolama, 6200-001 Covilhã, Portugal*
‡*Department of Computer Science, University of Nicosia, 46 Makedonitissa Avenue, 1700 Nicosia, Cyprus*
§*University of Politehnica of Bucharest, Bucharest, Romania*
¶*Technological Educational Institute of Crete, Heraklion, Crete, Greece*
‖*Applied Computer Science Department, University of Applied Sciences, Leipziger Strasse 123, D-36037, Fulda, Germany*
***Faculty of Computer Science and Engineering, University Ss Cyril and Methodius, ul. Rugjer Boskovic 16, Karpos 2, 1000 Skopje, The former Yugoslav Republic of Macedonia*

8.1 INTRODUCTION

The usefulness of the Ambient Assisted Living/Enhanced Living Environment (AAL/ELE) platform is a matter of intensive discussions nowadays (Garcia and Rodrigues, 2015). The emerging market of different devices and tools for medical, physical, emotional, environmental, activity, position, condition detection, proprietary and open source solutions, interconnected and processing data locally and in the cloud are driving the platform analyses and design. In this chapter, a generic open hierarchical model that is used as a reference model for existing and new systems is defined (Kryftis et al., 2015). The model is open and not limited to all new technologies like sensor networks, disruptive tolerant networks, dew, fog, cloud computing, web-based solutions, proprietary solutions with common interfaces. It includes definitions of end-to-end connections, data models, virtualization at different levels (Batalla et al., 2015).

The diversity of the market, stakeholders, end-users, providers, producers, integrators, investors is one of the main drivers of the technology development (Skourletopoulos et al., 2014). There is a need of small, medium

Ambient Assisted Living and Enhanced Living Environments.
DOI: http://dx.doi.org/10.1016/B978-0-12-805195-5.00008-9

and big scale solutions serving different end–users and stakeholders. There is a need of scaling the functionality and the price of the platform and adapting it to specific customers (Papanikolaou et al., 2014). The active aging technological segment is growing rapidly and will be more than 15 trillion in 2020 (http://ec.europa.eu/research/innovation-union/index_en.cfm?section=active-healthy-ageing&pg=silvereconomy). In addition to this, the smart home automation business in Europe will increase from 44 billion to 140 billion in 2020. The consideration of smart living environment, smart cars, and smart cities are also pushing the development of the market segment.

In next sections, we try to highlight most of the important issues on the subject without pretending to show a complete solution.

Platform users and stakeholders are quite different by means of aim and usage of the system. From one hand, elder demonstrate resistance from the use of electronic devices and software components. Young generations are behaving in an opposite way. They expect to have full control on the hardware and software of the system and expect to have open solution that is interoperable with the other systems. On the other hand, professionals use the system in everyday work and need to save time and efforts. This diversity in requirements needs to be solved by offering gentle automated solution for elders and professional system to the others that is scalable and easy to integrate. The solution should propose AAL as a Service (AALaaS) and ELE as a Service (ELEaaS).

The chapter consists of the state of the art positioning the work appropriately on the market. There are sections on AAL/ELE service definitions, end-user identification, traffic patterns, Quality of Service (QoS) and Quality of Experience (QoE) analyses including data modeling. The hierarchical model design presenting the layered model of open protocol stack and framework, sensor networks or other local area technologies for access part of the platform, dew, fog and cloud computing expressing the data models and data processing are next parts of the chapter. Applied technologies already existing on the market, implementations, use-case scenarios related to the platform testing at different levels of design, platform verification and validation through examples, use-cases and protocols that demonstrate the vitality of the proposed solution are mentioned in chapter 16.

8.2 STATE OF THE ART

There are many AAL and ELE Platforms already existing on the market nowadays. Parts of them are still under development. Parts of them are completed as solutions. The biggest part of the platforms is proprietary and this attracts issues as interoperability and integrity of the proposals. Many platforms are developed under support of different organizations including European Commission, AAL Forum, ZigBee Alliance, etc. In this chapter, we try to create an open reference model that could be an umbrella for many existing solutions including support for AALaaS and ELEaaS.

One of the well-known and developed AAL platform is UniversAAL (UniversAAL, 2014; Tazari et al., 2012). It defines framework starting from sensor networks and defining AAL services at application layer over different operating systems and middleware. The project analyzed and consolidated several FP6/7 platforms like PERSONA (2010, 2016) that consider scientific, technical, economical, psychological, political aspects of AAL. Another interesting project is MPOWER (2009) that created a middleware-based open source solution for homecare and performed pilots. A very interesting project is SOPRANO (2010) that developed object-oriented tool for context-aware services. An interesting approach for intelligent connected home for AAL is demonstrated in AMIGO (2009). Ontology-based open architecture is shown in OASIS (2011). Stress on validation was made by Vaalid project (VAALID, 2011). All the mentioned projects aim to provide:

- A reference model for AAL that is abstract and presents the level of relationships between subjects and objects in the environment as well as highlights important concepts related to AAL.
- An AAL reference architecture with abstract architectural elements that might be irrelevant to the building technologies and protocols in the domain. This approach is common in software development and not so common in hardware development where interoperability is based on the standard interfaces and protocols. An integrated approach is presented in this chapter.
- Basic set of API specifications, patterns, and models of high-level shared data. We develop the data models further.

Very interesting approach taking into account human-machine interfaces and psychological condition of the people is demonstrated in AsTeRICS (2013). It provides a flexible construction set for AAL service development combining sensor technology and computer vision. The project aimed helping people with reduced motor capabilities through new Human Machine Interfaces (HMI) at the standard desktop. The solution works on mobile phones or smart home devices. AsTeRICS platform consists of building blocks. Sensors are used for body activity control. Actuators are used for interface to the environment. The Embedded Computing Platform is customizable to the needs of the individual user.

Connected Health Lab (2016) (http://www-cps.hb.dfki.de/research/baall) is dealing with home care system by clustering end-users and nurses, hospital care systems, including hospital administration and Electronic Patient Records (EPR), national hospitals authority, interoperability between health labs.

The system defined in KIT (2016) is interesting and designed for telemedicine. It is modular and customized to specific diseases and needs of the end-users. Modules are pluggable and allow adaptable configuration (Drobics and Hager, 2014). The system works as a core platform supporting management, administration, utilities, and security. The focus is to collect data from sensors and analyze data statistically offline using a database. The system has three layers (Kreiner et al., 2013a, 2013b; Von der Heidt et al., 2014; Kastner et al., 2011; Drobics and Hager, 2014). The basic layer is for sensor data collection. The sensor data aimed health-based observations according to the predefined standard through web interface using profiles. Observed profiles are transferred for further analyses through Application Programmer's Interface (API). The data is processed and used for sensor management, rule-based decision support, data visualization using interactive charts, organizer and calendar functions, search engine for observations, data export, notification engine for reminder sending using SMS, email and Google Cloud Messaging, document management for PDF and word files. All generic services are connected to specific care plugins providing customized user interfaces for different stakeholders, health domains, specific diseases, and preventive telemedicine for elderly people.

The intLIFE PHR Platform (Vogiatzaki and Krukowski, 2015, 2014; Krukowski et al., 2013, 2014) is a system for diverse health applications funded by European Commission such as ICT-PSP-NEXES, AAL-PAMAP, FP7-StrokeBack, Artemis-CHIRON, FP7-ARMOR. It is open source solution helping people to take medicines on time, to be monitored remotely, to check preventively their status, to send possible warning to the responsible personnel, to see electronic health records etc.

8.3 AAL/ELE SERVICES

AAL/ELE services are a matter of many different definitions last decade. Wherever the AAL services are more related to the health of the people aiming more preventive and ill people support including different stakeholders in hospitals and at home, the ELE services are more related to the living environment of the people taking into account not only ill or elder but also other generations and people who like to have active and healthy way of living (Pires et al., 2016; Garcia et al., 2014). Identification of AAL and ELE services is too broad (Garcia and Rodrigues, 2015). Some sources consider services at sensor level with specific use-case scenarios (Ferro et al., 2015; Goleva et al., 2015c; UniversAAL, 2014). Other sources consider cloud or web-based services (Active aging, 2016). There are sources considering services end-to-end (Tazari et al., 2012; KIT, 2016; Goleva et al., 2015a, 2015b, 2015c).

Generic definitions of the AALaaS and ELEaaS could be seen in Stainov et al. (2016) and Mirtchev et al. (2016). The authors tried to define the services at cloud or fog level. Service virtualization at fog or cloud level allows service mobility and makes the service independent from access network technologies.

We try to highlight here the most important AAL/ELE services taking into account the end-user place where raw data is obtained. We also consider network at access, edge and core parts where data is collected and might be partially processed. The data centers for data collection and processing are either at core part of the network where the data is stored and analyzed or at dew and fog level. The services are virtualized at different levels in terms of time and place.

In this section, we try to explain necessary generic services that allowed definition of an open platform. The AAL/ELE service classification is based on many different criteria like end-user perspective, service provider views, technological constraints.

Most of the raw data is obtained from Home Care System that is based on sensors, home server, home controller and optional gateway to the Internet (Goleva et al., 2015a, 2015b, 2015c; Drobics and Hager, 2014). The data could be partially pre-processed, i.e. data acquisition could be performed locally at home server like it is proposed in Pires et al. (2016). Other authors like Skourletopoulos et al. (2016), Bessis and Dobre (2014) and Batalla et al. (2016) consider data processing at cloud level. We believe that in AAL/ELE platform there is a need of both approaches because the network of cared and caring people is heterogeneous and open by definition. There are many papers related to the personal and electronic health records storage, management and security of the transferred data (Porumb et al., 2015; Porambage et al., 2015). Many authors like Lamine et al. (2014) work at ontology and middleware level and stress on data processing issues.

Part of the basic services is activity identification and recognition. It is applicable for elder and young generations, healthy or ill. Some of the people concern privacy. In many cases when there is a need to monitor people 24/7 hours this service is valuable (Garcia and Rodrigues, 2015). Activity recognition is related to the digital signal processing (Gehrig et al., 2015; Kahveci et al., 2015; Díaz Rodríguez et al., 2013a, 2013b, 2014a, 2014b, 2014c; Díaz Rodríguez, 2015). Part of the work is defined using ontology. Some of the works consider real-time recognition via home experiments with multiple residents. The experiments could be done explicitly using people identification with RFID tags, bracelets, cameras. Experiments without explicit identification, using only ambient sensors in real-time are more interesting because of the applications like fall detection, activity-tracking, movement monitoring (Alemdar et al., 2015; Tunca et al., 2014; Durmaz Incel et al., 2013; Alemdar at al., 2014; Ustev et al., 2013; Ertan et al., 2012).

Identification could be cooperative when the system has access to the facilities like house, office, building, etc. It could be non-cooperative and used to recognize who is in the house (Garcia and Rodrigues, 2015). Services could be classified as context-related like ambient services (Monekosso et al., 2015a, 2015b, 2015c) and individual services (Pires et al., 2016).

Technology mapping to services could be done through interoperability via standard interfaces, protocols, standard formats for data exchange and virtualization at different levels (Monekosso et al., 2015a, 2015b, 2015c; Goleva et al., 2015a, 2015b, 2015c, 2016).

Another interesting part of services is connection to outdoor assistance. Most of outdoor services are mobile for the vehicles, end-users and environmental services in the parks, shops, club, school, hospital, etc. In many cases, there is a lack of continuous connectivity (Lambrinos, 2015; Jara et al., 2015; Chaudet and Haddad, 2013). The use of outdoor and indoor femtocells at the edge of the network could support the service. Sensor networks are capable to forward messages (Kreiner et al., 2013a; Goleva et al., 2015a, 2015b, 2015c). The access network is ubiquitous by nature. The use of gateways using standard interfaces allows abstraction and virtualization of the existing technologies and exhibits them to the application as a transparent single infrastructure layer. It could be done via Software Defined Network, middleware (Lamine et al., 2014), virtual machines, peer-to-peer connectivity (Papanikolaou and Mavromoustakis, 2013; Stainov et al., 2016, 2014).

While having services in Body Area Network (BAN) it should be also taken into account that part of them are personal. In order to make services healthy related there are also special regulations in different countries to be considered. Well known IEEE 802.15.4 standard combined with ZigBee is widely used for Personal Area Networks (PAN) and home/environment automation. It is well combined by all IEEE 802.11 standards. In this sense home/personal controllers use IEEE 802.15.4 to 802.11 gateways for network interconnection. BAN standard 802.15.6 is created long time ago and is widely used (Kurunathan, 2015). There are technological limitations, delays, interoperability issues, security considerations, lack of sustainable technologies, lack of coordinated legislation and standards (Riazul Islam et al., 2015; Al-Fuqaha et al., 2015). Many topics in this technology are still open like monitoring, security of the well-being devices, end-user customization, dynamics of the BAN network based on the wearable sensors, sensor power supply, psychological aspects related to the wearable sensors, the trust to the technology.

There are also many different cloud-based solutions on the market for BAN services. The tradeoff between data storing algorithms and placement of data in the cloud or locally is discussed in many projects and papers. The virtualization process requires further analyses where to store and process, what type of data to consider private or public, lack or presence of feedback channels or customized feedback data, data sharing, big data analyses. We introduce in the chapter a dew/fog/cloud computing approach using peer port and client/server computing defining further ELEaaS and AALaaS (Stainov et al., 2016; Mirtchev et al., 2016).

8.4 REQUIREMENTS ANALYSIS

AAL/ELE platform analyses is an important development phase in order to define correctly the main parts of the system, end-users and performance requirements. The main difficulty in this analyses phase is the distributed and mobile nature of the network and systems.

The requirement analyses starts with end-users and stakeholders. The end-users could be active working people who like to measure their daily activity. End-users could be children at their first three years when they might need additional health monitoring and concern. End-users could be children between 3 and 18 years when the daily activity, appropriate sport and studying sequences are essential for their well-being. It is especially valid for children at risk. End-users are retired people at home who like to continue their active aging keeping preventive monitoring of the health condition.

The technology is invaluable for ill people who like to stay at home and have a normal life. It is valuable for disabled people helping them to overcome their physical limitations (Kirchner et al., 2015). Patients in hospitals could benefit from the technology that monitors medications, rehabilitation, medical history, basic condition, previous and future prescriptions (http://robotik.dfki-bremen.de/en/research/robot-systems/aila.html).

The system will never be useful without all the people taking care of the patients. There is a need of proper interfaces to nurses, medical doctors, rehabilitation personnel, insurance companies, pharmacists, pharmaceutical companies, sport centers, family members. Because the roles of the end-users and stakeholders are different, they need to be profiled properly during system analyses phase.

The profile of the active people consists of few to many sensors for daily and occasionally weekend activities. Only part of the data should be processed and shown in real-time. Most of the data could be processed in near-real-time (Stainov et al., 2016) or in non-real-time. The scenarios between sensors and controller device are mostly by means of simplex or duplex messages. In some special cases voice, pictures, videos and multimedia could be also transmitted. In some environmental services, the messages are simplex. Usually in end-user to system interaction, the communication channel is duplex. The connection between controller and cloud application is duplex and could work in real-time in case of cloud-based real-time processing. Most of the operations are performed there in non-real-time. The cloud application could be enriched by many different additional functionalities based on the needs and requests from the end-users. When applied dew and fog computing will make the cloud-based application distributed and more complicated. The distribution of the data in the dew/fog/cloud and distribution of its processing, the capability of the data to move freely will be described in more details in the section on cloud computing.

The profile and the interfaces to the patients should be customizable to their needs. There are many divertive requirements here starting with small kids, elder with Alzheimer and finishing with hearth disease and diabetics. We need to point also that some of the patients would like to control the process whereas for some of the patients it is not possible. The interfaces to the patients that control their status alone are the same as the interface set for the active living people. The interfaces to the patients that control their status with the support of the medical and rehabilitating personnel are richer by means of restrictions and availability of the data as well as by means of interconnection between different system modules. Communication protocol set between the patient and caring people should be duplex, gentle, allowing negotiation, corrections, restrictions, chatting, mails, data sharing, data connection, voice and video support, file exchange in addition to the already defined functionality for the self-controlling people. Some of the functions should be a matter of configuration being enabled or disabled depending on the case. For example, people with dementia do not need all functionality related to the messaging. Additionally there is a need to consider different levels of alarms in all cases. Some of the alarms could be warnings only. Some of the alarms could call emergency team automatically.

The profile of the family members is the same as the profile of the active living people. Additionally they need to be capable to support the family member's data partially upon medical personnel demand or family member demand. A typical case is a baby or a person with dementia that have a family member taking care of her/him.

Caregivers' profile is similar to the profile of the family member. The main difference is in security functions related to the personal data of the patients. This functionality could be optional.

Nurses and rehabilitation personnel need access to the medical data of the patients and active living people for keeping prescriptions, everyday activity tracking, and necessary manipulations (Mandel and Autexier, 2016). The communication channels should be duplex and cover all functionality of the family members. Additionally, there might be a need of additional communication channel to the family members without the presence of end-user like in the case of kids under 18 or disabled elder.

Medical doctors need all kind of communication of the family members and nurses as well as priority function to the activities and manipulations including calling for emergency. Prescriptions and recommendations should be well archived and support functions for historical records, personal health record. Medical records analyses are necessary in addition. In all cases when the medical doctors taking care of the person are many, the prioritization should follow the rules set in the national and international level legislation. Medical doctors and nurses need also to have a separate communication channel that will allow cooperative consultations and decision making using all kind of possible data exchange like voice, picture, data, video, and multimedia.

Special case of stakeholders is hospital administration and insurance companies. They will need all kind of interfaces to the patient or her/his family member and be capable to keep track automatically on the activities and the events related to the administration and financial management of the end-user's health. The process depends very much on the national level legislations and customs. In this sense, the service should be customizable.

Finally, the service needs to be managed. One option is to allow the end-user and family members doing so. The other option is to offer managed services to the end-users. The last approach is going to be offered by utility companies. The role of Information and Communication Technologies (ICT) in managed services is essential. There is a need of the interface to the system administrator who will take care of the service by means of monitoring, upgrading, configuration, archiving, technology migrating, enhancements. Virtualization process at all levels will allow technological migration without active participation of the end-users. The end-user devices and activities affect only access network.

Additionally to the identified AAL services, there are many different options for ELE services. End-users in ELE services should control and manage the sensors at home, car, garden, school, shop, nearest cafeteria, kindergarten, i.e. all environmental sensors related to the personal network connected through different technologies and interfaces (Georgieva, 2016). Part of these sensors could be accessed by many end-users at ones. In this sense, there is a need of prioritization and conflict resolution. Notification service could be also useful.

Family members could control the personal network when the end-user is disabled or not allowed (Georgieva and Markova, 2015). They will need the same level of access. Other body like utility provider could control the network. There is a need to change legislation concerning the governance of the personal services, level of access and priorities.

There is a necessity to support different levels of warnings and alarms like gas leaking, CO_2 level increase, flooding, temperature raising, smoke detection etc.

There will be a need of AAL and ELE service cooperation, overlapping and possible unification. The level of cardinality could be a function of end-user cooperation.

After identification of the main users of the system, we may define the most generic requirements like:

- Open — the system should allow enhancements to the new and already existing technologies. It should allow internetworking of the heterogeneous segments on a single platform.
- Virtualized — the system should be virtual at different levels in order to allow better administration and service development.
- Generic — the system needs to be as generic as possible. This means that the services should not be specific but more natural for the end-users.
- Scalable — the system should be scalable and should be capable to be installed in very small scale and be expandable to the large scale at later stage.
- Module-based — the system should be proposed in modules that are interoperable and could map, merge, split, interact using standard interfaces and protocols.
- Customizable — the system should be as customizable as possible. Based on the generic nature of the initial set of functions the end-users and stakeholders should be capable to change configuration in order to serve their personal preferences, peculiarities, local customs, religion specific requirements etc.
- Granular — the system should work at different levels of granularity allowing the end-user to change it.
- Security and privacy should be included in all layers and components of the system.

The main system functions that reflect to the test scenarios, verification and validation plans could be summarized as:

- Sensor installation, uninstallation, configuration, monitoring
- Sensor transmission in simplex and/or duplex mode
- Communication channel between sensor and controller installation and management
- Sensor network controller installation and management
- Sensor network gateway installation and management
- Communication channel and protocols between controller and network gateway establishment
- Optional home server installation

- Communication channels between home server, controller and gateway establishment
- Configuration tool for home users
- Device/user authentication mechanisms
- End-user capability to install sensors
- End-user capability to change sensor, home server, controller, gateway configuration and access control rules
- End-user capability to read historical data
- End-user capability to set an alarm to the caring personnel or family members
- End-user capability to negotiate, explain, ask the data with care givers and stakeholders using different communication channels and applications
- Access network technologies supported and changeable to others like Bluetooth, WiFi, LAN, 3G, 4G, etc.
- Family members capability to check the status of the sensors and to manage them
- Cooperative management of some of sensors, rules for prioritization and customization
- Historical data accessibility
- Alarms management
- Data presented to the administration and insurance companies trackable by end-users, family members, hospital administration and other authorized bodies
- Additional services through home server availability
- Additional services through the dew/fog/cloud accessibility
- Nurses, rehabilitation personnel and medical doctors capability to prescribe and track details
- Medical doctors capability to set priorities
- Medical doctors and nurses capability to consult locally and remotely
- System administrators capability to monitor, manage and configure locally and remotely
- Administration staff and insurance companies data track ability and communication support to the end-users and family members
- Data collection
- Data storage
- Different stakeholders data processing at different levels
- Data processing locally at home server
- Data processing locally at service provider server

- Data processing in the cloud
- Statistical analyses availability
- Historical data availability
- Real-time data presentation availability
- Remote access availability

There are many other functions that might exist in the system and many of them are described during analyses and design of the system, data modeling, and traffic analyses. Use-case scenarios and tests for verification and validation are shown in Chapter 16.

8.5 HIERARCHICAL MODEL DESIGN

The AAL/ELE system is a distributed heterogeneous network by nature and has access, edge and core parts partially virtualized. We create a typical open hierarchical model.

The dew/fog/cloud based parts of the system are also hierarchical and distributed. The hierarchical modeling is also applied there for data location, storage, processing and mobility. Part of the services will be clear client/server based. Other part will be peer-to-peer. Due to the peer-to-peer overlays part the data will move freely in the network. Peer port will be applied for the data resilience (Stainov et al., 2014).

The proposed hierarchical model defines the rules in a similar way to the Open System Interconnection Reference Model (OSI RM). The AAL/ELE hierarchical model is extended by all new software technologies working in the dew/fog/cloud computing level (Figure 8.1).

The physical layer takes into account all existing technologies and possibility of interoperability. The channel layer is also based on the existing technologies. The door to the future technologies is open. The network layer explains details on connectivity that is essential in the future cloud and mobile networks. Transport layer is based on well-known IP but could incorporate other existing technologies through existing interfaces. Session layer is interesting by additional possibilities like sessions, associations from signaling systems, exchanges from storage area networks, bundle sessions from delay tolerant network etc. The ad hoc connectivity at access part of the network allows even more options here with multihoming, multisession, and multipoint connections. This part of the OSI RM is not developed well and it is expected also to play main role at secure and performance management layer. Security issues are mentioned in the model at presentation layer but should be considered at all levels. It is not shown

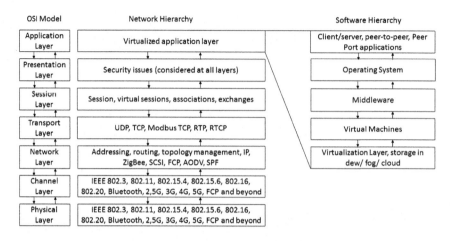

Figure 8.1 Open AAL/ELE system hierarchy.

on the Figure 8.1 for simplicity. Application layer is expanded and fully virtualized. It is possible to have applications running on a single machine and under single operating system. It is possible to have a client/server or peer-to-peer applications. Part of the services could use peer port for data resilience. It is also possible to have a virtual machine running on a single hardware item. Middleware allows distributed services. It is possible to have server farm with application servers and/or router servers that are fully virtualized. More options could be seen in the section on dew/fog/cloud computing on the open AAL/ELE platform.

The system interconnects subnetworks, domains, modules, and tools. In addition, it can be added that fusion of several infrastructures and more efficient management of the physical network including possibility for device-to-device communication is seriously considered as part of the so-called 5G cellular technology (Chávez-Santiago et al., 2015a, 2015b). Connections are not only heterogeneous but also could be slightly overlapped. For example, Wi-Fi and ZigBee sensors work at the same frequency band (Goleva et al., 2015a, 2015b, 2015c; Chávez-Santiago et al., 2015a, 2015b). While preparing the open platform many different already existing implementations are taken into account. Part of the implementations are quite specific like solutions for head tracking, remote health monitoring (Spinsante and Gambi, 2012a, 2012b, 2012c), digital TV integration, and remote monitoring through interactive television. A special issue in multi-sensor environment is synchronization and data fusion in PAN especially in cases when the data is used to take important decisions (Wåhslén et al.,

2012, 2011a, 2011b). Time synchronization is taken into account also in interactive television applications and activity recognition (Cippitelli et al., 2015). One universal way is applying Network Time Protocol (NTP) at application layer. Such approach seems to be easy during data collection to the cloud. The best way to synchronize data is to perform it at any layer of the system. Up to now, this is technologically constrained.

Special attention is paid to the medium access and routing protocols. Medium access should cover all existing technologies in IEEE 802.x and all mobile technologies like $2\frac{1}{2}$G, 3G, 4G, 5G, 6G and beyond. Bluetooth is also taking into account. At cloud level, the access technology is Fiber Channel Protocol or IP.

Routing is also quite diverse in the access and in the core parts of the network. AODV and shortest paths coexist with static routing (Reina et al., 2016). Ad hoc topologies require additional algorithms for network segmentation, network unification, load balancing, bundle layer routing etc.

Sensor collaboration, data sharing, topology dynamics are presented in Hansen et al. (2016) as well as in Rolla and Curado (2015). Application-awareness is shown in Wamser et al. (2014a, 2014b). The work also developed an idea for home router and middleware services. Context-awareness of the network at application layer is presented in Chilipirea et al. (2016). Performance monitoring defined at network and transport layers in wireless sensor networks could be seen in Wåhslén et al. (2011a, 2011b).

Preliminary work on the open platform for AAL services is demonstrated in Garcia and Rodrigues (2015), Goleva et al. (2015a, 2015b, 2015c). The idea is developed further in Goleva et al. (2016). The network consists of sensor networks at access parts and application servers in the cloud. The network is used not only for measurement of the AAL/ELE data but also for management of the end-devices and proactive behavior towards the end-users and stakeholders when necessary. In this sense, the network is supporting duplex communication channels in all cases when it is possible and applicable. The important parts of the network has no single point of failure.

In Dupont et al. (2015) there is analyses that makes the cloud part of the network green explaining how the performance is adapted to the available energy from renewable sources. Sensor network as well as network at dew level also could be green using new technologies like EnOcean (Goleva et al., 2016).

The system is capable to collect data from home/personal/car/city/village environment and to build personal cloud architecture based on the

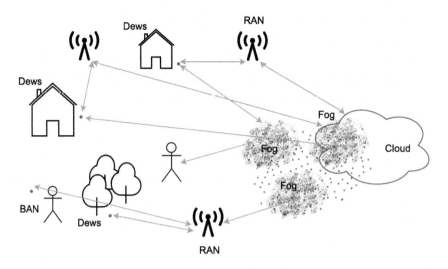

Figure 8.2 Personal AAL/ELE services.

existing infrastructure (Figure 8.2). The personal cloud information is redundant. The redundancy of the sensors is a matter of personal decision. The redundancy of the data after the home server is a matter of cloud AAL/ELE services. We demonstrated on Figure 8.2 possible cloud-based AAL/ELE services platform that looks like a fishnet. Small dots on the picture represent home servers that are dews in the cloud-based network. They collect data from local sensors. Part of the data could be relayed by other sensors. Home servers or dews are connected to small clouds called fogs. Fog could be based on private or public clouds using client/server and peer-to-peer technologies. Radio Access Network (RAN) on the figure represents mobile network fog level. Fog islands are connected to the cloud. The data collected locally is transferred to the fog/cloud after its verification and is a matter of additional analyses (Goleva et al., 2015a, 2015b, 2015c). Dews, fogs or clouds could be private or public. The interconnection between dews, fogs and clouds will be a matter of service level agreements (SLA).

8.6 SENSOR NETWORKS. DEW, FOG AND CLOUD COMPUTING

Sensor technologies and 5G are vastly developing area used to interconnect many different devices into a single network. Sensor network could be clus-

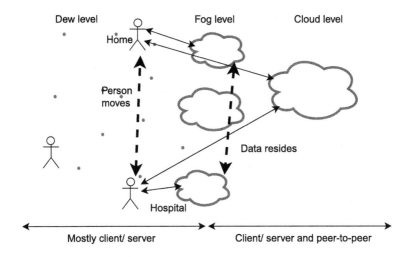

Figure 8.3 Dew/fog/cloud computing platform.

tered. It could be static of dynamic, i.e. ad hoc. Part of the sensors could relay the information. The data from sensor network is collected by one or more controllers and transferred to the gateway and home server. There might be more than one gateways for redundancy. The home server could perform preliminary data filtering and acquisition (Figure 8.3). Personal data could freely reside at dew/fog/cloud level following the mobility pattern of the person. Data residing procedure may not be always in real-time. Only critical data could be moved in real-time.

The home server collects data from the entire home, garden, cars, and all people in the home. It is also responsible for the connectivity to the fog-computing servers and neighboring dew-computing servers for redundancy and data exchange. The same could be applied for shop, school, office, cinema servers and so on. Data from local home/school/hospital/office/shop servers related to the personal computing is transferred either at dew computing level or at fog computing level. In a typical peer-to-peer environment, the data is indexed in a distributed manner and data residing could follow the indexed information as background services.

Fog-computing level contains regional servers and storage systems that compute and store personal information. Fog-computing facilities could be a matter of subscription to the company like Google, local Internet Service Provider, Health Insurance Company, local mobile service provider, municipality service, or third party provider. At fog-computing level it is easy to accumulate all necessary data for personal computing network and

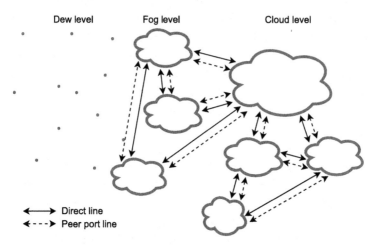

Figure 8.4 Peer port application at fog level.

facilities because people tends to move in a specific region (Figure 8.4). The solid lines on the figure represent direct communication channels. The dashed line presents peer port channels. Both channels are interchangeable, i.e. backup each other in almost real-time. The same approach could be applied at dew level and at cloud level as well as between levels.

Accommodation of the data in the cloud in interconnected data centers served by many servers in server farms and distributed in a big region allows globalization of the AAL/ELE services including performance and storage optimization. Future smart cities, smart homes, smart offices, smart hospitals, smart schools, smart shops, i.e. smart environment as well as development of the municipality services will allow rapid AAL/ELE service development, mobility of the users, devices, remote access and monitoring, better resource management.

In the sense of dew/fog/cloud computing the personal computing network that provides AAL/ELE services are a virtual overlay (like a fishnet) item that is spread in the area the end-user is active and connected. The development of the personal computing is going to open additionally other sensitive issues to the end-users like security, sensitivity of the information, privacy. All of them will reflect to the law and legislation locally, regionally and globally. Connectivity of any single sensor to the network in a reliable way will allow mobility of the data and resources processing throughout the platform. In this sense, data processing could be done at different level and it could depend on the technology, green energy requirements, end-user preferences, local law etc. The new dew/fog/cloud computing paradigm changes the performance, QoS and QoE characteristics significantly.

8.7 TRAFFIC PATTERNS, QOS AND QOE REQUIREMENTS

Traffic patterns generated by the AAL/ELE services are going to be quite different from the typical voice, video, CCTV, TV, audio, data, multimedia flows. The pattern generated by the single sensor might be question/answer or unidirectional repeated single data item. The patters after TV camera are video flows with different encoding, rates and frame sizes. The pattern after the smart phone is voice flow or picture or messages, or aggregated multimedia. The patterns after the home controller or home server are aggregated data from many questions and answers as well as cameras, smart phones, tablets, as well as backups of the data. The home servers may receive the data from fog servers and cloud servers that is accumulated from offices, shops, schools, streets neighboring home servers where the end-user had been. The same is applicable for the traffic between fog computing and cloud computing facilities. The use of multipathing, multihoming and peer port (Stainov et al., 2014) could make the traffic even more complicated.

Generally speaking, the irregularity or regularity of the traffic is so high that it is difficult to predict what will be the necessary channel capacity requirements and perform proper network planning. In our previous works (Goleva et al., 2015a, 2015b, 2015c, 2016) we have already shown that the traffic patterns are not exponential by nature but tend to change their probability distribution function depending on the codecs, data aggregated, performance tasks, protocols, etc. Attempt for traffic flow classification is made on Table 8.1. Some of the services and flows are compound. The dynamic nature of the AAL/ELE services is the main reason for difficulties in network design. After measurements of the traffic in real environment and in the labs we estimated traffic changes based on the inter-arrival time and servicing time starting from well-known exponential, towards Gaussian, gamma, Pearson, deterministic, Pareto, lon-normal distributions.

The QoS and QoE requirements follow traffic patterns and application specific features. Whereas QoS is more related to the traffic flows, performance of the system, resource configuration and design, the QoE is more related to the perception of the service by the end-user. It is very important for specific users like disabled or elder people to develop easy and comfortable Human Machine Interfaces (HMI) that will allow better use of the service. Very often, the HMI is developed by the support

Table 8.1 Traffic flows specification in AAL/ELE platform

No	Media type/AAL/ELE service type	Item size	Inter-arrival time	Servicing time
1	Sensor to controller and controller to sensor	From 10 bytes to 200 or more bytes	Any distribution. Usually multimodal or deterministic	Depending on the technology in the range of milliseconds
2	Sensor to sensor relay	Composition of many sources like 1	Any distribution. Usually multimodal, gamma or Pareto	Depending on the technology in the range of milliseconds or even seconds
3	Controller to gateway in sensor network	TCP or MODBUS TCP segments with accumulated data from services 1 and 2	Any distribution with high irregularity. Usually gamma or Pareto	Usually in milliseconds
4.	Gateway or controller to home server	TCP segments with accumulated data from services 1 and 2	Any distribution with high irregularity. Usually Pareto or gamma	Usually in milliseconds for real-time services and seconds for non-real-time services
5.	WiFi or 3G/4G gateway to the home or fog server	TCP segments with accumulated data from sensors and devices	Any distribution. Usually Pareto of gamma	It depends on the technology and data. It is usually in milliseconds or seconds
6.	Home or fog server to WiFi or 3G/4G gateway	TCP segments with accumulated data from PAN and other entities	Any distribution. Could be approximated to gamma and Pareto	It depends on the technology and data. It is usually in milliseconds or seconds
7.	Home server to the fog server	TCP segments with aggregated traffic from home, cars, garden, BANs	It could be deterministic for short time if the data is not in real-time. It is not deterministic if the data is in real-time	In milliseconds for the real-time data and in seconds for non-real-time data like archiving. Could be also in minutes

continued on next page

Table 8.1 (*continued*)

No	Media type/AAL/ELE service type	Item size	Inter-arrival time	Servicing time
8.	Fog server to home server	TCP segments with files, raw data from other servers, pictures, video files, reports	It could be deterministic for short time if the data is not in real-time. It is not deterministic if the data is in real-time	In milliseconds for the real-time data and in seconds for non-real-time data like archiving. Could be also in minutes
9.	Fog server to cloud server	Fiber channel protocol over IP data with read and write requests	Any distribution. During the session could be partially deterministic. Usually in milliseconds	Depends on the length of read/write requests. Usually in milliseconds. Could be also in minutes in database replications
10.	Cloud server to fog server	Fiber channel protocol over IP data with read and write requests	Any distribution. During the session could be partially deterministic. Usually in milliseconds	Depends on the length of read/write requests. Usually in milliseconds. Could be also in minutes

of psychologists in order to navigate the interface to the necessities of the patients. HMIs should be also highly adaptable being prepared in modules and parametrized.

Special attention to the patient's requirements could be found in De Santis et al. (2015). It is related to the smart home and active ageing of the end-users. Technologies for disabled patients could be seen also in Gasparrini et al. (2015). In Spinsante et al. (2015a, 2015b) a multimodal interaction with elderly-friendly interfaces is demonstrated. Interoperability between systems is presented in Rossi et al. (2014). Raffaeli et al. (2014) has shown smart TV application for personal e-health services. The system architecture and its zero configuration for 6LowPan networks are presented in Carlier et al. (2014). Synchronization and data fusion issues are explained in Cippitelli et al. (2015). A very specific cases are considered in Kirchner et al. (2015), Mandel and Autexier (2016).

Quality of Service and Quality of Experience requirements could be mapped to the traffic patterns in cases of the known probability density function (pdf) of the inter-arrival times and packet lengths. An analyses of the smart home patterns and lab-based Voice over IP and Video over IP pdfs could be seen in Goleva et al. (2015a, 2015b, 2015c, 2016) in details. Because the pdfs are changing dynamically from deterministic, through multimodal, geometric, gamma, Pearson, Pareto, exponential, etc. we propose to map pdfs of the inter-arrival time to the gamma distribution and resource management to the Polya distribution. The appropriate mapping will allow better design and performance analyses. This is part of the future research plans of the authors.

8.8 APPLIED TECHNOLOGIES

The AAL/ELE platform is open to many different and even diverse technologies. We may start description with different sensor platforms like EnOcean, ZigBee, ZigBee Green Power. Then, we may go through Bluetooth, WiFi and mixed options and continue with Azure, VMware, Common Object Request Broker Architecture (CORBA), Storage Area Network (SAN), Delay/Disruptive Tolerant Networks (DTN), software-defined networks (SDN) (Lange et al., 2015), representational state transfer (REST) architecture (Spinsante et al., 2015a, 2015b), social middleware for cloud-based resource intensive applications (Mavromoustakis et al., 2016, 2014).

The context — aware social networks are a special case for AAL service implementation in the cloud. The opportunistic nature of the AAL services and applications and its special requirements to the connectivity is shown in Lukowicz et al. (2012). The resources of socially connected peers can be used in order to extend the capabilities of the mobile devices, by providing extra computing, storage resources, as well as the execution guarantee within a predefined time frame. Resource offload scheduling scheme is proposed in Mavromoustakis et al. (2016) and Mousicou et al. (2013) towards enabling offloading of the resources from a mobile device to another, based on their social collaboration.

Users that are communicating on a regular basis (as a part of the social-interaction) can be part of the device-oriented context awareness. This is a result of the device-based triggering mechanisms for devices data exchange. As a result, each device knows the activity of the other device that it communicates with, which in terms of resources can be used for offloading memory-hungry tasks. As devices exchange system-level messages, it can be used for optimal utilization of redundant resources of other devices that are critically in-need of processing resources. Comparative performance evaluations, in the presence of "critical-process executions", as well as in the sense of meeting the required deadlines, were performed for the comparison with other similar schemes to prove the validity and the efficiency of the proposed framework, in contrast to the nodes' lifetime extensibility.

Nowadays the semantic technology research allows for the communication between these two district words namely, the heterogeneous data and applications, as well as the different processes, providing an abstraction through cross-layering (Lukowicz et al., 2012). In this respect, there is a great need to develop operating systems that are capable in adjusting changes to connectivity variations, whereas at the same time, provide high performance process and context execution within the mobile processing environment. Mobile devices are inclined to energy consumption limitations that usually aggravate the reliability and the on-time proper execution of the underlying applications that are running. In this direction, we propose a mechanism that considers the energy consumption of the mobile devices, exploiting several context applications (processing power hungry applications). The proposed approach adopts an offloading process, which is using the social context of the each mobile user, in order to allow offloading of the processing resources in partitions. The latter targets the minimization of the energy consumption through the social collaboration model, which enables each mobile user to offload resources for conserving energy. The offloading mechanism is adopted as a part of each mobile application start-up, in order to minimize the GPU/CPU efforts and in turn the energy that is consumed by each mobile device, when it runs out of memory or processing resources.

Resource-poor implementations like the mobile-peer-to-peer communicating devices in AAL aim to leverage the advantages from the mobile cloud computing paradigm. Mobile Cloud oriented devices allow the exploitation of the cloud paradigm into mobile oriented framework (Papanikolaou and Mavromoustakis, 2013). It seems that one of the greatest challenges for mobile cloud paradigm is the energy conservation, while the resources are being manipulated on the cloud. Limitations in battery lifetime host many restrictions for media-oriented implementations of the mobile cloud and are a barrier for exchanging resources through mobile networks (Dimitriou et al., 2013). The QoS offered by these devices is significantly aggravated by the energy-hungry applications (e.g. video services, interactive gaming, etc.). Applications requiring processing power leave an open gap, in achieving the desired level of mobility within a certain context and it becomes particularly acute in service degradations, when bandwidth and temporal characteristics, as well as related network limitations restrict such applications to run. Most of these processing power-hungry applications heavily depend on the underlying device-oriented infrastructure, limiting the usage for devices, where the needed processing power is not supported (Mousicou et al., 2013). Authors in Abolfazli et al. (2014) refer to the impact of Cloud-based Mobile Augmentation (CMA) to various issues of the communication diversities with regards to the performance limitation factors that influence on the adoption of CMA, whereas in Satyanarayanan (2010) they expose the processing deficiencies of mobile clients due to slow processing speed and limited RAM in wireless devices. Mobile devices are anticipated to support high demands in memory with high processing capabilities similar to desktop computing for adequately performing computing-intensive tasks. However, the most mobile devices tend to become smaller in size, whereas they are incapable of hosting high requirements in memory, which can only be supported by powerful processors, being able to perform large number of instructions set through time (Park and Choi, 2012).

In this context, harvesting the processing power by measuring the capabilities of each mobile device based on temporal characteristics (Mousicou et al., 2013) and social parameters can be used as an input for lifetime extensibility of such devices. Efficient resource management has to be achieved within the context of the mobile cloud-computing paradigm. This should include effective allocation issues of the processing power, in collaboration with the memory resources, as well as the network bandwidth. Different dynamic resource allocation policies have been explored in Slegers et al. (2009), towards elaborating on the enhancement of the application execution performance and the efficient utilization of the resources. Other research approaches dealing with the performance of dynamic resource allocation policies, had led to the development of a computing framework (Warneke and Kao, 2009) that takes into account the countable and measureable parameters affecting the task allocation. Authors in Chun and Maniatis (2009) statically partition the service tasks and resources between the client and the server portions. The service is then reassembled on the mobile device at a later stage. This approach has many weaknesses, since it considers the resources of each cloud rack, depending on the expected workload and execution conditions (CPU speed, network performance). In addition, a computation offloading scheme is proposed in Chun and Maniatis (2009) that is exploited in cloud computing infrastructures to minimize the energy consumption of a mobile device. Work of Chun enables the execution of specified applications that do not have executions time limitations. However, such criteria do not include the communication characteristics (i.e. transmission diversities, servers processing overhead) during the communication process with the mobile users' claims. In addition, the approach does not take into account the available processing resources (Barbera et al., 2013), the utilization of the device memory, the remaining energy and the available capacity with the communication of each of the device with the closest — in terms of latency — cloud terminal. Finally, the research approach in Kumar and Lu (2010) proposes different analytical models, towards addressing offloading computation and elaborating on offload mechanism to offer energy conservation. Within this context, this section presents an offloading resource mechanism, which is incorporated into an energy-efficient model that encompasses offloading context applications within the social collaboration.

The strength of social interaction will also affect the offloading process, which will positively impact the energy conservation mechanism. This is performed, in order to define the set of nodes that according to social parameters, can host any measured offloading executable task.

The concept of "friendship" denotes that each "friend" peer is aware of the communication context (Lukowicz et al., 2012) and the related characteristics of their "friends". This is determined by the amount of direct or indirect social interaction among the different users, belonging to the network defined as the direct friendship evaluation from node to node with given probability of a node, being connected to k other nodes at time t (Mavromoustakis and Dimitriou, 2012). This results in a large number of nodes have a small node degree and therefore very few neighbors, but a very small number of nodes having a large node degree and therefore becoming hubs in the system. Taking into consideration the social interactions resource intensive applications as depicted in Satyanarayanan et al. (2009), we may investigate a novel architectural prototype supported by a middleware in the system.

To this end the middleware shows the different layers in the architecture and the elements derived from Dimitriou et al. (2013) and Mousicou et al. (2013), indicating the likelihood of an individual to move from given point to a certain direction and having associated measured 'capability' of each device to handle offloaded power intensive application (Satyanarayanan et al., 2009).

8.9 IMPLEMENTATIONS, USE-CASE SCENARIOS

Open AAL/ELE service implementation starts with isolated islands of AAL labs, special implementations in the hospitals, special institutions for elderly, grows towards clusters and domains ending to private and public clouds. The deployment, interoperability and aggregation of the solutions will follow the market demands and protocol conversion. The use-case scenarios may differ from implementation to implementation due to the specific needs and customization. The data model should follow the direction from sensors to the cloud. Backward data direction needs very careful analyses. For example, there is a need of interaction between medical doctors and patients. This does not apply visibility of all the available information to the doctors and to the end-users (Table 8.2).

Some of the sensors need polling in order to send data. Others need to be woken up. Some of the sensors report regularly on timer. Part of the sensors report few times like EnOcean technology without any acknowledgment. After the home/car/tram/school/park controller the data could be directly forwarded or partially processed. Recent annals on big data claim that sending raw data to the cloud is better than local acquisition from performance point of view (Cooklev et al., 2015). We have to say that at dew level there is a need of data processing at least for collection of personal data and data mapping between many dews and the fog nodes. The similar

Table 8.2 Types of data processed at different platform levels

Place	Type of data and possible processing

Sensor
	Sensors might be sources, proxies and sinks. Acknowledgment is optional. Also:
	  – Sources usually could keep data from a single measurement.
	  – Proxies could keep own data and data that needs forwarding. They support input and output queues for forwarding.
	  – Sinks are behaving like proxies. They support input and output queues for forwarding.

	Sensors in BAN might be sources and sinks. Acknowledgment is optional. There is also a gateway to the mobile dew/fog level where:
	  – Sources usually could keep data from a single measurement.
	  – Sinks could keep own data and data that needs forwarding. They support input and output queues for forwarding. Sinks and gateway could be a single device.
	  – Optional body server keeps current and historical data for the body.
	  – Optional body server supports input buffers for all sensors.
	  – Optional body server supports output buffers to all sensors.
	  – Optional body server supports end-user interfaces.
	  – Optional body server supports interface to/from the dew and fog including multipathing and multihoming.
	  – Optional body server is capable to understand and react to the messages from dew and fog level including data analyses.

	Home controllers, servers and gateways:
	  – Keep current and historical data for the house and all people around.
	  – Support input buffers for all sensors.
	  – Support output buffers to all sensors.
	  – Optionally perform data acquisition.
	  – Support end-user interfaces.
	  – Support interface to/from the fog and cloud including multipathing and multihoming.
	  – Are capable to understand and react to the messages from fog and cloud including data analyses.
	  – Support peer port to the neighbor dews and fogs.
	  – Support AAL/ELE services at dew level.
	  – Support data mobility.

continued on next page

Table 8.2 (*continued*)

Place	Type of data and possible processing
	Base station and base station controllers: - Support communication to/from dews via different paths, load balanced, and using peer port. - Support communication to/from fog servers via different paths, load balanced, and using peer port. - Support communication to/from clouds via different paths, load balanced, and using peer port. - Store raw data at fog level. - Support data mobility. - Perform data analyses at fog level. - Provide AAL/ELE services at fog and dew level.
	Shop, hospital, school, cinema controllers, servers and gateways: - Keep current and historical data for the building and all people around upon subscription. - Support input buffers for all sensors. - Support output buffers to all sensors. - Perform data acquisition that allows correct personal data to be sent in real-time and near-real-time to the home or fog servers. - Support end-user interfaces. - Support interface to/from the dew/fog/cloud including multipathing and multihoming. - Are capable to understand and react to the messages from dew/fog/cloud including data analyses. - Support peer port to the neighbor entities. - Support AAL/ELE services at dew/fog level. - Support data mobility.
	Tram, car, train, bus controllers, servers and gateways: - Keep current and historical data for the device and all people around upon subscription. - Support input buffers for all sensors. - Support output buffers to all sensors. - Perform data acquisition that allows correct personal data to be send in real-time and near real time to the home or fog servers. - Support end-user interfaces. - Support interface to/from the dew/fog/cloud including multipathing and multihoming. - Are capable to understand and react to the messages from dew/fog/cloud including data analyses.

continued on next page

Table 8.2 (*continued*)

Place	Type of data and possible processing

	- Support peer port to the neighbor entities.
	- Support AAL/ELE services at dew/fog level.
	- Support data mobility.
	
	Fog level servers and data centers support:
	  - interfaces to/from the dew/fog and cloud including multipathing, multihoming, peer port, and load balancing
	  - data storage
	  - data analyses
	  - data history
	  - server mobility
	  - data mobility
	  - end-user interfaces
	  - all AAL/ELE services

	Cloud level servers and data centers support:
	  - interfaces to/from the dew/fog and cloud including multipathing, multihoming, peer port, and load balancing
	  - data storage
	  - data analyses
	  - data history
	  - server mobility
	  - data mobility
	  - end-user interfaces
	  - all AAL/ELE services

situation exists at fog level where the data should be locally stored/processed, shared, managed and appropriate data should be sent back to the dew or to the cloud level. Data processing facilities at fog level could be essential in case we are speaking of local elders that do not travel. On the contrary, moving the data and AAL/ELE services to the cloud will allow easy globalization of the service helping the people to use the capability of the telemedicine, telemonitoring, teleworking to support end-users remotely. In this sense cloud-based AAL/ELE services makes them generic and open and will allow service spreading along The Globe.

Data analyses tools are another interesting and important topic in the cloud (Chilipirea et al., 2015; Pop et al., 2015). The big data analyses will save efforts and resources locally and could be made in a more professional way at global level (Bessis and Dobre, 2014; Ionescu et al., 2015). Furthermore the AAL/ELE services will become not only a matter of personal subscription for users, could be offered at municipality level, be egovernmentally based, will be continental, union, global (Schweitzer et al., 2016). Globalization of the service is addressing also security and privacy issues, political differences, religion requirements, local customs (Dinsoreanu and Potolea, 2014; Stulman et al., 2012, 2013). In this sense, the AAL/ELE services need to be adaptable and this reflects to the globally based data processing.

In a typical AAL lab, the main services are related to the work inside the lab, their remote management and work that needs interfaces outside the lab (http://www-cps.hb.dfki.de/research/baall, http://robotik.dfki-bremen. de/en/research/robot-systems/aila.html). The data flow from the lab to the rest of the world is quite limited. There is no or quite limited way to aggregate the work of AAL labs in a cluster. Therefore, cloud based approach to or simply virtualization of the AAL/ELE services is a coming thing with specific requirements and design limits.

8.10 CONCLUSION AND FUTURE WORK

This chapter presents the basic structure and hierarchy of the open AAL/ELE platform that is virtualized at different levels. There are already many existing platforms on the market and some of them pretend to be global. Most of the platforms are proprietary and do not show capability to be aggregated and integrated in an interoperable way to the possible global solution. In this chapter, we show such possibility without solving all open issues, especially the ethical and privacy concerns.

The open AAL/ELE generic platform has access parts based mostly on sensor networks without being limited to the sensors only. Access networks accumulate the data from AAL/ELE services in small cloud called dew at home/office/car level. Then, the data from every dew is carefully shared with other dews and with fog computing centers. Feedback is also expected from the fog. The fog computing centers process the data carefully and take decisions what, when, how to share with dews, peering fogs and higher-level clouds. The cloud level is the global level that allows better spreading of the services and high-level data processing.

Our future work plans continue with better standardization of the platform, better interface and protocol definitions, more precise data models, data processing tools identification. We are also looking for the solutions to solve the privacy and security issues. Furthermore the AAL/ELE services need to be enriched by adaptable end-user and stakeholder interfaces.

ACKNOWLEDGEMENTS

Our thanks to ICT COST Action IC1303: Algorithms, Architectures and Platforms for Enhanced Living Environments (AAPELE), ICT COST Action IC1406: High-Performance Modeling and Simulation for Big Data Applications (cHiPSet) and TD COST Action TD1405: European Network for the Joint Evaluation of Connected Health Technologies (ENJECT).

REFERENCES

Abolfazli, S., Sanaei, Z., Ahmed, E., Gani, A., Buyya, R., 2014. Cloud-based augmentation for mobile devices: motivation, taxonomies and open challenges. IEEE Commun. Surv. Tutor. 16 (1), 337–368.

Active aging, 2016. Innovation Union, A Europe 2020 initiation, The Silver Economy. http://ec.europa.eu/research/innovation-union/index_en.cfm?section=active-healthy-ageing&pg=silvereconomy.

Alemdar, H., Kasteren, T.V., Niessen, M.E., Merentitis, A., Ersoy, C., 2014. A unified model for human behavior modeling using a hierarchy with a variable number of states. In: IEEE International Conference on Pattern Recognition. ICPR 2014, Stockholm, August 2014.

Alemdar, H., Tunca, C., Ersoy, C., 2015. Daily life behaviour monitoring for health assessment using machine learning: bridging the gap between domains. Pers. Ubiquitous Comput. 19 (2), 303–315. http://link.springer.com/article/10.1007/s00779-014-0823-y.

Al-Fuqaha, A., Guizani, M., Mohammadi, M., Aledhari, M., Ayyash, M., 2015. Internet of things: a survey on enabling technologies, protocols, and applications. IEEE Commun. Surv. Tutor. 17 (4), 2347–2376.

AMIGO, 2009. Amigo ambient intelligence for the networked home environment. EU funded project, http://cordis.europa.eu/project/rcn/71920_en.html.

AsTeRICS, 2013. Assistive technology rapid integration & construction set. EU funded project, http://www.asterics.eu/index.php?id=88.

Barbera, M.V., Kosta, S., Mei, A., Stefa, J., 2013. To offload or not to offload? The bandwidth and energy costs of mobile cloud computing. In: Proc. of INFOCOM 2013. April 2013, Turin, Italy.

Batalla, J.M., Kantor, M., Mavromoustakis, C.X., Skourletopoulos, G., Mastorakis, G., 2015. A Novel methodology for efficient throughput evaluation in virtualized routers. In: Proc. of IEEE International Conference on Communications 2015. IEEE ICC 2015, London, UK, 08–12 June.

Batalla, J.M., Mastorakis, G., Mavromoustakis, C.X., 2016. International Journal on Network Management (IJNM), Special Issue on "Management of the Internet of Things and Big Data". Wiley, April 2016.

Bessis, N., Dobre, C. (Eds.), 2014. Big Data and Internet of Things: A Roadmap for Smart Environments. Springer. (First published March 11th 2014). ISBN 3319050281 (ISBN13: 9783319050287), 470 pages.

Carlier, M., Braeken, A., Lemmens, B., Smeets, R., Segers, L., Steenhaut, K., Touhafi, K., Aerts, A., Mentens, N., 2014. 6LowPan — towards zero-configuration for wireless building automation: system architecture. In: Proceedings of the 1st Int. Electron. Conf. Sens. Appl., 1–16 June. In: Sciforum Electronic Conference Series, vol. 1.

Chaudet, C., Haddad, Y., 2013. Wireless software defined networks: challenges and opportunities. In: Proc. of IEEE International Conference on Microwaves, Communications, Antennas and Electronics. IEEE COMCAS, Tel-Aviv, Israel, Oct.

Chávez-Santiago, R., Szydelko, M., Kliks, A., Foukalas, F., Haddad, Y., Nolan, K., Kelly, M., Masonta, M., Balasingham, I., 2015a. 5G: the convergence of wireless communications. Wirel. Pers. Commun. 83 (3), 1617–1642.

Chávez-Santiago, R., Haddad, Y., Lyandres, V., Balasingham, I., 2015b. VoIP transmission in Wi-Fi networks with partially-overlapped channels. In: Proc. of IEEE Wireless Communications and Networking Conference. WCNC, March 2015, New Orleans, USA.

Chilipirea, C., Petre, A.C., Dobre, C., Pop, F., Xhafa, F., 2015. Enabling vehicular data with distributed machine learning. In: Trans. Computational Collective Intelligence XIX, pp. 89–102.

Chilipirea, C., Laurentiu, G., Popescu, M., Radoveneanu, S., Cernov, V., Dobre, C., 2016. A comparison of private cloud systems. In: 30th International Conference on Advanced Information Networking and Applications Workshops. WAINA, pp. 139–143.

Chun, B.G., Maniatis, P., 2009. Augmented smartphone applications through clone cloud execution. In: HotOS.

Cippitelli, E., Gasparrini, S., Gambi, E., Spinsante, S., Wåhslén, S., Orhan, I., Lindh, T., 2015. Time synchronization and data fusion for RGB-depth cameras and inertial sensors in AAL application. In: IEEE International Conference on Communications (ICC) Workshop on ICT-Enabled Services and Technologies for eHealth and Ambient Assisted Living. London.

Connected Health Lab, 2016. Connected Health Lab. Castres, France, http://chl.univ-jfc.fr/?q=en. http://www.irit.fr/e-Health-Platform.

Cooklev, T., Darabi, J., Mcintosh, C., Mosaheb, M., 2015. A cloud-based approach to spectrum monitoring. IEEE Instrum. Meas. Mag. 18 (2), 33–37. http://dx.doi.org/10.1109/MIM.2015.7066682.

De Santis, A., Gambi, E., Montanini, L., Pelliccioni, G., Raffaeli, L., Rascioni, G., Spinsante, S., 2015. Smart homes for independent and active ageing: outcomes from the TRASPARENTE project. In: Italian Forum on Ambient Assisted Living. Lecco (IT), May 2015.

Díaz Rodríguez, N., 2015. Semantic and fuzzy modelling for human behaviour recognition in smart spaces: a case study on ambient assisted living. TUCS Dissertation Series 186. Åbo Akademi University (Finland) and University of Granada (Spain).

Díaz Rodríguez, N., Lilius, J., Pegalajar Cuéllar, M., Delgado Calvo-Flores, M., 2013a. Rapid prototyping of semantic applications in smart spaces with a visual rule language. In: Häkkila, J., Whitehouse Kamin, K., Krüger, A., Tobe, Y., Hilliges, O., Yatani, K., Dey, A.K., Gellersen, H.W., Huang, E.M., Ojala, T., Santini, S., Mattern, F. (Eds.), Proceedings of the 2013 ACM Conference on Pervasive and Ubiquitous Computing Adjunct Publication. UbiComp '13 Adjunct ACM.

Díaz Rodríguez, N., Pegalajar Cuéllar, M., Lilius, J., Delgado Calvo-Flores, M., 2014a. A fuzzy ontology for semantic modelling and recognition of human behaviour. In: Knowledge-Based Systems. Elsevier.

Díaz Rodríguez, N., Pegalajar Cuéllar, M., Lilius, J., Delgado Calvo-Flores, M., 2014b. A survey on ontologies for human behaviour recognition. In: ACM Computing Surveys. ACM.

Díaz Rodríguez, N., León Cadahía, O., Cuéllar, M., Lilius, J., Delgado-Calvo-Flores, M., 2014c. Handling Real-world context-awareness, uncertainty and vagueness in real-time human activity tracking and recognition with a fuzzy ontology-based hybrid method. In: Sensors. MDPI.

Díaz Rodríguez, N., Wikström, R., Lilius, J., Pegalajar Cuéllar, M., Delgado Calvo-Flores, M., 2013b. Understanding movement and interaction: an ontology for Kinect-based 3D depth sensors. In: Bravo, J., Ochoa, S., Urzaiz, G., Chen, L.L., Oliveira, J. (Eds.), Ubiquitous Computing and Ambient Intelligence. Context-Awareness and Context-Driven Interaction. In: Lecture Notes in Computer Science. Springer.

Dimitriou, C., Mavromoustakis, C.X., Mastorakis, G., Pallis, E., 2013. On the performance response of delay-bounded energy-aware bandwidth allocation scheme in wireless networks. In: IEEE ICC 2013 Int. Workshop on Immersive & Interactive Multimedia Comm. over Future Internet. Budapest, Hungary, 9–13 June.

Dinsoreanu, M., Potolea, R., 2014. Opinion-driven communities' detection. Int. J. Web Inf. Syst. 10 (4), 324–342.

Drobics, M., Hager, M., 2014. KIT-Aktiv — Fit & Aktiv im Alter. In: Deutschen AAL Kongress 2014. VDE Verlag, ISBN 978-3-8007-3574-7, p. 5.

Dupont, Corentin, Hermenier, Fabien, Schulze, Thomas, Basmadjian, Robert, Somov, Andrey, Giuliani, Giovanni, 2015. Plug4Green: a flexible energy-aware VM manager to fit data centre particularities. Ad Hoc Netw. 25, 505–519.

Durmaz Incel, O., Kose, M., Ersoy, C., 2013. A review and taxonomy of activity recognition on mobile phones. In: Special Issue on Personal Health Systems for Well-Being and Lifestyle Change. BioNanoScience 3 (2), 145–171. http://dx.doi.org/10.1007/s12668-013-0088-3.

Ertan, H., Alemdar, H., Incel, O.D., Ersoy, C., 2012. Designing a wireless sensing system for continuous behavior and health monitoring. In: Ubihealth'2012. New Castle, UK, June 2012.

Ferro, E., Girolami, M., Salvi, D., Mayer, C., Gorman, J., Grguric, A., Ram, R., Sadat, R., Giannoutakis, K.M., Stocklöw, C., 2015. The universAAL platform for AAL (Ambient Assisted Living). J. Intell. Syst. http://dx.doi.org/10.1515/jisys-2014-0127. ISSN print 0334-1860, ISSN web 2191-026X, web: http://www.degruyter.com/view/j/jisys.ahead-of-print/jisys-2014-0127/jisys-2014-0127.xml.

Garcia, N.M., Rodrigues, J., 2015. Ambient Assisted Living. Rehabilitation Science in Practice Series. CRC Press, ISBN 9781439869857.

Garcia, N.M., Garcia, N.C., Sousa, P., Oliveira, D., Alexandre, C., Felizardo, V., 2014. TICE.Healthy: a perspective on medical information integration. In: 2014 IEEE-EMBS International Conference on Biomedical and Health Informatics. 1–4 June, pp. 464–467.

Gasparrini, S., Cippitelli, E., Spinsante, S., Gambi, E., 2015. Depth cameras in AAL environments: technology and real-world applications. Chapter 2. In: Theng, Lau Bee (Ed.), Assistive Technologies for Physical and Cognitive Disabilities. IGI Global.

Gehrig, T., Al-Halah, Z., Ekenel, H.K., Stiefelhagen, R., 2015. Action unit intensity estimation using hierarchical partial least squares. In: 2015 11th IEEE International Conference and Workshops on Automatic Face and Gesture Recognition. FG, Ljubljana, pp. 1–6.

Georgieva, T., Markova, V., 2015. Some aspects of the development of high-speed Wi-Fi networks. Proc. Univ. Ruse 54 (11).

Georgieva, T., 2016. Experimental investigation of radio performance in wireless access network. Int. Sci. J. "SCIENCE. BUSINESS. SOCIETY" (ISSN 2367-8380) Year I (4), 14–17.

Goleva, R., Atamian, D., Mirtchev, S., Dimitrova, D., Grigorova, L., Rangelov, R., Ivanova, A., 2015a. Traffic analyses and measurements: technological dependability. In: Mastorakis, G., Mavromoustakis, C., Pallis, E. (Eds.), Resource Management of Mobile Cloud Computing Networks and Environments. Information Science Reference, Hershey, PA, pp. 122–173.

Goleva, R., Stainov, R., Savov, A., Draganov, P., 2015b. Reliable platform for Enhanced Living Environment. In: First COST Action IC1303 AAPELE Workshop Element 2014, in Conjunction with MONAMI 2014 Conference. Wurzburg, 24 Sept., 2014. Springer International Publishing, ISBN 978-3-319-16291-1, pp. 315–328.

Goleva, R., Stainov, R., Savov, A., Draganov, P., Nikolov, N., Dimitrova, D., Chorbev, I., 2015c. Automated ambient open platform for Enhanced Living Environment. In: Loshkovska, S., Koceski, S. (Eds.), ICT Innovations 2015, ELEMENT 2015 Workshop. Ohrid, FyROM, 1 Oct. 2015. In: Advances in Intelligent Systems and Computing. Springer International Publishing, Switzerland, pp. 255–264.

Goleva, R., Stainov, R., Wagenknecht-Dimitrova, D., Mirtchev, S., Atamian, D., Mavromoustakis, C.X., Mastorakis, G., Dobre, C., Savov, A., Draganov, P., 2016. Data and traffic models in 5G network. In: Mavromoustakis, C.X., Mastorakis, G., Batalla, J.M. (Eds.), Internet of Things (IoT) in 5G Mobile Technologies, 2016. Springer International Publishing, Cham, ISBN 978-3-319-30913-2, pp. 485–499.

Hansen, H.V., Goebel, V., Plagemann, T., 2016. DevCom: device communities for user-friendly and trustworthy communication, sharing, and collaboration. Comput. Commun. (ISSN 0140-3664) 85, 14–27. http://dx.doi.org/10.1016/j.comcom.2016.02.001.

http://robotik.dfki-bremen.de/en/research/robot-systems/aila.html. AILA Mobile Dual-Arm-Manipulation, University of Bremen, Germany.

http://www-cps.hb.dfki.de/research/baall. BAALL: The Bremen Ambient Assisted Living Lab, University of Bremen, Germany.

IEEE 802.15 WPAN™, Task Group 6 (TG6), Body Area Networks. http://www.ieee802.org/15/pub/TG6.html.

Ionescu, V., Potolea, R., Dinsoreanu, M., 2015. Data driven structural similarity. In: KDIR 2015, pp. 67–74.

Jara, A.J., Bocchi, Y., Genoud, D., Thomas, I., Lambrinos, L., 2015. Enabling federated emergencies and Public Safety Answering Points with wearable and mobile Internet of Things support: an approach based on EENA and OMA LWM2M emerging standards. In: 2015 IEEE International Conference on Communications. ICC, London, 2015, pp. 679–684.

Kahveci, A.Y., Alemdar, H., Ersoy, C., 2015. Sleep quality monitoring with ambient and mobile sensing. In: 2015 23rd Signal Processing and Communications Applications Conference. SIU, Malatya, 2015, pp. 507–510.

Kastner, P., Lischnig, M., Tritscher, J., Eckmann, H., Schreier, G., 2011. DiabMemory — Proof of Concept für mHealth bei Patienten mit Diabetes Mellitus. In: Schreier, G., Hayn, D., Ammenwerth, E. (Eds.), Poster: eHealth 2011. Wien, 26.05.2011–27.05.2011. OCG, Wien, ISBN 978-3-85403-279-3, pp. 275–280.

Kirchner, E., de Gea Fernandez, J., Kampmann, P., Schröer, M., Metzen, J.H., Kirchner, F., 2015. Intuitive interaction with robots — technical approaches and challenges. In: SyDe Summer School, pp. 224–248.

KIT, 2016. Keep in Touch. Telemedical Solution. https://kit.ait.ac.at/technology/plattform/.

Kreiner, K., Gossy, C., Drobics, M., 2013a. Towards a light-weight query engine for accessing health sensor data in a fall prevention system. Stud. Health Technol. Inform. 205, 1055–1059.

Kreiner, K., De Rosario, H., Gossy, Ch., Ejupi, A., Drobics, M., 2013b. Play up! A smart knowledge-based system using games for preventing falls in elderly people. In: Proceedings of the eHealth 2013. Vortrag: eHealth 2013 — Health Informatics Meets eHealth. Wien, 23.05.2013–24.05.2013. OCG, ISBN 978-3-85403-293-9, pp. 243–248.

Krukowski, A., Vogiatzaki, E., et al., 2013. Patient Health Record (PHR) system. In: Maharatna, K., Bonfiglio, S. (Eds.), Next-Generation Remote Healthcare: A Practical System Design Perspective. Springer Science and Business Media, New York, ISBN 978-1-4614-8842-2. Chapter 6.

Krukowski, A., Charalambides, M., Chouchoulis, M., 2014. Supporting medical research on chronic diseases using integrated health monitoring platform. In: 4th International Conference on Wireless Mobile Communication and Healthcare — "Transforming Healthcare Through Innovations in Mobile and Wireless Technologies", Special Session on "eHealth@Home — Infrastructure and Services for Remote Multi-parametric Monitoring, Analysis and Support". 3–5th November, Athens, Greece.

Kryftis, Y., Mavromoustakis, C.X., Mastorakis, G., Pallis, E., Batalla, J.M., Rodrigues, J., Dobre, C., Kormentzas, G., 2015. Resource usage prediction algorithms for optimal selection of multimedia content delivery methods. In: Proc. of IEEE International Conference on Communications 2015. IEEE ICC 2015, London, UK, 08–12 June.

Kumar, K. Lu, Y., 2010. Cloud computing for mobile users: can offloading computation save energy? IEEE Comput. Soc., April, 38–45.

Kurunathan, J.H., 2015. Study and overview on WBAN under IEEE 802.15.6. U. Porto J. Eng. (ISSN 2183-6493) 1, 11–21.

Lambrinos, L., 2015. On combining the Internet of Things with crowdsourcing in managing emergency situations. In: 2015 IEEE International Conference on Communications. ICC, London, pp. 598–603.

Lamine, E., Tawil, A.H., Bastide, R., Pingaud, H., 2014. Ontology-based workflow design for the coordination of homecare interventions. In: Camarinha-Matos, L., Afsarmanesh, H. (Eds.), Collaborative Systems for Smart Networked Environments, 15th IFIP WG 5.5 Working Conference on Virtual Enterprises, Proceedings. PRO-VE 2014, Amsterdam, The Netherlands, October 6–8. Springer, Berlin, Heidelberg, pp. 683–690.

Lange, S., Gebert, S., Spoerhase, J., Rygielski, P., Zinner, T., Kounev, S., Tran-Gia, P., 2015. Specialized heuristics for the controller placement problem in large scale SDN networks. In: 2015 27th International Teletraffic Congress. ITC 27, Ghent, pp. 210–218.

Lukowicz, P., Pentland, S., Ferscha, A., 2012. From context awareness to socially aware computing. IEEE Pervasive Comput. 11 (1), 32–41.

Mandel, C., Autexier, S., 2016. People tracking in ambient assisted living environments using low-cost thermal image cameras. In: Chang, C., Chiari, L., Cao, Y., Jin, H., Mokhtari, M., Aloulou, H. (Eds.), Inclusive Smart Cities and Digital Health: 14th International Conference on Smart Homes and Health Telematics, Proceedings. ICOST 2016, Wuhan, China, May 25–27. Springer International Publishing, pp. 14–26.

Mavromoustakis, C., Mastorakis, G., Batalla, J.M., 2016. Energy harvesting and sustainable M2M communication in 5G mobile technologies. In: Internet of Things (IoT) in 5G Mobile Technologies. In: Modeling and Optimization in Science and Technologies, vol. 8, p. 99. Chapter 10.

Mavromoustakis, C.X., Dimitriou, C.D., 2012. Using social interactions for opportunistic resource sharing using mobility-enabled contact-oriented replication. In: 2012 International Conference on Collaboration Technologies and Systems (CTS 2012), in Cooperation with IoT 2012. Denver, Colorado, USA, pp. 195–202.

Mavromoustakis, C.X., Mastorakis, G., Papadakis, S., Andreou, A., Bourdena, A., Stratakis, D., 2014. Energy consumption optimization through pre-scheduled opportunistic offloading in wireless devices. In: The 6th International Conference on Emerging Network Intelligence. EMERGING 2014, August 24–28, Rome, Italy.

Mirtchev, S., Goleva, R., Atamian, D., Mirtchev, M., Ganchev, I., Stainov, R., 2016. A generalized Erlang-C model for the Enhanced Living Environment as a Service (ELEaaS). Cybern. Inf. Technol. 16 (3), 104–121. http://dx.doi.org/10.1515/cait-2016-0037.

Monekosso, D., Flórez-Revuelta, F., Remagnino, P., 2015a. Guest editorial. In: Special Issue on Ambient-Assisted Living: Sensors, Methods, and Applications. IEEE Trans. Human-Mach. Syst. 45 (5), 545–549.

Monekosso, D., Flórez-Revuelta, F., Remagnino, P., 2015b. Ambient Assisted Living [Guest editors' introduction]. IEEE Intell. Syst. 30 (4), 2–6.

Monekosso, D., Flórez-Revuelta, F., Remagnino, P., 2015c. Guest editorial. In: Special Issue on Ambient-Assisted Living: Sensors, Methods, and Applications. IEEE Trans. Human-Mach. Syst. 45 (5), 545–549.

Mousicou, P., Mavromoustakis, C.X., Bourdena, A., Mastorakis, G., Pallis, E., 2013. Performance evaluation of dynamic cloud resource migration based on temporal and capacity-aware policy for efficient resource sharing. In: Proceedings of the 2nd ACM Workshop on High Performance Mobile Opportunistic Systems. HP-MOSys '13. ACM, New York, NY, USA, pp. 59–66.

MPOWER, 2009. SINTEF ICT, 034707: Middleware platform for eMPOWERing cognitive disabled and elderly (MPOWER). http://www.sintef.no/projectweb/mpower/.

OASIS, 2011. Quality of life for the elderly. EU funded project. http://www.oasis-project.eu/index.php/lang-en/component/content/133?task=view&cat=21.

Papanikolaou, K., Mavromoustakis, C.X., 2013. Resource and scheduling management in cloud computing application paradigm. In: Zaigham, Mahmood (Ed.), Cloud Computing: Methods and Practical Approaches. Springer, ISBN 978-1-4471-5106-7, pp. 107–132.

Papanikolaou, K., Mavromoustakis, C.X., Mastorakis, G., Bourdena, A., Dobre, C., 2014. Energy consumption optimization using social interaction in the mobile cloud. In: Proc. of International Workshop on Enhanced Living EnvironMENTs (ELEMENT 2014), 6th International Conference on Mobile Networks and Management (MON-AMI 2014). Wuerzburg, Germany, September.

Park, J., Choi, B., 2012. Automated memory leakage detection in Android based systems. Int. J. Control. Autom. Syst. 5 (2).

PERSONA, 2010. Perceptive spaces promoting independent aging. EU funded project. http://www.telemed.no/persona-perceptive-spaces-promoting-independent-aging. 541667-247950.html.

PERSONA, 2016. PERSONAlised Health Monitoring System. EU funded project. http://cordis.europa.eu/project/rcn/108629_en.html.

Pires, I.M., Garcia, N.M., Pombo, N., Flórez-Revuelta, F., 2016. From data acquisition to data fusion: a comprehensive review and a roadmap for the identification of activities of daily living using mobile devices. Sensors 16 (2), 184.

Pop, F., Dobre, C., Cristea, V., Bessis, N., Xhafa, F., Barolli, L., 2015. Deadline scheduling for aperiodic tasks in inter-Cloud environments: a new approach to resource management. J. Supercomput. 71 (5), 1754–1765.

Porambage, P., Braeken, A., Gurtov, A., Ylianttila, M., Spinsante, S., 2015. Secure end-to-end communication for constrained devices in IoT-enabled Ambient Assisted Living systems. In: 2015 IEEE 2nd World Forum on Internet of Things (WF-IoT), pp. 711–714.

Porumb, M., Barbantan, I., Lemnaru, C., Potolea, R., 2015. REMed: automatic relation extraction from medical documents. In: Proceedings of the 17th International Conference on Information Integration and Web-based Applications & Services. iiWAS '15. ACM, New York, NY, USA. Article 19.

Raffaeli, L., Gambi, E., Spinsante, S., 2014. Smart TV based ecosystem for personal e-Health services. In: Proc. 8th International Symposium on Medical Information and Communication Technology. ISMICT, Florence, 2–4 April.

Reina, D.G., Ruiz, P., Ciobanu, R., Toral, S.L., Dorronsoro, B., Dobre, C., 2016. A survey on the application of evolutionary algorithms for mobile multihop ad hoc network optimization problems. Int. J. Distrib. Sens. Netw. 2016, Article ID 2082496. http://dx.doi.org/10.1155/2016/2082496.

Riazul Islam, S.M., Daehan, K., Humaun, K., Mahmud, H., Kyung-Sup, K., 2015. The Internet of Things for health care: a comprehensive survey. IEEE Access 3, 678–708.

Rolla, V.G., Curado, M., 2015. Enabling wireless cooperation in delay tolerant networks. Inf. Sci. (ISSN 0020-0255) 290, 120–133. http://dx.doi.org/10.1016/j.ins.2014.08.035.

Rossi, L., Belli, A., De Santis, A., Diamantini, C., Frontoni, E., Gambi, E., Palma, L., Pernini, L., Pierleoni, P., Potena, D., Raffaeli, L., Spinsante, S., Zingaretti, P., Cacciagrano, D., Corradini, F., Culmone, R., De Angelis, F., Merelli, E., Re, B., 2014. Interoperability issues among smart home technological frameworks. In: Proc. of 10th IEEE/ASME International Conference on Mechatronic and Embedded Systems and Applications — Workshop on Advances in AAL: From Research to Industrial Development. Senigallia, Italy, 10–12 September.

Satyanarayanan, M., 2010. Mobile computing: the next decade. In: Proc. ACM MCS '10. San Francisco, USA.

Satyanarayanan, M., Bahl, R.C.P., Davies, N., 2009. The case for VM-based cloudlets in mobile computing. IEEE Pervasive Comput. 8 (4), 14–23.

Schweitzer, N., Stulman, A., Shabtai, A., Margalit, R.D., 2016. Mitigating denial of service attacks in OLSR protocol using fictitious nodes. IEEE Trans. Mob. Comput. 15 (1), 163–172. http://dx.doi.org/10.1109/TMC.2015.2409877.

Skourletopoulos, G., Mavromoustakis, C.X., Mastorakis, G., Pallis, E., Chatzimisios, P., Batalla, J.M., 2016. Towards the evaluation of a big data-as-a-service model: a decision theoretic approach. In: IEEE INFOCOM Session on Big Data Sciences, Technologies and Applications (BDSTA 2016) — 2016 IEEE Infocom BDSTA Workshop, IEEE International Conference on Computer Communications. 10–15 April 2016, San Francisco, CA, USA.

Skourletopoulos, G., Bahsoon, R., Mavromoustakis, C.X., Mastorakis, G., Pallis, E., 2014. Predicting and quantifying the technical debt in cloud software engineering. In: Proceedings of the 19th IEEE International Workshop on Computer-Aided Modeling Analysis and Design of Communication Links and Networks. IEEE CAMAD 2014, Athens, Greece, 1–3 December.

Slegers, J., Mitriani, I., Thomas, N., 2009. Evaluating the optimal server allocation policy for clusters with on/off sources. Perform. Eval. 66 (8), 453–467.

SOPRANO, 2010. Service oriented programmable smart environments for older Europeans. EU funded project. http://cordis.europa.eu/project/rcn/80527_en.html.

Spinsante, S., Gambi, E., 2012a. Home automation systems control by head tracking in AAL applications. In: 2012 IEEE First AESS European Conference on Satellite Telecommunications. ESTEL, Rome, 2012, pp. 1–6.

Spinsante, S., Gambi, E., 2012b. Remote health monitoring by OSGi technology and digital TV integration. IEEE Trans. Consum. Electron. 58 (4), 1434–1441. http://dx.doi.org/10.1109/TCE.2012.6415017.

Spinsante, S., Gambi, E., 2012c. Remote health monitoring for elderly through interactive television. Biomed. Eng. Online 11 (1), 1–18. http://dx.doi.org/10.1186/1475-925X-11-54.

Spinsante, S., Cippitelli, E., De Santis, A., Gambi, E., Gasparrini, S., Montanini, L., Raffaeli, L., 2015a. Multimodal interaction in an elderly-friendly smart home: a case study. In: Agüero, R., Zinner, T., Goleva, R., Timm-Giel, A., Tran-Gia, P. (Eds.), Mobile Networks and Management, 6th International Conference. MONAMI 2014, Würzburg, Germany, September 22–26, 2014. In: Lecture Notes of the Institute for Computer Sciences, Social Informatics and Telecommunications Engineering, pp. 373–387. Revised Selected Papers.

Spinsante, S., Gambi, E., Montanini, L., Raffaeli, L., 2015b. Data management in ambient assisted living platforms approaching IoT: a case study. In: 2015 IEEE Globecom Workshops. GC Wkshps, San Diego, CA, pp. 1–7.

Stainov, R., Goleva, R., Demirova, M., 2014. Reliable transmission over disruptive cloud using peer port. In: 10th Annual International Conference on Computer Science and Education in Computer Science 2014. CSECS 2014, 4–7 July, Albena, Bulgaria, pp. 153–166. ISSN 1313-8624.

Stainov, R., Goleva, R., Mirtchev, S., Atamian, D., Mirchev, M., Savov, A., Draganov, P., 2016. AALaaS intelligent backhauls for P2P communication in 5G mobile networks. In: BlackSeaCom 2016. June 6–9, Varna, Bulgaria.

Stulman, A., Lahav, J., Shmueli, A., 2012. MANET secure key exchange using spraying Diffie-Hellman algorithm. In: 2012 International Conference for Internet Technology and Secured Transactions. 10–12 Dec, pp. 249–252.

Stulman, A., Lahav, J., Shmueli, A., 2013. Spraying Diffie-Hellman for secure key exchange in MANETs. In: Security Protocols XXI. Springer, Berlin, Heidelberg, pp. 202–212.

Tazari, M.R., Furfari, F., Fides-Valero, A., Hanke, S., Hoeftberger, O., Kehagias, D., Mosmondor, M., Wichert, R., Wolf, P., 2012. The universAAL reference model for AAL. In: Augusto, J.C., Huch, M., Kameas, A., Maitland, J., McCullagh, P., Roberts, J., Andrew, A., Wichert, R. (Eds.), Handbook of Ambient Assisted Living: Technology for Healthcare, Rehabilitation and Well-being. In: Ambient Intelligence and Smart Environments, vol. 11. IOS Press, Amsterdam, pp. 612–625. ISBN: 978-1-60750-836-6 (Print), ISBN: 978-1-60750-837-3 (Online).

Tunca, C., Alemdar, H., Ertan, H., Incel, O.D., Ersoy, C., 2014. Multimodal wireless sensor network based ambient assisted living in real homes with multiple residents. Sensors 14, 9692–9719. http://www.mdpi.com/1424-8220/14/6/9692.

UniversAAL, 2014. Universal open architecture and platform for ambient assisted living. http://www.universaal.org/index.php/en/.

Ustev, Y.E., Durmaz, O., Ersoy, C., 2013. User, orientation, device independent activity recognition using smartphones. In: ACM UbiMI'13, Ubiquitous Mobile Instrumentation. Zurich, September 2013.

VAALID, 2011. Accessibility and usability validation framework for AAL interaction design process. EU funded project. http://cordis.europa.eu/project/rcn/86723_en.html.

Vogiatzaki, E., Krukowski, A., 2014. Serious games for stroke rehabilitation employing immersive user interfaces in 3D virtual environment. In: Special Issue. J. Health Inform. (ISSN 2175-4411) 6, 105–113.

Vogiatzaki, E., Krukowski, A., 2015. Modern Stroke Rehabilitation Through e-Health-based Entertainment. Springer International Publishing, ISBN 9783319212937. https://books.google.bg/books?id=-UqGCgAAQBAJ.

Von der Heidt, A., et al., 2014. HerzMobil Tirol network: rationale for and design of a collaborative heart failure disease management program in Austria. Wien. Klin. Wochenschr. 126 (21–22), 734–741.

Wåhslén, J., Orhan, I., Sturm, D., Lindh, T., 2012. Performance evaluation of time synchronization and clock drift compensation in wireless personal area network. In: Bodynets 2012. Oslo, Norway, September, 2012.

Wåhslén, J., Orhan, I., Lindh, T., 2011a. Local time synchronization in Bluetooth piconet for data fusion using mobile phones. In: 8th International Workshop on Wearable and Implantable Body Sensor Networks. BSN 2011, Dallas, US, May 2011.

Wåhslén, J., Orhan, I., Lindh, T., Eriksson, M., 2011b. A novel approach to multisensor data synchronization using mobile phones. Int. J. Auton. Adapt. Commun. Syst. (IJAACS) 6 (3).

Wamser, F., Ifländer, L., Zinner, T., Tran-Gia, P., 2014a. Implementing application-aware resource allocation on a home gateway for the example of YouTube. In: Mobile Networks and Management. Würzburg, Germany, September 2014. In: Lecture Notes of the Institute for Computer Sciences, Social Informatics and Telecommunications Engineering.

Wamser, F., Zinner, T., Ifländer, L., Tran-Gia, P., 2014b. Demonstrating the prospects of dynamic application-aware networking in a home environment. In: ACM SIGCOMM. Chicago, IL, USA, August.

Warneke, D., Kao, O., 2009. Efficient parallel data processing in the cloud. In: Proceedings of the 2nd Workshop Many-Task Computing on Grids and Supercomputers. Nov. 14–20. ACM, Portland, OR, USA, ISBN 978-1-60558-714-1, pp. 1–10.

CHAPTER 9

Developing Embedded Platforms for Ambient Assisted Living

Cristian-Győző Haba Ph.D. Professor, Liviu Breniuc Ph.D. Professor, Valeriu David Ph.D. Professor
"Gheorghe Asachi" Technical University of Iasi, Faculty of Electrical Engineering, Bd. D. Mangeron, 23, 700050, Iasi, Romania

9.1 INTRODUCTION

Rapid development in sensor technology, wireless communications and health information systems, based mainly on miniaturization and integration, resulted in the proliferation of new devices and applications in the medical domain. This formed the ground base for advanced wireless healthcare, mobile health and assisted living applications (Kulkarni and Ozturk, 2011; Kuroda et al., 2012; Mukhopadhyay and Postolache, 2013).

The problem of population aging is affecting most of the developed countries which recognize it to be one of the most important issues of the next years, with direct influence on health and social sectors (increased budgets and number of employees, new required services and skills), work productivity and job market. The European Commission in its Europe 2020 strategy identifies that population ageing is one of the societal challenges. Also, the World Health Organization estimates that the people with age over 60 will be about 1.2 billion by 2025 and about 2 billion in 2050 (Hossain and Ahmed, 2012). On the other hand the European Commission recognizes that finding solutions to this problem can result in developing new technologies and services that can increase competitiveness and create new highly skilled jobs (EU, 2014).

New trends in providing healthcare services for older people or patients are based on relocating the services from clinics and hospitals to people's homes, especially taking into account that 90% of the older persons prefer the comfort of their own home (Rashidi and Mihailidis, 2013). Measuring vital signals, initiating interaction between patient and doctor, monitoring and verifying patient medication can now be done outside the clinic, as a result of successful tentative for integration of these activities with the em-

Ambient Assisted Living and Enhanced Living Environments.
DOI: http://dx.doi.org/10.1016/B978-0-12-805195-5.00009-0

bedded systems implementing the home automation. Many of these home automation systems are starting to include apart from usual functions such as control and monitoring of household systems (e.g. light, door, fan and air conditioning), support for elderly and disabled people with special attention taken to those living alone.

The advent of reach featured embedded systems combined with new intelligent sensors opened innovative directions in remote diagnostic and treatment, survey and patient continuous monitoring in different surroundings and situations, with real time processing of information and rapid intervention. The resulting effects are reduction of energy consumption, time savings, increased comfort, security and an enhanced ambient assisted living.

In the chapter we study the perspectives opened by the ambient assisted living systems, considering especially some applications related to intelligent elderly and patient monitoring in different environment conditions.

We will present a system for developing ambient assisted living systems monitoring some biological signals in association with signals provided by home automation component in order to identify the possible interactions that can significantly add support for an enhanced living environment.

The proposed system may be used for the long term survey of biomedical signals in different environments resulting from body temperature, biological impedance (Breniuc et al., 2014) and human physical activity. Thus, there are considered different types of movement, free fall (Breniuc et al., in press), sleeping activity and cardiac activity (electrocardiography, photoplethysmography) with heart rate variability (HRV) monitoring in correlation with home automation signals that can have an influence on the first ones: room temperature, lighting, air conditioning, noise, level of illumination and other outside weather conditions (Breniuc and Haba, 2014) or electromagnetic interferences (David et al., 2013; Nica et al., 2011). Different protocols for communication between central unit and client modules as well as central unit and cloud services have been tested and implemented (Breniuc et al., 2010) and will be discussed in the chapter.

Finally, we will discuss how modularity and flexibility built in the system can be used to further develop the platform and to implement customized products with various levels of functionality and complexity.

9.2 ONGOING RESEARCH

The development of embedded platforms for proactive health as well as for the intelligent monitoring, supporting and improving lives of the persons (older adults, disabled, chronically ills, etc.), facilitating the rapid intervention in emergency medical conditions or in difficult situations occurred in daily activities, are under the focus of many research groups all over the world.

There are many endeavors to connect and harmonize different domains such as e-health, remote monitoring, telemedicine, mobile health, ambient intelligence (innovative human computer interactions), smart rooms or even smart cities and ambient assisted living in order to create the versatile, safe and more user friendly systems for increasing the quality of life, simultaneous with decreasing the healthcare costs and efficient exploitation of environments.

Moreover, the concept "Internet of Things" allows the connection of anything, anyplace, anytime, any network, and any service — opening the door to many different applications with a great technological, economic and social impact (Riazul Islam et al., 2015).

These intelligent monitoring and supporting platforms are distributed systems implemented using different architectures. They contain wireless sensors networks (including mobile and wearable sensors) with multi-sensor nodes for physiological, medical or environment data collection, advanced signal processing devices, different communication infrastructure and protocols, as well as monitoring equipment (personal digital assistants — PDAs, smartphones, tablets, notebooks and computers).

There are many such systems based on middleware architecture which integrate medical health devices and use different communication protocols (Pereira et al., 2014), systems composed of wearable sensors/devices (smart shoe with GPS, chest band with sensors/possibilities for electrocardiogram — ECG, electromyogram — EMG, electro dermal activity — EDA recordings, fall detection, etc.) that work in conjunction with personal mobile devices such as the smartphone or tablet (Silva et al., 2014).

Of great importance is the development of the smart multi-sensory systems capable of real time and continuous assistance of people with sensory disability, especially of visual impaired people (Ando, 2012). Thus, in the case of blind people it is proposed a system (Ando et al., 2014) for user localization (based on ultrasound transmitter-receiver transducers), and user-environment interaction, which permits a real time assistance with a high spatial resolution (an accuracy of about 4 cm for the localization system).

Because the embedded platforms permit the individual persons monitoring at hospital, at home or in any other type of environment — with the possibility to send information and alerts in difficult situations — their versatility, ability to be configured and customized become very important features. There are personal health systems (PHSs) with additional decision-support capabilities which can adapt to changes of the user (patient/person), features that are very useful in cardiology (Fayn and Rubel, 2010) or other diseases.

In addition, due to cognitive decline of the older person during the time, as well owing to the diversity and dynamism of environments, assistive living systems are designed with adaptability and extendible characteristics in order to follow the changes of people abilities and of the surrounding ambient. The dynamic integration of assistive services with sensing and interaction devices is accomplished with middleware and semantic mechanisms. Thus, the system modularity (e.g. based on service-oriented approach), semantic web technologies, the Distributed Open Service Gateway initiative (DOSGi) allow the selection of the suitable assistive services and devices for patients in their environment, giving more flexibility, transparency, scalability and simple reconfiguration (Aloulou et al., 2014). Such a system (tested for three assistive services, which are very common and thus most required by old persons and caregivers), may be developed taking into account many other activities. In addition, the developments of a "Virtual Caregiver (ViCG)", that provides real time assistance to the older persons, constitutes a real support for human caregivers, which intervene only for additional support (Hossain and Ahmed, 2012).

Considering the progress made in sensor technologies (mobile and wearable sensors based on human-tissue compatible materials); assistive robotic technologies; security domain (noninvasive user authentication methods based on biometric and physiological features) as well as in smart home, there are many opportunities to transfer the results with good outcomes also in ambient assisted living.

The levels of both acquired physiological signals and in general the responses of used sensors/devices are very low, what involves special precautions even against used mobile or wireless communications systems. Moreover, medical devices (Nica et al., 2011), user support devices or electrical appliances (David et al., 2009) may be important electromagnetic field sources. Because of the very small distances between electromagnetic fields sources and the user, special precaution must be taken in the case of body-worn devices, smart garments or E-textiles.

Generally, the studies and research concerning the biological and health effects of electromagnetic fields consider: the identification of some biological signal modification in case of person exposed to electromagnetic fields (Andritoi et al., 2013; Branzila and David, 2013); the determination of the fields and currents induced in human body by external fields or the specific absorption rate — SAR (Nica et al., 2012).

Some of the main problems that arise and must be solved in the future are:

* miniaturization of systems and making them ubiquitous;
* developing low-power systems or which have energy harvesting capability;
* converging multitude of actual standards and increasing system interoperability;
* providing security for systems and for the sensitive data transmitted between system components.

In this chapter we consider both the monitoring of biological signals and environmental parameters with the possibility of their integration and control. A particular focus is set on versatility, flexibility and adaptability to different situations and applications, which gives our proposed system the ability to be configured and tuned to user's special environment and health conditions.

9.3 AMBIENT ASSISTED LIVING CHALLENGES AND APPLICATIONS

Ambient assisted living systems use the information, communication and generally new technologies to monitor, support and socially connect persons (especially old adults) in their daily living and working environment (Rashidi and Mihailidis, 2013; Queiros et al., 2015; Monekosso et al., 2015a, 2015b; Geman et al., 2015).

As shown in Figure 9.1 the AAL systems *assist the user in their environment, survey their physical health* (remote monitoring) and *perform a user behavior analysis.*

The aim of AAL is to improve user health conditions and wellness, as well as to ensure user social interaction and safety/security. Beside continuous assistance of a large category of people, ambient assisted living also constitutes a real support for caregivers.

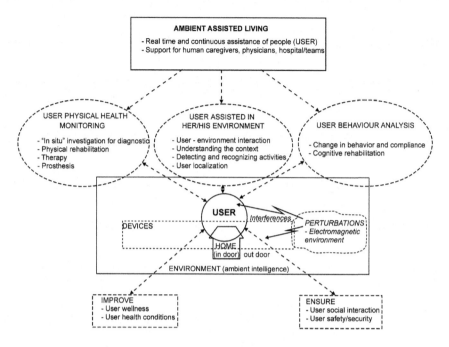

Figure 9.1 General view of ambient assisted living.

The devices used for gathering and processing information or for responding adequately to different situations are embedded on the user's body and environment (ambient intelligence). Thus, the miniaturizations, compatibility, connectivity as well as data accuracy and security are all very important.

One of the basic problems in the design and operation of the AAL system is the analysis, reduction and control of disturbing factors (perturbations). Due to multiple and diverse electromagnetic fields sources, their wide spatial and timely spreading, there is an "electromagnetic environment" (David et al., 2013), which acts both on electronic equipment (electromagnetic compatibility problems) and on user health (bioelectromagnetic compatibility problems).

There are many aspects of ambient assisted living:

- Activity recognition (sitting, walking, running) in smart homes (intelligent home automation) (Sun et al., 2010, Fernandez-Luque et al., 2014).
- Objects or persons localization and identification (e.g. identification of the person, based on biometric characteristics). The accuracy of local-

ization systems are in the range of meters (1 m ÷ 5 m) in the case of Wi-Fi technologies and in the range of centimeters (about 4 cm) in the case of ultrasound based systems (Ando et al., 2014).

- Event detection, classification and localization. One of the wide-spread such task is the fall detection using sensors or vision-based devices, with the possibility to identify, classify and discern falls from other activities of daily living (Ando et al., 2015).
- Location/situation awareness.
- Behavior analysis, e.g. behavioral patterns based on a statistical predictive algorithm that models circadian activity (Virone et al., 2008).
- Interaction with users (human-human, or human-object interaction) resulting in system ability to be aware of the context, model it to predict future situation evolution (Reerink-Boulanger, 2012).

Besides their basic goal for increasing and monitoring the quality of peoples' life (support mobility and autonomy of the frail users), the ambient assisting living may be useful:

- to estimate the general health status of the person (long term or continuous health status monitoring);
- for emergency detection and intervention (fall detection, medical emergency or hazard detection);
- for emotional wellbeing and to assist the caring of elderly people (social connectedness and communication);
- to detect disease conditions and/or progression;
- to evaluate a medical intervention (therapy, prosthesis, rehabilitation, diagnosis);
- to encourage the older persons in the achievement of their daily activities offering safety and independence.

In Figure 9.2 are outlined the involved elements in Ambient Assisted Living, namely different *users*, *environments*, used *technologies/devices* and some *applications* or *assistive solutions*.

By means of an appropriate data processing and analysis, it is possible *to support persons* in everyday activities, *to monitor patients* ("in situ" health investigations) and *to intervene rapidly* in emergency cases.

9.4 PROPOSAL APPROACH

AAL systems rely on a wide range of technologies, devices and services starting from the simplest sensors to complex systems for monitoring,

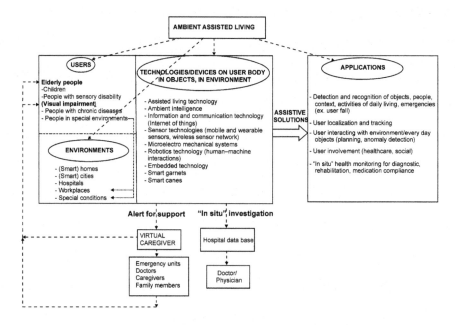

Figure 9.2 Involved elements and applications of AAL.

surveillance, alerting and remote nursing. The AAL system is a link between various users (including assisted persons) and the environment in
which the assisted person is living. This link includes multi directional paths
and involves multiple levels of interaction.

9.4.1 Sensors and Data Acquisition Systems

An AAL system is an automated system which implements a sense-process-
act cycle. The activities of this cycle are the following:

- Continuous or periodic reading of data set to determine the status
  of the assisted person, the environment in which the person and the
  system are located and the possible interactions between them. This
  is accomplished through a set of sensors or data acquisition systems
  (Postolache et al., 2012). Data acquisition is performed in different locations in the residence, some of which can be with well defined fixed
  positions while others can be mobile as they are fixed (mounted) on
  the assisted person or on objects that the person carries on (eye glasses,
  sticks, shoes, clothes etc.).

- The acquired data must be processed in order to be used by the AAL system components. The system may use primary data obtained directly from sensors or secondary data resulting from the fusion of data from several sensors or from local data processing.

AAL system acquired data will be interpreted and together with existing information will be used to elaborate decisions and take certain actions.

- The actions will be carried out using actuators which are in direct or indirect interaction with the person in care, the environment or other participants in the system. These actions can range from very simple, such as turning on or off a light bulb, to more complex ones such as calling a care giver, nurse or physician.

9.4.2 Platforms

As we can see, the implementation of an AAL system will require a platform covering multiple technologies, device and service types. Key features will refer to:

- data acquisition,
- data processing and implementation of algorithms for decisions,
- delivering actions.

While basic functionality is important, other features must be taken into consideration for the system as they are important from the user point of view. Some of these additional features are:

- easy to use interface for all users (assisted person, care givers, doctors);
- low power consumption resulting from two system requirements:
  o the system must not overload the bills for energy and communications services (telephony, Internet, security) of the users,
  o extended battery life for mobile elements or devices which are not powered from the main power supply;
- customization of system hardware and functionality so they can be adapted to people with:
  o different health conditions or abilities,
  o living in different homes/environments,
  o interacting with different users,
  o having different requirements for services.

To meet these requirements, a system that is flexible, adaptable and configurable must be considered.

9.4.3 Connectivity

An AAL system is by its purpose a distributed system whose operation cannot be conceived without the exchange of information between its different components. Information exchange is done using either wired or wireless communication channels with a trend oriented to the extension of wireless solutions. As the complexity of these systems continues to augment, the amount of data exchanged and communicated within and beyond the system is likely to increase. Therefore, communication is an important feature to be considered when designing and implementing an AAL system.

The rapid advance of microelectronics and communications technologies has led to an unprecedented increase in connectivity. This is supported by mobile and wireless communications such as WiFi, Bluetooth and ZigBee (Breniuc et al., 2010). This growth increases the possibility of connection and supports designers to create systems and devices more mobile and interconnected. This provides greater flexibility in system architecture development, installation in the operating area, in ensuring the accessibility to system components, in data acquisition and processing and in transmission of commands and data to and from the system. Software that can run on these systems further increases the flexibility and connectivity and eases the use of these systems. Increased connectivity takes into account not only the system itself but also the connections, wired or wireless, with other systems or applications running on dedicated computers or in the cloud.

Increasing connectivity brings a lot of advantages to applications, the most important being:
- Ability to update the software application running on different AAL subsystems directly in the field. In such systems we can consider:
  o software updating due to adding new sensors, actuators or introducing new derived quantities that must be computed,
  o changing the type, amount or precision of the collected data,
  o updating the algorithms used in system operation.
- Ability to update the hardware structure (if using programmable integrated circuits). Updating this structure will have the same purpose as that for the software update, only here it is envisaged that those features will be implemented in hardware and not in software.
- The possibility of taking data samples by a moving operator using a dedicated device or using a tablet or mobile phone that connects directly to the system via mobile communication or through cloud services to which the two parties, system and operator are connected.

These samplings are required at system installation, configuration and customization, when system operation is periodically checked and for recalibration of sensors providing sensitive data.

- Realization of a flexible plug-and-play system. In such a system the components can be inserted/plugged or unplugged automatically with the system identifying each situation and making necessary changes. This feature is important either for the measurement system used in the initial stage of system customization or for replacing malfunctioning and deprecated components. Such an intelligent system is expected to identify the connection/disconnection of devices and perform their intelligent control in order to have an overall image of the entire system.

- Receiving a continuous stream of data from measuring and monitoring nodes (to characterize environment and person activity) that can be temporarily stored in system memory. Data can then be transferred at regular intervals to dedicated processing units or to analysis and diagnostic applications running on dedicated computers or in the cloud.

- Transformation of mobile devices (tablets, smartphones) into components of the ALL system or, transforming them into user interfaces to the system by taking advantage of their increasing performance: processing power, memory capacity, screen and video camera resolution, multiple communication possibilities (GSM, GPRS, 3G, Wi-Fi, Bluetooth, near field connectivity — NFC), and availability of integrated sensors (acceleration, temperature, GPS). Directly or by adding accessories, these devices can be converted into meters for distance, angle, geographic location, humidity, noise or nuisance, into surfaces inspection elements, into barcode or RF tags readers or can be used to capture images or videos that enhance information about persons and their environment.

9.4.4 System Architecture

A general architecture for an ALL system is depicted in Figure 9.3. It is well understood that this picture can only give a schematic image of a real ALL. As we can imagine, there are a lot of factors that influence the final structure and functionality of such a complex system configured and customized for a certain site, assisted person and system users. Some of the most important factors are the following:

- the type, size and shape of the assisted person's household,
- assisted person number and needs,

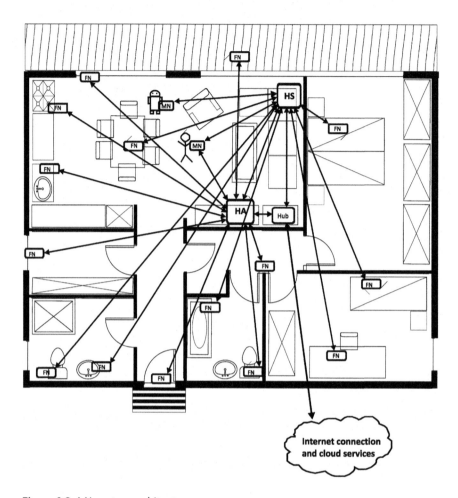

Figure 9.3 AAL system architecture.

- available wired and wireless connection to LANs and WANs,
- number and type of system users (care givers, nurses, doctors),
- assisted person's social and cultural background,
- neighborhood.

In order to cope with the implied level of complexity we are looking to create the overall system as a composition of several categories of subsystems or devices (Figure 9.3):

- Health System (HS) — will be dedicated for collecting, storing, processing and displaying data related to assisted person's physiological and health parameters. Data can be collected directly from the assisted person's wearable devices and from other mobile and fix nodes.

- Home Automation System (HA) — dedicated to implementing control and monitoring of household goods like lighting system, windows and doors, fans, air conditioning (AC) systems, relay type controls, measuring environment parameters and characterizing environment conditions, performing home surveillance and granting home security.

- Mobile Nodes (MN) — placed on monitored persons or objects that move along with the assisted persons (stick, clothes, wheel chair etc.) or even artificial personal assistants (robots).

- Fixed Nodes (FN) — are measuring and monitoring nodes which are placed in the home and testify human interaction with the environment (e.g. sensors in chairs, floor, bed and kitchen, bathroom or other room objects) or that monitor or characterize human activity (radars, position sensors, microphones, light barriers, CO_2 sensors).

In this set-up HS and HA are two important components of the AAL system. They collect data, do the processing and take actions based on the implemented algorithms. Depending on the implementation we can consider three types of operation modes:

- Health System (HS) and Home Automation System (HA) are working independently and are communicating directly using a wired or wireless connection;

- HS and HA are working independently but the communication and cooperation between the two is implemented using an application (cloud application) outside the system. The connection to Internet is implemented using a hub;

- HS and HA are embedded in the same system and implemented as different software components. Communication between the two is implemented at application level.

9.5 IMPLEMENTATION AND EVALUATION

Based on the architecture depicted in Figure 9.3 we have developed several prototypes of the system components. We will refer here to the Home automation system, Health System as well as to the Fixed and Mobile Nodes.

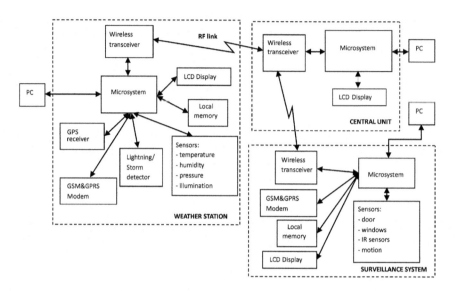

Figure 9.4 Home automation system block diagram.

9.5.1 Home Automation

Home automation (HA) system was designed to increase home comfort and security. The HA is made of three modules: Central Unit, Weather Station and Surveillance System. The block diagram of the system is given in Figure 9.4. The Weather Station module contains the sensors that will collect information on different weather parameters (temperature, humidity, pressure, illumination level). Based on temperature and humidity the system computes the thermal comfort index. In addition this module contains a Lighting/Storm detector which will warn, if it is the case, the distance and level of a storm with lighting activity. Based on this information, the user can take the appropriate measures for ensuring an adequate thermal comfort and the measures to avoid damages related to a heavy storm (inside rain, broken curtains and windows). The Surveillance System will provide information regarding the status of windows and doors (open or closed), movement detection in areas of interest etc. The information will be used to identify unwanted intrusions and illegitimate presence in the home.

The system can operate as a whole with Central Unit being the part which coordinates the whole operation. Communication between the modules is done through the RF link. The maximum distance between the modules depends on the type of RF link and also on home architecture and component materials. Actual RF technologies ensure the coverage of distances and wall materials within a normal house. Information from the two modules is collected, processed and displayed by the Central Unit. The data can also be transmitted to a PC in order to further analysis, processing and storing.

Weather Station and Surveillance System can also operate in a standalone mode. Weather Station module can be placed also outside the house. For applications that involves distribution of stations over a wider area (ex. multi building nursing homes), the position is determined with a GPS receiver. The use of the GPS is helpful if we want to create a map that characterizes a geographical area from the point of view of environment parameters like temperature, humidity, noise or chemical pollution. This implies connection to the Weather Station of appropriate sensors, movement of the station in the area of interest, collection and storage of data in local memory along with geographical coordinates, data analysis and building the map using an application such as Google Earth or similar.

Each of the two modules is build around a microcontroller based microsystem that offer the following features:
- coordinator of the module for standalone operation;
- data collection from sensors and processing;
- connectivity of the module with a PC for data transfer, debugging and software updating;
- local data display and temporary local data storage;
- RF connectivity and GSM/GPRS communication.

In case of a special situation the modules have the possibility to send an automated SMS or GPRS message to a preset phone number. For the Weather Station a special situation is considered when a weather parameter is over a preset threshold, comfort index outside normal range or if a storm is nearer than the preset distance. The special situations for the Surveillance System are considered the alteration of the status of a window (broken or opposite normal case — open instead of closed), detection of an intrusion etc. Generation of automated SMS or GPRS messages is possible using a GSM & GPRS modem module. The sent message will contain information about the special situation and if needed, the geographical coordinates of the Weather Station.

When choosing the components used to build the HA it is useful to keep in mind a few rules:
- The microsystems used should be based on microcontrollers from the same family. This will result in the use of the same programming envi-

ronment (with the consequence of better exploiting the programmer's expertise), using the same simulation, implementation, debugging and software updating methods, thus decreasing the overall development effort. In addition it allows the smooth transition to another member of the microcontroller family (with a direct consequence on microsystem hardware and software optimization in terms of efficient use of its resources: flash memory, the number of Input/Output — I/O lines while reducing microsystem size and energy consumption etc.);

- System should include several microsystem power supply solutions (USB — Universal Serial Bus, from batteries to solar cells etc.) and voltage stabilizers 5/3.3 V to accommodate different connecting modules;
- Micro connectors to output voltages must be provided in order to power peripheral modules, together with digital and analog I/Os where peripheral modules can easily be attached or removed. Serial communication lines for easy and rapid interconnection of peripherals or similar microsystems must also be available;
- Use peripheral modules which meet the specifications requirements and that have same electrical and mechanical compatible connectors with those existing on the controlling microsystems;
- Use peripheral modules that can be controlled through standard serial interfaces (UART, SPI, I2C etc.).

In order to develop a prototype for the HA we have used different modules manufactured by Digilent Inc. dedicated to build microcontroller based embedded systems. Products manufactured by this company generally correspond to the requirements listed before. From their large offer we found chipKIT microsystems very suitable for our purposes. These microsystems are based on the Microchip 32 bit PIC32 microcontroller which can be used to develop applications either using the MPLABX design environment or the MPIDE which is an Arduino type development environment tailored for the PIC32 microcontrollers. The chipKIT modules have two form factors, one compatible with Pmod peripheral modules and the other compatible with Arduino "Uno" and "Mega" form factor shields.

For the HA prototype we have used chipKIT modules compatible with Pmod modules that can be connected directly into the chipKIT connectors and which implement different peripheral functions. The microcontroller based modules available in this category are chipKIT MX3, chipKIT Pro MX4 and chipKIT Pro MX7. They differ in the type of microcontroller used, which imposes restrictions on available hardware resources, and the number of available connectors for the Pmod modules. For example, the main components and features of chipKIT MX3 are: one PIC32MX320F128H microcontroller (128K Flash, 16K RAM, 80 MHz operating frequency, 42 I/O pins, 12 analog inputs etc.), 5 connectors with 12 pins for Pmods modules, one I2C connector and one USB connector for programming and debugging. The 42 I/O pins support different communication protocols I2C, SPI and UART and other functions such as PWM (Pulse Width Modulation), external interrupt lines etc.

The most performing chipKIT microsystem, chipKIT Pro MX7, has the following components and features: one PIC32MX795F512L microcontroller (512KB Flash, 128KB RAM, 80 MHz operating frequency, 52 I/O pins, 16 analog inputs etc.), 6 connectors with 12 pins compatible with Pmod modules, 2 USB connectors, 2 I2C connectors, 2 CAN (Control Area Network) and 1 Ethernet connector. The 52 I/O pins allow communication using I2C, SPI, UART, CAN, USB and Ethernet interfaces, as well as implementation of PWM signals, PMP/PSP (Port Master Parallel/Port Slave Parallel) mode, external interrupt signals etc.

The selection of the microsystem to be used in developing the HA prototype will be determined by the complexity and amount of hardware resources that can cover the needs of the application (Flash and RAM memory, number of I/O pins, number of Pmod compatible connectors, types and number of communication interfaces). The advantage of using these modules relays in the flexibility of replacing the initial microsystem with one with more resources without changing the application software.

For implementing the RF communication we have used a PmodRF1 module. This module is based on a AT86RF212 transceiver with the following parameters: IEEE 802.15.4 certification, 700/800/900 MHz ISM Frequency Band, 250 kbit/s transfer rate, 6 km maximum distance communicating range (direct sight), controlled using the SPI communication interface. For communication we have implemented the ZigBee protocol where the modules are connected in a mesh network with Central Unit having a FFD — Full Function Device role while Weather Station and Surveillance System having a RFD — Reduced Function Devices role. The use of ZigBee protocol enables the connection between HA subsystems and also connection of HA to other compatible systems. The mesh network allows data exchange also between RFD modules, resulting in the reduction of application complexity (ex. the use of a single temperature sensor that provides temperature data both for the Weather Station and for the Surveillance System).

In order to reduce power consumption, communication between modules is restricted to take place with a certain frequency and for a specific amount of time. Suspension of communication can also occur due to degrading of communication conditions, temporary unavailability of central unit because of power failure, maintenance or upgrading actions. For these reasons, temporary local storage of data is an important feature of the Weather station and Surveillance system modules.

For local storage of data two solutions are widely used: one using an EEPROM memory with I2C/SPI compatible interface and the other which is using an SD card type memory.

The SD card solution is more used because of its high speed SPI interface, large storing capacity and easiness of data transferring to a PC. Data is stored on the SD card using the CVS (Comma Separated Values) format which can be easily imported and processed in Excel or other data processing software.

In the case of Weather Station, a record would contain data from sensors (temperature, humidity, pressure, illumination level), computed comfort index, geographical data received from the GPS receiver and time and date stamp. In our prototype, for local data storage was used a PmodSD module that provides the interface with an SD card.

For geographical localization we have used a PmodGPS module based on a MediaTek GPS MT3329 circuit. The module is controlled using a simple UART interface. The data can be output using different sentence types which are based on the National Marine Electronics Association (NMEA) protocols. Our prototype is using a RMC (Recommended Minimum Data for GPS) output sentence type which contains most of the GPS information in a concentrated form.

The GSM/GPRS modem was built using a SIMCOM SIM900 circuit which supports communication in 850/900/1800/1900 MHz bands and can be controlled using AT commands transmitted using an UART interface.

Detection of storms accompanied by lighting phenomena was implemented using a Lighting Storm Detector based on the AS3935 Lightning Sensor integrated circuit. The AS3935 can be programmed via a 4-wire standard SPI or an I2C interface. The sensor outputs information about the presence of a storm located at a distance of maximum 40 km indicating also the strength of the storm.

Figure 9.5 Prototype for the Weather Station.

Local data display was implemented using a PmodOLED OLED display with 128x32 pixel resolution and controlled using the SPI interface. In applications for elderly people the use of a bigger display would be a better accepted solution. We have therefore also tested a PmodOLED2 OLED display which is larger and has a 256x64 pixel resolution with 16 levels of grey.

Regarding the choice of sensors required for Weather Station and Surveillance System modules actual market provides a multitude of solutions. In Weather Station we have used Sensirion SHT21 for temperature and humidity measurements, Epcos T5400 sensors for pressure measurements and Maxim MAX44009 for measurements of illumination levels. All these sensors can be controlled using an I2C interface.

The microsystem can be powered from the USB interface but in mobile applications it was powered from a set of four 1.2 V/1000 mAh batteries. The microsystem includes the power stabilizing circuitry for providing 5 V and 3.3 V used to supply the microcontroller on the board and also the modules connected to the Pmod headers. A prototype of the Weather Station is given in Figure 9.5.

Application running on the Weather Station was written using MPIDE. The compatibility with the Arduino platforms makes initial prototyping very fast as a large set of libraries for controlling sensors, communication interfaces, displays and other devices are freely available.

9.5.2 Health System

This system is designed to be used in combination with other components of the AAL system, allowing the collection, processing, storage and transfer of medical data in order to increase health monitoring and a preliminary or "in situ" medical diagnosis.

A current problem encountered in the use of a health system is setting optimal parameters of the monitoring process in conjunction with assisted person and monitoring protocol. Depending on the assisted person and its house, the health system must be placed in a location that won't interfere with person's activity. The existence of power cords and communication cables for the transmission of information (commands and data) between the health system and other components of the AAL system makes difficult the acceptance from the part of the assisted person.

The use of wireless communication improves usability of the health system. It allows increased mobility of the wearable nodes and therefore, a better acceptance from the assisted person.

A storage system can be sometimes useful to extend capabilities of AAL system by collecting and accumulating information from all AAL system components for archiving purposes or for further data analysis, with the aim of optimizing the use of the entire AAL system and the process of elaborating a better AAL functionality.

Following the design steps presented in Haba (2012) we have created a prototype for the HS made of three node types, one for controlling the HS, one for mobile node with the interface with the user and one node for collection and storage of physiological or medical data and operation data.

A minimal network will consist of three nodes, one of each type that can communicate with each other based on an established communication protocol. This minimal network can be tested and in a later phase can be extended with other nodes. For example there can be two Mobile (Wearable) Nodes (one on assisted person and one on his stick) and one Storage Unit that can serve them for storage purposes.

In Figure 9.6 is depicted the Health system structure containing the three node types and the modules used for their implementation. The nodes have different functionalities and all can communicate via wireless communication modules.

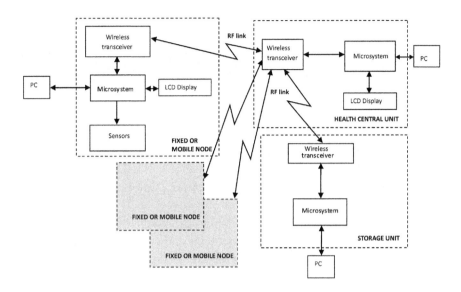

Figure 9.6 The Health system block diagram.

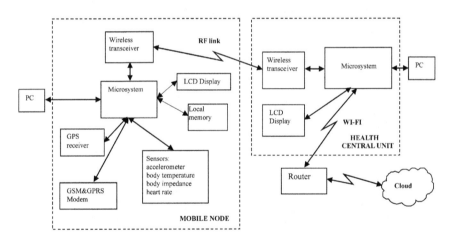

Figure 9.7 The SDMA block diagram including a Mobile Node.

9.5.3 Fixed and Mobile Nodes

In Figure 9.7 we present the architecture of a System for Developing Medical Applications (SDMA) that was used to develop and implement the fixed and mobile nodes of a health monitoring system (Breniuc et al., in press).

A SDMA includes at least two modules: Health Central Unit (HCU) and a Fixed Node (FN) or a Mobile Node (MN). The Health Central Unit and the Fixed Nodes are placed in the patient surveillance and monitoring area while the Mobile Nodes are usually worn or carried by the patient. The modules communicate using RF links which determines also the longest distance between the modules. Health Central Unit controls the operation of the FN and MN via the RF link and also uses this link for data transfer. In our prototype the RF communication was implemented using the ZigBee protocol. This protocol allows connection of HCU or FN and MN to other existing complex systems which support ZigBee protocol. An example is the ability to establish a connection between the FNs or MNs with the Home Automation System described in paragraph 9.5.1.

Data are processed by the Health Central Unit and can also be sent for storage and display to a PC. Data considered important is displayed on a local display. A greater versatility of the Health Central Unit is obtained using a microsystem that can connect using a wired or wireless connection to the Internet and thus access cloud services. Collected health data can be sent to be stored in patient record where it can be also accessed by authorized persons (doctors, caregivers, family members etc.).

The FN or MN has a flexible and reconfigurable structure so it can cover a great number of use cases. The number and type of sensors available on the FN or MN are established based on the number and type of physiological signals measured for the monitored patient. For example, in the case of free fall detection, three axis accelerometers are used. Temperature is measured with appropriate temperature sensors, respiration is measured using thoracic impedance while for heart rate are used ECG modules. Normally, the operation of the FNs and MNs is controlled by the HCU but in certain applications the FN or the MN can work also as standalone system where all control and processing is performed locally in the node. For the standalone operation the FN or MN should include the following components:

- local display for displaying important data collected from patient;
- local memory for temporary storage of data to be transferred to a PC or over the Internet for further processing;
- GPS receiver for patient localization;
- wireless communication link for data exchange but also for sending warning messages in critical situations. By critical situation we understand the case when the patient's health or life is in danger (ex. freefall, increased heart rate, abnormal body temperature etc.).

For maintenance purposes, debugging or software updating a serial connection to a PC is desirable.

Component selection for building the SDMA must follow the same rules used for selecting the modules for the HA system. In order to reduce costs, maintain compatibility and reuse code, for the SDMA prototype we have used modules manufactured by Digilent Inc. similar to those used in prototyping the HA. Some modules unavailable from Digilent or other manufacturers were designed and prototyped by authors with a special care to maintain compatibility with other components of the system (Pmod or Arduino connectors, voltage levels etc.).

As wireless transceiver we have used PmodRF1 module but we also made tests with the PmodRF2 module (based on Microchip MRF24J40 transceiver, IEEE 802.15.4 certified, ISM band 2.405–2.48 GHz operation, 250 kbit/s or 625 kbit/s transfer rate, about 300 m maximum distance communicating range, simple SPI communication interface). For local display, local storage, localization using GPS system and GSM&GPRS messaging we have used the same modules as for the HA implementation.

When choosing the microsystem to implement the FN or MN we must consider which are the patient's physiological measurements that must be collected and what are the already available modules that can do these measurements. In this regard, the modules to be used can be compatible with Pmods modules or can be chipKIT modules compatible with Arduino chipKIT shields.

For our prototypes we have used both type of modules and when not available, we have built measurement modules compatible with either of the types. For example, for the Heart Rate monitoring we have used chipKIT Max32 microsystem compatible with Arduino shield 'Mega' form factor. The chipKIT Max32 board is based on a PIC32MX795F512L microcontroller, the same as for the chipKIT Pro Mx7. The advantage of using this form factor was the easy connection of various ECG Arduino shields available on the market.

The Health Central Unit was built using the chipKIT WF32 microsystem with the following features: Microchip PIC32MX695F512L microcontroller (80 MHz operating frequency, 512K Flash, 128K SRAM, 43 I/O pins, 12 analog inputs etc.), micro SD card connector, USB 2.0 OTG controller with A and micro-AB, Microchip MRF24WG0MA Wi-Fi module etc. These features offer an integrated local storage of data on a Micro SD card and connection of the system to a Wi-Fi router so no additional modules are needed to implement these functionalities. This type of microsystem can be used also for building more complex FN or MN that need more memory and computing resources.

Similar to the HA system, the supply of Health Central Unit, FNs and MNs can be done via the microsystem USB interface or from batteries.

The application running on the Health Central Unit was written and tested using the MPIDE environment with all the advantages deriving from the use of an Arduino programming environment.

In Breniuc et al. (in press) is presented a Mobile Node prototype configured for detection of free fall in elderly people. FreeFall detection was implemented using a PmodACL module based on an Analog Devices ADXL345 3-axis accelerometer. The resolution of the accelerometer (specialized for Free Fall detection) is programmable in the range $\pm 2/\pm 4/\pm 8/\pm 16$ g, and can be controlled through either an SPI or an I2C interface. The Mobile Node is built using the chipKIT MX3 microsystem and PmodRF1 transceiver. In Figure 9.8 is presented the prototype for the Mobile Node that detects freefall.

In Breniuc et al. (2014) is presented a system for impedance measurements based on an Analog Devices AD5933 circuit (Figure 9.9). The system uses a chipKIT MX4 microsystem, a PmodOLED2 display, a PmodSD SD card interface and a PmodENC encoder for function selection. The system may be used for respiration measurements.

A similar system was prototyped for measurements of heart rate as well as for body temperature and humidity. The Mobile Node included an Olimex ECG/EMG module for heart rate measurement and a Sensirion SHT21 sensor for temperature and humidity measurement. The module is built around the chipKIT WF32 microsystem which can connect directly to a wireless router. A PmodOLED display is used to show data locally, RF communication is done using the PmodRF1 transceiver and function selection is performed using a PmodENC encoder. The prototype of the mobile node designed for heart rate measurement is presented in Figure 9.10.

9.6 DESIGN CONSIDERATIONS

Software development depends primarily on the chosen microcontroller system.

Lately, most of the microcontroller manufacturers are increasing support for developers by providing them access to a full range of design resources. These resources are divided into the following categories:

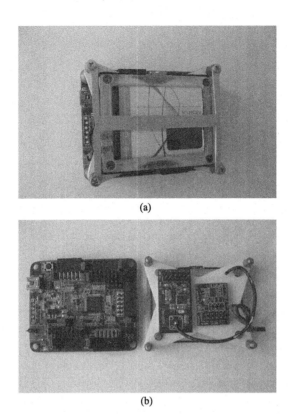

(a)

(b)

Figure 9.8 Mobile node prototype for freefall detection a) packed and b) unpacked.

Figure 9.9 Prototype of the impedance meter used for respiration measurements.

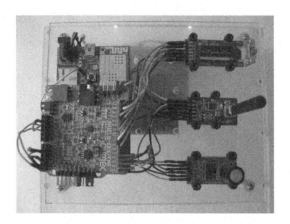

Figure 9.10 Mobile node prototype for heart rate measurement.

- Evaluation boards. Built around a key component and including other devices and circuitry that support the evaluation process;
- Reference designs. These are developed for different application areas including reference home automation and healthcare;
- Development boards enabling the creation of prototypes for AAL systems. These boards include already some of the needed hardware resources such as: power circuitry, sensors, LCD display, communications modules (Bluetooth, Wi-Fi, Ethernet), I/O elements such as micro buttons and micro switches, LEDs. On newest development boards we can find sensors, touch buttons, sliders and panels. Existing connectors on the board providing access to the microcontroller pins, enable insertion of standardized peripherals modules (Pmods, Arduino shields, etc.) or user made ones. Availability of designs, schematic diagrams and design tips eases the development of the initial prototype of the AAL system;
- Rich documentation including datasheets, white papers, reference designs, code examples, user guides, reference manuals;
- Code libraries to implement most common embedded system functionalities;
- Application platforms offering usable code implementing general functionalities but also for implementing specific functions such as sensor data fusion, standard or proprietary communication protocol stacks (USB, Bluetooth, Ethernet, Wi-Fi), motor control, display control etc.

AAL system development will always involve at least two steps. The first is the prototyping step, when the product concept is developed from initial specifications and a concept proof is obtained. The second step starts with the product concept and takes it the long way into the hands of the users.

In order to reduce the time consumed with the first step, the designer will look to use already existing components whether it involves development of the hardware or software part. In this stage the effort will be focused on shaping and developing the concept, namely the basic functionalities that will characterize future product. It is a recent practice, experienced mainly in small companies to use development boards and software components that are on the borderline between hobby and entry-level boards for system prototyping. Demonstration and development boards using these hardware and software modules may not be optimal for the future product but they allow the creation of a functional prototype in a very short time. In our case, for prototyping purposes we used peripheral modules and development boards from Digilent and MPIDE programming environment compatible with Arduino libraries. We have thus benefit from the large set of Arduino libraries and code examples available on the Internet to rapidly develop the functionality of our system components.

Following the prototype development, the second stage of the design will focus on making a product that is not only functional but also satisfies a number of optimality criteria such as low cost, small size, low power consumption, pleasant design, ubiquitous and usability. In addition, it will have to comply with specific guides, standards and regulations in the area of healthcare systems (EEC, 1993; ANSI, 2009; AAMI, 2004; IEC, 2014) and personal data security (Zaidi, 2003; COM, 2012).

For both FNs and MNs the most important optimizations will take into consideration system miniaturization, microcontroller resource management, reduction of energy consumption and limiting the radiation emissions of wireless communication modules.

In this stage for software development will be used professional programming environments (MPLAB or MPLABX instead of MPIDE) which include enhanced development tools such as compilers with advanced code optimization, debugging and memory tracing tools, code profiling and code versioning tools.

The aim of building the platform was to provide a support for multiple applications based on established or emergent hardware and software technologies. The architecture of the platform evolved from different other platforms that have been used for smaller and more dedicated applications which were aimed at measuring only one parameter such as temperature or humidity.

The platform was conceived to support the development of AAL applications but it can also be used for other areas that can take advantages of the platform reach set of features. The proposed platform can spread over multiple applications in the area of home automation and AAL.

The different platform features can be associated to specific needed features for an ALL system. Table 9.1 presents a covering matrix where an x sign in cell (pf, af) indicates that platform feature pf covers ALL feature af and thus can be used for the implementation of that functionality.

This matrix can be used for the design and configuration of a specific application. The matrix can be customized for a particular development board whose hardware and software resources are known. Establishing the features we need for the Healthcare or Home Automation system we want to build, we can see whether a particular platform allows implementation of the system.

An important feature of the proposed system is sharing collected and processed data between the two subsystems. Information resulting from data collected by HA system can be used by HS in order to take specific actions related to the assisted person's condition or environment. These actions can range from issuing simple visual or audio alerts to more complex actions like changing acquisition rate for specific vital signals, start recording data to SD card or sending alert messages to caregivers or doctors.

The type of data shared between the two subsystems and the action types to be taken in certain situations strongly depends on the assisted person and the caregiver system in which the person is integrated.

Based on the features of our proposed implemented system we have identified the possibility to implement the following interactions:

- Deviation from normal and rapid modifications of atmospheric pressure values can have negative effects on health and daily activity of persons with health problems. Numerous studies have shown that change in atmospheric pressure, both in the ascending and descending direction may, influence the health of the patient with negative effect on the respiratory, renal and cardiovascular systems. Changes can also produce reduction of reaction time and modification of attention that can lead to accidents, and can induce and increase anxiety to persons with mental illnesses.

- The influence of temperature, humidity and comfort index on assisted person's health. Though people can adapt to variation of temperature and humidity conditions, elderly people can be more severely affected due to their weakened heath conditions. Modification in temperature

Table 9.1 Platform features per ALL features covering matrix

Platform features	AAL features														
	Data acquisition	Data fusion	Signal processing	Data processing	Data storage	Wired communication	Wireless communication	Time stamp	User interface	Digital controls	Analog controls	Health monitoring	Home automation	Cloud services	Localization
Microcontrollers		×	×	×	×	×	×	×	×				×		
Digital I/Os	×					×			×	×			×		
Analog I/Os	×		×								×		×		
Timers	×		×					×	×				×		
Real Time Clock	×			×	×			×	×				×		
PWM									×			×	×		
UART	×								×						
I2C	×				×	×				×	×				
SPI	×				×	×				×	×				
Ethernet						×						×	×	×	×
Custom RF	×						×								
RF ZigBee							×		×						
RF Bluetooth							×		×			×	×		
RF WiFi							×		×			×	×	×	×
Mobile (GSM/GPRS)							×					×	×	×	×

continued on next page

Table 9.1 (*continued*)

Platform features	AAL features														
	Data acquisition	Data fusion	Signal processing	Data processing	Data storage	Wired communication	Wireless communication	Time stamp	User interface	Digital controls	Analog controls	Health monitoring	Home automation	Cloud services	Localization
Communication protocols	×					×	×							×	
Display									×						
Temperature sensor	×	×										×	×	×	
Humidity sensor	×	×										×	×	×	
Position sensor	×	×										×	×	×	
Acceleration sensor	×	×										×	×	×	
Illumination sensor	×											×	×		
Storm detector	×												×		
Microphone	×								×			×	×	×	
Gas(es) sensor	×											×		×	
Impedance meter	×					×	×					×		×	
Electrocardiograph	×					×	×					×		×	
Encryption				×	×							×	×	×	
Power saving modes	×								×			×	×	×	
Power supply (multiple)									×			×	×	×	

and humidity can result in problems with respiratory and cardiovascular systems as well as can result in worsening conditions for persons suffering from dermatitis.

- HA sensors for presence, lighting and noise can be used to identify unusual activity of assisted person which can either be characterized by an atypical behavior of the assisted person (multiple visits to the toilet during night) or an unusual level of activity related to the time of day (long sleeps during day or high activity during night) that can suggest the modification of some health conditions.

- Occurrence of unexpected events in the home or surroundings such as home violation, window brake, failure of one or several utility systems (water supply cut off, power outage, phone or Internet failure) can induce a state of anxiety, insecurity and panic with direct influence on assisted person's status or health condition. These conditions can be captured by the HA system, shared with HS and take preventive actions for the benefit of assisted person.

In Table 9.2 are presented the situations when HA and HS interaction should occur, which are the elements used to detect these situations (sensors, detectors, meters) and the actions to be taken by each of the subsystems.

Implementation is based on data exchange between the two systems, storage and patient specific tuning of threshold values for signals involved in the detection of special situations, local processing and fusion of two data types, running appropriate detection algorithms resulting in the issue of associated actions. The availability of wireless communication modules and powerful 32-bit microcontrollers in both HA and HS allowed us to implement some of these types of interactions in our prototypes presented in the previous section (Breniuc and Haba, 2014; Breniuc et al., in press).

In developing an AAL system is of a great importance to characterize the environment where the system will operate and also the influence of the system to this environment and user. Recordings of the biological signals (ECG, EEG, EMG) in environments with high magnetic and/or electric fields are important. We developed a methodology to record and process ECG signals from persons during the magneto therapy procedures and also modification of heart rate variability (HRV) due to a high level electromagnetic environment which is described in Andritoi et al. (2013).

For determining the heart rate variability (HRV), we used two methods, one based on Wigner function and another one based on Wavelet functions

Table 9.2 Home automation and Health systems interaction

	Detection	HS action	HA action
Weather conditions	pressure sensor, storm detector, anemometer	alert caregiver, increase measurement rate, local data recording	display alert on local display, increase visibility of locally displayed data, window closing
Comfort index	temperature sensor, humidity sensor	compare body with room temperature	regulate temperature, humidity of HVAC
Unusual activity	proximity sensor, switches, optical barriers	alert caregiver, find pattern	adjust lighting
Unusual events	motion detector, environmental sensor, door/window alarm, noise sensor, utility meters	alert caregiver, increase measurement rate, local data recording	alert security company, alert utility company

for detection of R peak from perturbed ECG signal and the results are presented in Branzila and David (2013).

The characterization of the electromagnetic interference and protection against it is a very important step in the design of AAL imposing the application of developed methodologies for subsystem validation.

9.7 CONCLUSION

The development of an ALL system implementing intelligent monitoring, preventive medicine, support and improvement of life conditions of disabled, ill or older adults is a complex task that must rely on a plethora of technologies, devices and services.

Flexibility, modularity and interconnectivity are the vital features of a system trying to cover so many characteristics implied by the operation of an ALL system, at the same time balancing as much as possible the different aspects which are influenced by technology, price and human psychology.

The presented platform includes a large number of devices, modules, technologies and features that allow designers to select various implementation solutions to be customized for the needs of a specific application. Based on the particularity of the specification, designers can decide which modules should be included in the implementation and which ones should discarded.

The proposed platform is modular, flexible and easily adaptable to various applications, with limited resources (low cost and simplicity in operation). Functionality of the platform has been demonstrated by the implementation of several use cases such as of human free fall detection, body temperature and humidity measurement and heart rate monitoring.

The authors are currently working on defining different metrics that will support a better search in the solution space and the development of a methodology for the selection of optimum set of fixed and mobile nodes for each AAL feature included in the covering matrix.

REFERENCES

AAMI, 2004. AAMI TIR32:2004. Medical device software risk management.

ANSI, 2009. ANSI/AAMI/IEC TIR80002-1:2009. Medical device software — Part 1: Guidance on the application of ISO 14971 to medical device software. (Identical adoption of IEC/TR 80002-1:2009).

Aloulou, H., Mokhtari, M., Tiberghien, T., Biswas, J., Yap, P., 2014. An adaptable and flexible framework for assistive living of cognitively impaired people. IEEE J. Biomed. Health Inform. 18 (1), 353–360.

Ando, B., 2012. Measurement technologies to sense "user in the environment" for ambient assisted living. IEEE Instrum. Meas. Mag., December 2012, 45–49.

Ando, B., Baglio, S., Lombardo, C.O., 2014. RESIMA: an assistive paradigm to support weak people in indoor environments. IEEE Trans. Instrum. Meas. 63 (11), 2512–2528.

Ando, B., Baglio, S., Lombardo, C.O., Marletta, V., 2015. An event polarized paradigm for ADL detection in AAL context. IEEE Trans. Instrum. Meas. 64 (7), 1814–1825.

Andritoi, D., David, V., Ciorap, R., Branzila, M., 2013. Recording and processing electrocardiography signals during magnetotherapy procedures. Environ. Eng. Manag. J. 12 (6), 1231–1238.

Branzila, M., David, V., 2013. Real time electrocardiogram signal processing for R peak detection using Wigner and Wavelet functions. Environ. Eng. Manag. J. 12 (6), 1207–1214.

Breniuc, L., Haba, C.G., 2014. Embedded system for increasing home comfort and security. In: 2014 International Conference and Exposition on Electrical and Power Engineering (EPE). 16–18 October 2014, Iasi, Romania, pp. 881–886.

Breniuc, L., Haba, C.G., David, V., 2010. Wireless network for the development of measurement applications based on ZigBee protocol. Bul. Inst. Politeh. Iași LVI (LX), 41–54, f.1.

Breniuc, L., David, V., Haba, C.G., 2014. Wearable impedance analyzer based on AD5933. In: 2014 International Conference and Exposition on Electrical and Power Engineering (EPE). 16–18 October 2014, Iasi, Romania, pp. 585–590.

Breniuc, L., David, V., Haba, C.G., in press. Embedded system for developing medical applications. Environ. Eng. Manag. J. 16 (1).

COM, 2012. 11 final, 2012/0011 (COD), Proposal for a Regulation of the European Parliament and of the Council on the protection of individuals with regard to the processing of personal data and on the free movement of such data (General Data Protection Regulation), 2012.

David, V., Nica, I., Salceanu, A., Breniuc, L., 2009. Monitoring of environmental low frequency magnetic fields. Environ. Eng. Manag. J. 8 (5), 1253–1261.

David, V., Sălceanu, A., Ciorap, R.G., 2013. Acquisition and analysis of biomedical signals in case of peoples exposed to electromagnetic fields in pervasive and mobile sensing and computing for healthcare. In: Mukhopadhyay, Subhas Chandra, Postolache, Octavian A. (Eds.), Technological and Social Issues. Springer, Berlin, Heidelberg.

EEC, 1993. Council Directive 93/42/EEC of 14 June 1993 concerning medical devices, 1993.

EU, 2014. EU Decision No 554/2014/EU of The European Parliament and of the Council, of 15 May 2014, on the participation of the Union in the Active and Assisted Living Research and Development Programme jointly undertaken by several Member States, 2014.

Fayn, J., Rubel, P., 2010. Toward a personal health society in cardiology. IEEE Trans. Inf. Technol. Biomed. 14 (4), 401–409.

Fernandez-Luque, F.J., Martinez, F.L., Domenech, G., Zapata, J., Ruiz, R., 2014. Ambient assisted living system with capacitive occupancy sensor. Expert Syst. 31 (4), 378–388.

Geman, O., Sanei, S., Costin, H., Eftaxias, K., Vyšata, O., Procházka, A., Lhotská, L., 2015. Challenges and trends in ambient assisted living and intelligent tools for disabled and elderly people. In: Proc. of IWCIM — Computational Intelligence for Multimedia Understanding Conference. October 29–30, 2015, Prague.

Haba, C.G., 2012. Initial development of wireless applications using FPGAs. In: 2012 International Conference on Electrical and Power Engineering (EPE 2012). October 25–27, 2012, pp. 741–746.

Hossain, M.A., Ahmed, D.T., 2012. Virtual caregiver: an ambient-aware elderly monitoring system. IEEE Trans. Inf. Technol. Biomed. 16 (6), 1024–1031.

IEC (International Electrotechnical Commission), 2014. Application of usability engineering to medical devices, International IEC Standard 62366 edition 1.1 2014-01. International Electrotechnical Commission.

Kulkarni, P., Ozturk, Y., 2011. mPHASIS: mobile patient healthcare and sensor information system. J. Netw. Comput. Appl. 34 (1), 402–417.

Kuroda, T., Sasaki, H., Suenaga, T., Masuda, Y., Yasumuro, Y., Hori, K., Ohboshi, N., Takemura, T., Chihara, K., Yoshihara, H., 2012. Embeded ubiquitous services on hospital information system. IEEE Trans. Inf. Technol. Biomed. 16 (6), 1216–1223.

Monekosso, D., Florez-Revuelta, F., Remagnino, P., 2015a. Guest editors' introduction, ambient assisted living. IEEE Intell. Syst. 30 (4), 2–6.

Monekosso, D., Florez-Revuelta, F., Remagnino, P., 2015b. Special issue on ambient-assisted living: sensors, methods, and applications. IEEE Trans. Human-Mach. Syst. 45 (5).

Mukhopadhyay, S.C., Postolache, O.A., 2013. Pervasive and Mobile Sensing and Computing for Healthcare. Technological and Social Issues. Springer, Berlin, Heidelberg.

Nica, I., David, V., Dafinescu, V., Salceanu, A., Haba, C.G., 2011. Characterization of electromagnetic radiation from a patient monitor. Environ. Eng. Manag. J. 10 (4), 561–566.

Nica, I., David, V., Lăzărescu, R., Ciorap, M., 2012. Estimations for human body exposure in some electromagnetic environments. In: 2012 International Conference and Exposition on Electrical and Power Engineering (EPE 2012). 16–18 October 2012, Iasi, Romania, pp. 541–543.

Pereira, R., Barros, C., Pereira, S., Mendes, P.M., Silva, C.A., 2014. A Middleware for Intelligent Environments in Ambient Assisted Living. 978-1-4244-7929-0/14/$26.00 ©2014. IEEE, pp. 5924–5927.

Postolache, O., Costa Freire, J., Girão, P.M., Dias Pereira, J.M., 2012. Smart sensor architecture for vital signs and motor activity monitoring of wheelchair' users. In: International Conf. on Sensing Technology — ICST, vol. 1. Kolkata, India, pp. 1–6.

Queiros, A., Silva, A., Alvarelhao, J., Rocha, N.P., Teixeira, A., 2015. Usability, accessibility and ambient-assisted living: a systematic literature review. Univ. Access. Inf. Soc. 14, 57–66.

Rashidi, P., Mihailidis, A., 2013. A survey on ambient-assisted living tools for older adults. IEEE J. Biomed. Health Inform. 17 (3), 579–590.

Reerink-Boulanger, J., 2012. Services technologiques intégrées dans l'habitat des personnes âgées: examen des déterminants individuels, sociaux et organisationnels de leur acceptabilité. Sociology. Universite Rennes 2.

Riazul Islam, S.M., Kwak, D., Kabir, H., Hossain, M., Kwak, K.-S., 2015. The Internet of Things for health care: a comprehesive survey. IEEE Access 3, 678–707.

Silva, B.M.C., Rodrigues, J.J.P.C., Simoes, T.M.C., Sendra, S., Lloret, J., 2014. An Ambient Assisted Living Framework for Mobile Environments. 978-1-4799-2131-7/14/$31.00 ©2014. IEEE, pp. 448–451.

Sun, H., De Florio, V., Gui, N., Blondia, C., 2010. Building mutual assistance living community for elderly people. In: Soar, Jeffrey (Ed.), Intelligent Technologies for the Aged: The Grey Digital Divide, pp. 207–219.

Virone, G., Alwan, M., Dalal, S., Kell, S.W., Turner, B., Stankovic, J.A., Felder, R., 2008. Behavioral patterns of older adults in assisted living. IEEE Trans. Inf. Technol. Biomed. 12 (3), 387–398.

Zaidi, K., 2003. Harmonizing U.S.-EU online privacy laws: toward a U.S. comprehensive regime for the protection of personal data. 12 Mich. St. U. J. Int'l L. 169.

CHAPTER 10

Wearable Electronics for Elderly Health Monitoring and Active Living

Raluca Maria Aileni*, Alberto Carlos Valderrama†, Rodica Strungaru*
*Politehnica University of Bucharest, Faculty of Electronics, Telecommunication and Information Technology, Romania
†Polytechnic Faculty of Mons, Electronics and Microelectronics Department, Belgium

10.1 INTRODUCTION

The care services ecosystem for elderly involve healthcare infrastructure, assistance services and ambient assisted living, which means that medical, economic and social factors have huge importance in establishing a healthcare policy (Camarinha-Matos et al., 2015). The elderly assisted living by monitoring and analyses have the objective to improve the life quality and to reduce the hospitalization costs (Fontecha et al., 2015). For elderly, the continuous health monitoring is very important because a high cause of mortality is associated with comorbidity phenomena (association of diseases like cardiovascular problems (hypertension, hypotension), non-physical activities (obesity) and Alzheimer's disease). In addition to representing one of the worst health outcomes, comorbidity is associated to a complex clinical management and increased health care costs.

Wireless sensors, for Personal Area Network (PAN), were used for monitoring in many healthcare applications. The collected data can be stored locally, in aggregators (like tablets or smartphones), or remotely by using desktop software applications and Cloud Computing services. The wireless sensors present limitations in terms of memory, low power consumption, energy autonomy, communication and efficient data management. For remote monitoring it is needed a high-performance computing and massive storage infrastructures to deal with real-time sensors data. Patients can interact with hospitals or doctors online, but the collected data must be stored permanently. In this context, computing power is mandatory to allow the use of complex algorithms (on/off line) able to extract precious information from medical databases. Cloud Computing services can support healthcare

Ambient Assisted Living and Enhanced Living Environments.
DOI: http://dx.doi.org/10.1016/B978-0-12-805195-5.00010-7

application platform and software services. In addition, the Cloud Computing technology offers computing, storage and software services (SaaS) with customization possibilities and virtualization at low cost.

Elderly or patient surveillance in their living environment using a wearable sensors network represents a high interest for scientists involved in risk, accident and failure detection (Augustyniak et al., 2014; Dasios et al., 2015). In case of the elderly, the declining phenomenon of the physiological sensorial capabilities such as hearing, vision and the coordination in space occur because of the decreasing the biological functions (Terrose et al., 2013; Siciliano et al., 2009). In terms of accident detection, researchers generally combine two types of motion sensors: optical, for visual gesture recognition based on image analysis and accelerometers for posture and tracking. Data from wearable embedded sensor networks are synchronized using composed patterns for detection of danger and risk situations (Augustyniak et al., 2014). For instance, changes in the body posture of elderly diabetics and people with cardiovascular diseases may indicate critical situations like tremors, failure or heart attack.

The patient environment also provides information useful for evaluating, preventing or detecting health problems. For instance, the prototype UbiCare, developed for home care of elderly people, involves the ambient environment parameters tracking and the incidents associated with moving, sitting and sleeping by using web interfaces. Thus, caregivers can be alerted when unknown activity patterns are detected (Dasios et al., 2015).

The motivation of AAL technologies for safety, health and wellness conditions comes firstly from elderly behavior: they prefer to remain alone in their own homes, even if this involves high costs with caregivers (Terrose et al., 2013). Wearable computing comes helping direct sensing. Today, biomedical parameters, such as cardiac rhythm or breath, can be measured using sensors embedded into patient garments (Kang et al., 2006). For the elderly patients, the garments with embedded monitoring system will allow sending alerts to emergency stations, doctors and relatives, and will lead to hospitalization and caregivers' costs reduction. However, in a non-controlled environment, multiple unknown phenomena are the source of unexpected behaviors. Thus, the amount and consistency of the sensed data is fundamental to the system robustness. To deal with large amounts of medical and ambient data from multiple sensors and embedded devices, the CoCaMAAL model brings Cloud Computing to AAL systems (Forkan et al., 2014).

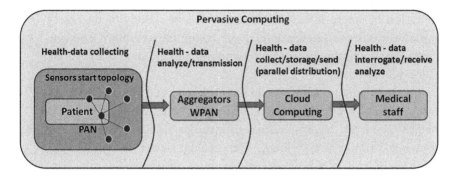

Figure 10.1 Pervasive computing (Kang et al., 2006).

For elderly patients, garments with embedded monitoring system (Kang et al., 2006) will lead to minimal cost for hospitalization and caregivers and will allow transmission of signal alerts to emergency stations, doctors and relatives.

10.1.1 Scientific Perspective

The continuous increasing of the healthcare costs and new advances in material science, miniaturization of the electronics devices and communications, create the opportunity for personalized health monitoring and management by ubiquitous/pervasive monitoring of a patient's or worker health condition.

The goal is to prevent the negative effect of the diseases by predictive modeling and analysis, based on data obtained using pervasive computing (Figure 10.1).

The wearable devices integrated into wireless personal or body networks (WPNs or WBNs) can sense, process and communicate vital signs through Internet for healthcare monitoring. These WPNs and WBNs (fitted for medical applications) are non-invasive and offer continuous ambulatory health monitoring.

Body Sensor Network (BSN) for medical applications consists on data fusion and Cloud Computing services (PaaS - Platform as a Service, SaaS - Software as a Service) for data storage and sharing solutions.

The data fusion includes preprocessing (filter the noise), feature extraction (data abstraction), data fusion computation (modeling different information types and fusion), and data compression (reducing the information stored in memory and transmitted by the transceiver).

The fusion between WWBSN (Wearable Wireless Body Sensor Networks), IoT (Internet of Things) and Cloud Computing will allow doctors, emergency stations or caregivers to track and receive data from BSNs about patients in different places.

By using biomedical sensors, the human behavior and physiology, associated to various physical and mental diseases, can be tracked and studied.

The WWBSN can cover monitoring for cardiovascular, diabetic problems or mental disorders (Alzheimer). The wearable devices are integrating for human body wear by using bracelets, garments or patches.

The development of embedded electronics into comfortable, washable and easy wearable textile by merging the electronics with textile or flexible structures is in development and represents a challenge for researchers from material science and electronics.

10.1.2 Future Market perspective

According to studies conducted by ABI Research in 2019 (Aileni and Iftene, 2014), market for wearable type devices will be 66 million users compared to 6 million users in 2013. According to IHS MEMS and Sensors report on wearable in 2014, the market dynamics for products intended for monitoring activity will increase in 2019. Popular devices will be watches/bracelets with integrated sensors, smart clothing and activity monitoring biomedical parameters for sports activities (whatsthebigdata).

The care and concern for the health of the elderly has always been and still is a challenge for researchers in the sense of help to streamline their care.

The European Commission classifies the care for elderly in formal/informal care and home care services provided by family members or professional caregivers.

According to the European Commission's 2009 Ageing Report, the number of people older than 79 will triple in 2060 and the persons older than 65 suffering from at least one disability in activities of daily living, more than double in 2060 (44.4 million), which will lead to the healthcare sector in resources allocation needs and medical staff requirements increasing.

The elderly dependent by medication and formal or informal healthcare services in hospital or at home will triple in the number.

The continuous monitoring of bio-signals for critical heath evaluation is a process specific for hospitals that can lead to hospitals costs increase.

The market for biomedical sensors is expected to increase at 14.58 billion dollars by 2018 at a CAGR of 6.05 percents over the period 2014–2020 and will be adaptive for each type of disease (memsjournal).

The comorbidity phenomena based on neurological and physiological changes that occurs to the older people involves attention at national and international level. Even if due to new medications and techniques of modern surgery, people are living longer, the human body transformation (between 50-80 years) and aging process, are influenced by genetic factors and by environment, lifestyle, geography and working conditions. The elderly are dealing with two or more of the following health disorders (cardiovascular diseases, depression, Alzheimer, arthritis, osteoporosis, incontinence, diabetes, cancer).

The health management for elderly consists in planning, coordinating and acting for health and life quality improvement by maintaining elderly' independence. Elderly care manages social and personal requirements for daily activities assistance and health care.

Elderly care management integrates health and physiological care. Elderly care can be different by cultural, economic and social perspective (for example, many countries in Asia and East Europe prefer the traditional methods of care by younger generations of family members). The costs allocated for elderly healthcare is higher than allocated for others.

In the past, the traditional way was providing the elderly care at home. Actually, the elderly healthcare is provided by medical — social insurance and charitable institutions because today the family size is reduced, the expectation of the elderly are higher and the geographical family dispersion due to the work places distribution is increasing.

According to the EC Ageing Report, the elderly care consists of home healthcare provided by caregivers at home, at a residential family care home or continuing care retirement communities (CCRCs). The old people should be monitored (targeted parameters like humidity, temperature, pulse) in a way that does not cause discomfort to the person.

10.2 BACKGROUND

The healthcare monitoring is specific for hospitals and consists of tracking bio-signals by using sophisticated devices with many cables and sensors attached to the patient and connected to the electronic display. The electronic equipment used in hospitals required qualified personnel for use. In addition, the patient mobility is reduced which lead to discomfort for

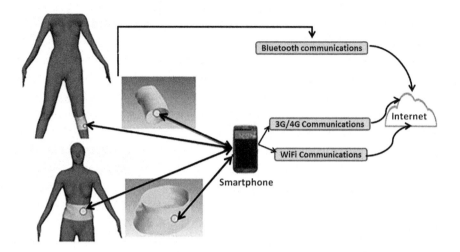

Figure 10.2 Wearable modules for health monitoring (Kang et al., 2006).

patient. By using wearable monitoring system integrated in cloth pieces (small wireless sensors), the patient can be monitored and the bio-signals can be tracked, conditioned, analyzed and sent to the medical staff. Today, the sensors used for environmental and health sensing are smart systems and consists of analog sensors, microcontrollers and communication system and a possibility is to integrate the smart sensor in a chip by using 3D advanced packaging technology.

The electronic devices for monitoring, integrated in textile structure, must be very small, easy to wear, use and comfortable for user.

The sensors integrated into wearable structures (cloth, bandage, bracelet) should be mapped in specific body areas for bio-signals accuracy tracking and in direct contact with human body skin (for temperature, humidity and pulse bio-signals tracking). The wearable sensors system should contain-elementary processing capabilities for signal pre-processing and data communication.

The wearable system is a modular system based on different body areas for the vital signs monitoring (Figure 10.2).

The monitoring system can be described by 3 layers:
-   Skin — first interface with sensors
-   Sensors — second interface with garment
-   Garment — last interface that contain conductive yarns or conductive coating for interconnection of the electronics device embedded

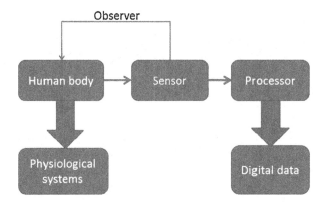

Figure 10.3 Health monitoring by using biomedical instrument.

10.3 HEALTH MONITORING BY USING WEARABLE DEVICES

10.3.1 Definitions

Bio-signals monitoring is a medical intervention defined as the act for collection and analysis of cardiovascular, respiratory, and body temperature data, in order to determine and prevent complications (businesswire). The bio-signals values situated in a range over normal values occur in case of diseases, and an alteration of vital signs is used to evaluate a patient's progress. The biomedical signals are measured using invasive and noninvasive sensors. Both types of sensors, invasive and noninvasive are wearable devices for health monitoring (Figure 10.3).

Health parameters monitoring, if the device is not complex and does not require medical qualification for use, can be achieved in hospitals or at home by qualified personnel or by patient.

The electronic device for noninvasive monitoring can be used for integration in textile structures and to be carried like a layer on the surface of human body skin.

E-textiles are fabrics with embedded electronic devices (sensors or actuators embedded). Terms used for e-textile definition are as smart clothing, smart garment, smart textile, electronic fabric and smart fabrics because the textile support is associated with smart components such as smart sensor or smart actuator. Wearable technology, wearable devices, fashion electronics are accessories or smart garment, which have integrated electronic devices.

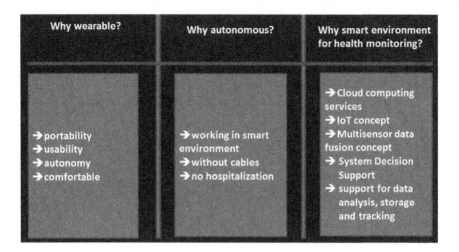

Why wearable?	Why autonomous?	Why smart environment for health monitoring?
→ portability → usability → autonomy → comfortable	→ working in smart environment → without cables → no hospitalization	→ Cloud computing services → IoT concept → Multisensor data fusion concept → System Decision Support → support for data analysis, storage and tracking

Figure 10.4 Wearable system justifications.

Wearable devices are "things" with embedded electronics and software, i.e. forming Internet of Things. Wearable embedded devices allow portability, usability, autonomy and comfort for patients (Figure 10.4).

10.3.2 Physiological Parameters Monitoring

Vital signs analyzed for elderly health are:
- pulse
- temperature
- humidity (skin moisture)
- breathing rhythm

According to medical dictionary, the pulse represents the dilation of an artery in discrete steps, by increasing the volume of blood into the vessel during the heart's contraction (medical-dictionary). Even if the bio-signal for pulse can be measured for different body regions (Figure 10.5), the most important are brachial, radial and carotid sites.

In the past, the arterial blood pressure was measured using a device called sphygmograph.

The measurement of the arterial and veins blood pressure pulse can be achieving using a photoplethysmogram, based on pulse oximeter. It illuminates the skin and measures the changes in light absorption.

The pulse tracking using photoplethysmogram is based on a gradual chart with numerous gradual elevations (businesswire).

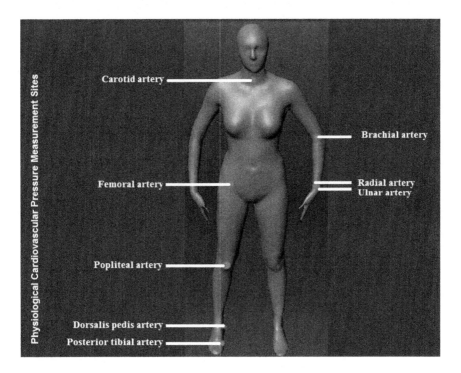

Figure 10.5 Regions on human body area for measuring pulse values.

The arterial blood pressure is the difference between systolic and diastolic values and it is an indicator for normal tension or for abnormal events such as hypotension or hypertension (businesswire). In the systolic phase the arterial pressure has maximum value because the aortic valve of the heart is open, the blood flows through arteries and the heart decreases the volume (businesswire). In the diastolic phase, the arterial blood pressure has minimum value because the aortic valve is closed and the heart increases in volume (businesswire).

The breathing is divided in chest breathing that is under voluntary control and the more rhythmic diaphragmatic breathing. Improper breathing can have adverse effects on the nervous system and can cause anxiety, stress, insomnia and exhaustion (businesswire).

There is a direct dependence between respiration and cardiac cycle because by varying the intrapleural pressure in expiration and inspiration the pressure on the vena cava and the filling of the right atrium will be influenced (normal sinus arrhythmia) (businesswire; medical-dictionary). When the frequency and depth of respiration increases, the venous return increases, leading to increased cardiac rhythm (businesswire).

Body temperature is a balance state result between heat rate production by metabolic processes and heat rate absorption or heat dissipation in environment through garment structure. The higher or lower body temperature is the human body answer to infections or harmful environment conditions, causing in the hypothalamus to set temperature to $37\,°C$, because less than $34\,°C$ is considered a hypothermic state caused by environmental condition or diseases and over $37\,°C$ is considered a hyperthermia that is caused by infections (businesswire).

The heat dissipation is based on 3D wall heat conduction (body-air) and if environmental condition presents low temperature, the body will lose the heat because the human body thermal resistance at the skin level is lower. Hypothermia is a result from human body exposure to the environment with low temperature and increased humidity. The report for heat production/heat dissipation is subunitary.

In addition, the hypothermia can be caused by a reduction of metabolic heat rate due to severe illness (medical-dictionary).

The temperature indicator is very important for elderly because in this case, the metabolic processes and skin thermal resistance are reduced.

10.3.3 Solution Analysis for Elderly Healthcare Monitoring

The diseases associated with ageing process require methods for improving the life quality and increase the surveillance for elderly. The public resources allocated in hospitals and for caregivers' remuneration are insufficient and not uniformly distributed geographically. The role of the wearable devices for healthcare system is to improve medical act quality by distance monitoring for preventing, and alerting (Figure 10.6).

The bio-signals (temperature, skin moisture, pulse) tracking by using biomedical sensors allows studies for human behavior and body physiology. The human body' bio-signals are modified when a stimulus or biological harmful effect provided by physical and mental diseases has occurred. The concept of data fusion refers to the usage of data synthesis from the wireless sensors network on the same level (similar sensors) and on different levels (different category sensors) for developing the decision systems (Aileni et al., 2015a).

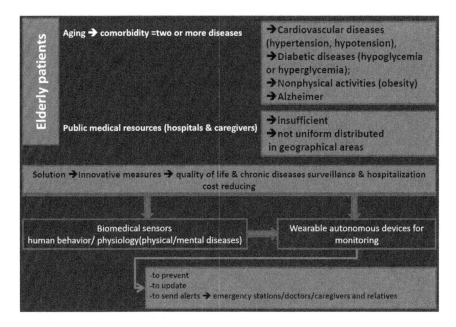

Figure 10.6 Ecosystem design for elderly health monitoring.

The Cloud Computing services such as PaaS or SaaS are used for medical applications oriented on storage, access, management of private health information (Aileni et al., 2015b).

The sensors integrated in textile are useful for monitoring the biosignals important for a large area of diseases:

- Cardiovascular diseases symptoms (hypertension, hypotension)
- Diabetic disease symptoms (hypoglycemia or hyperglycemia)
- Diseases associated with absence of the physical activities (obesity)
- Alzheimer

10.3.4 Wearable Electronic Devices Integrated in Garment

10.3.4.1 Smart Garment

The smart garment should offer comfort to wearer by flexibility, handling and design. Smart garments are formed of layers:

- The inner layer, the closest one to the skin, is intended for comfort and support, and is characterized by underwear products

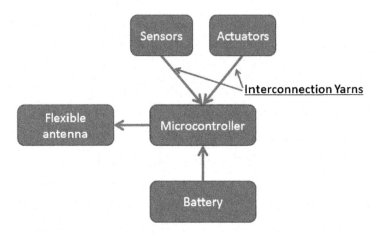

Figure 10.7 Wearable electronics model for integrating into garment by using conductive yarns.

- The outer layers, represented by outerwear products, are used to add comfort (e.g. warmth) and protection from the surrounding environment

A critical aspect when considering e-textiles is the integration and interconnection of textiles and electronics. Because textile and electronics are from materials with different characteristics, embedding electronic components is limited. The solutions consist of attaching electronic devices to textile substrates (Figure 10.7).

The electronic device can be integrated into clothing by using:
- Printing
- Attachable/detachable systems
- Adhesives
- Textile structures — yarns and fiber with which the electronics parts make common item

In general, textile structures are electrically inactive (Moore and Zouridakis, 2004) and can be converted into electrically active surface by using surface coating with metals in plasma, printed electronics or by using conductive yarns in weaving or knitted structures (Xiaoming, 2015).

10.3.4.2 Electrical Conductivity and Resistivily of the Fabric

The density of conductive yarns in the textile structure (woven, knitted or sewn) is important for modeling of textile conductive samples designed for wearable electronic devices (sensors and actuators). By using Energy

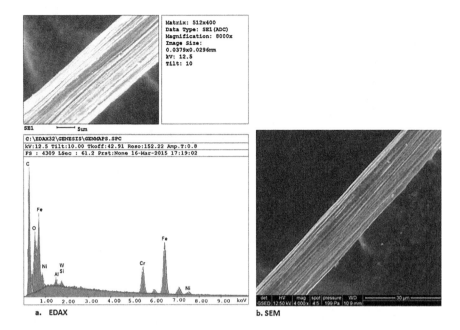

Figure 10.8 Stainless steel fiber analyses (Aileni and Dinca, 2015).

Dispersive X-ray analysis (EDAX) the chemical elements presented in the fiber structure are displayed (Figure 10.8a). In Figure 10.8, the pronounced peaks for chemical elements Fe, Ni, Al, W and Cr (Aileni and Dinca, 2015) are presented.

The electrical conductivity (or electrical resistivity) of a substrate (in surface of volume) is one of the main properties and requirements of smart garment and determines the applications of the e-textile. The resistivity defines the intrinsic nature of the material and is linear (yarns, surface (2D textile) or volume resistivity (3D textile)). Because of the textile attributes (softness and flexibility), electrical properties are being increasingly demanded in technical applications to perform functions such as heating, electromagnetic interference shielding, electrical data transportation or signal detection. However, in case of integration of the electronic components (sensors, actuators) on the textile surface (e-textile), constraints related to system design that requires high computational performance, low power consumption and fault tolerance may occur. The physical nature of the e-textile (discrete model) and the faults developed by open and short circuit can disconnect/drain the battery, affecting both battery life and the performance of the e-textile. Finally, it can affect the accuracy of signals from the e-textile (Aileni et al., 2015a).

10.3.4.3 Metallic Yarns Integration

For sensing systems, a textile material with electrical behavior and electric insulation is required. There are several constraints and incompatibilities with the available textile and clothing technologies related to the interconnection and integration of micro-electronic systems in the textile structures.

The production of an embedded network of conductive yarns within a textile structure may be obtained through textile fabrication techniques (weaving, knitting) and cloth manufacturing techniques (seaming, embroidery). Woven fabrics are used as anisotropic electrical conductive substrates with conductive yarns in weft or warp direction or as isotropic electrical conductive substrates by conductive yarns interwoven in warp and weft directions.

The electro-conductive textile substrate is obtained by integration of fiber in textile structure like:

-   Conductive fibers or yarns (metallic fibers or metallic alloys) — filament steel yarns, copper yarns
-   Polymer-based fibers or yarns (e.g. polyester — steel-grafted polyester yarns, polyamide) grafted with a conductive function
-   Treatment the of textile substrates with conductive products (metallic or conductor polymers)

Another interesting fiber is Nitinol, based on Shape Memory Alloys, and can be used to transmit discomfort signals (stress-strain) due to swelling and coupled with other electronic textile components to monitor bio-signals.

10.4 SOFTWARE SYSTEM ARCHITECTURE

The software and hardware system for elderly health monitoring are based on four sensors for measuring pulse, temperature, respiration, humidity and rate (Figure 10.9). The system will be used for elderly patients with diabetes.

The conductive yarns (semiconductors) used in textile structures for the integration of sensors/actuators with the motherboard can affect signal data accuracy because of the yarns resistivity changes with temperature variations, body thermal flow and due to the textile property to be thermal conductive (Aileni and Dinca, 2015). The electronic devices (sensors

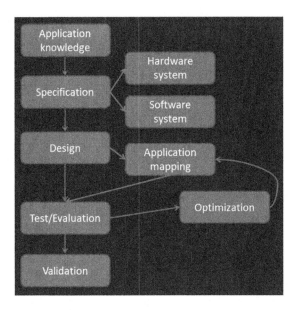

Figure 10.9 System architecture.

or actuators) integrated into the textile (e-textile) require a high computational performance, low power consumption and fault tolerance control (Figure 10.10).

In many cases, the sensors output may generate the errors that can be considered like fault events (Fontecha et al., 2015):

- Partial or total output loss
- Abrupt/continuous switching between modes of functioning
- Nonlinear aberrations

We considered hyperglycemia (10.1) and hypoglycemia (10.2) as the critical events for diabetic elderly patients. The events are function of the values for temperature, pulse, breathing rhythm and skin moisture.

$$\text{Hyperglycemia} = f(\text{temperature, pulse, breath rhythm, humidity}) \quad (10.1)$$

$$\text{Hypoglycemia} = f(\text{temperature, pulse, breath rhythm, humidity}) \quad (10.2)$$

In a mainboard, ATmega328 there were integrated two sensors for temperature (MCP9700) and skin humidity (SEN01400P) monitoring. The sensor data processing has been described in Figure 10.11.

We consider the mathematical expression (10.3) for sensors (Aileni et al., 2015a):

$$S(V, e) = \{V(t), e(t)\}, \quad (10.3)$$

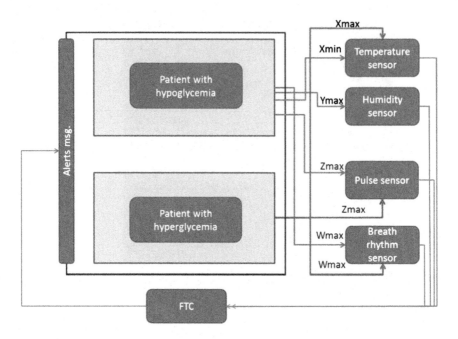

Figure 10.10 Fault tolerance control for monitoring system (Aileni and Dinca, 2015).

where:

S is the mapping function for the environmental numerical values V (Aileni et al., 2015a)

$e(t)$ are values for environment (temperature, humidity)

t is the time (Aileni et al., 2015a)

After tests with 1000 milliseconds delay, we obtained numerical data for signal processing (Figure 10.11).

We find that there is no linear dependence (10.6), (10.7) between humidity and temperature by signal processing (Figure 10.12), covariance (10.4), (10.5) and correlation coefficient analysis (10.6), (10.7), (10.8) for data from sensors. These values are in inverse proportionality (10.8). It explains the thermoregulatory human system that uses the moisture process to normalize the body temperature.

The Pearson coefficient (10.6) from statistics, was used for analyzes and correlation the data sets (humidity and temperature vectors).

$$\text{cov}\left(humidity,\ temperature\right) = \begin{vmatrix} 2.5957 & 0.0623 \\ 0.0623 & 0.9059 \end{vmatrix} \tag{10.4}$$

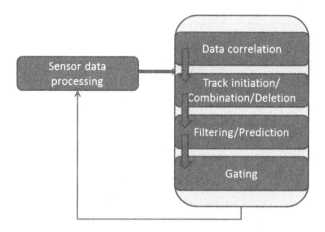

Figure 10.11 Data processing (Aileni and Dinca, 2015).

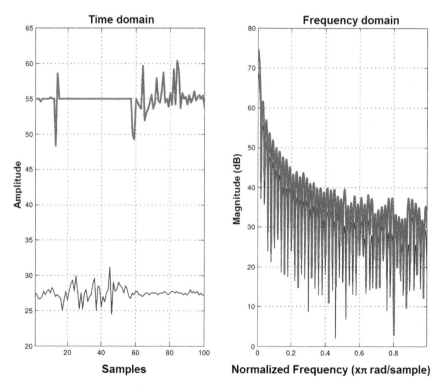

Figure 10.12 Temperature (blue line, bottom on the graphs) and skin moisture (green line, upper parts of the graphs) representation in time and frequency. (For interpretation of the references to color in this figure legend, the reader is referred to the web version of this chapter.)

$$\text{cov}(humidity,\ temperature)_{1,2} = \text{cov}(humidity,\ temperature) = 0.0623 \tag{10.5}$$

$$R_{xy} = \frac{n\sum_{i=1}^{n} x_i y_i - \sum_{i=1}^{n} x_i \sum_{i=1}^{n} y_i}{\sqrt{[n\sum_{i=1}^{n} x_i^2 - (\sum_{i=1}^{n} x_i)^2][n\sum_{i=1}^{n} y_i^2 - (\sum_{i=1}^{n} y_i)^2]}} \tag{10.6}$$

$$R_{xy} = \text{corrcoef}(x,\ y) <=> R_{x,y} = \begin{vmatrix} 1.000 & 0.0406 \\ 0.0406 & 1.000 \end{vmatrix} \tag{10.7}$$

$$R_{xy(1,2)} = R_{xy(2,1)} = 0.0406 \neq 1 \tag{10.8}$$

where:

$x = $ humidity

$y = $ temperature

$R_{x,y} = $ correlation coefficient (Pearson coefficient) (10.6).

The big data received from biomedical sensors requires the design of new technologies, complex algorithms and software for collecting, storing and managing the huge data volume by using support decision systems and Cloud Computing.

For analyzing the parameters from patients, we propose a support decision system (Figure 10.13).

The system architecture consists of 5 levels (Aileni et al., 2015a):

- Level 1 — data transmission (biomedical sensors aggregators)
- Level 2 — big data (data collecting, sampling and storage)
- Level 3 — medical information (data mining)
- Level 4 — diseases knowledge (data synthesis)
- Level 5 — decision support system

Big data processing from sensors requires a data fusion (multisensory data fusion) with preprocessing (filtering the noise), feature extraction (data abstraction), data fusion computation (modeling different information type and fusion), and data compression (reducing the information stored in memory and transmitted by the transceiver).

Multisensory data fusion (Figure 10.14) can be defined as fusion of the different data sources in order to obtain final information with less uncertainty. Data fusion and big data integration in complex systems for security or healthcare can lead to system failure, fault or errors. Multisensory data fusion appeared due to our society's tendency into ubiquitous computing environments with robotic services everywhere.

It is important for a medical system to have data fusion for information from different stakeholders involved in medical domain (Figure 10.15).

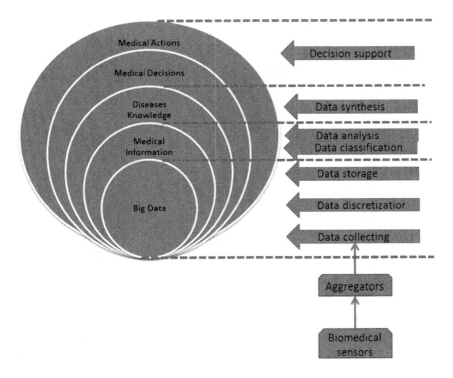

Figure 10.13 Support decision systems for medical staff (Aileni et al., 2015a).

Cloud Computing services are required for storage, access and big data analyzes. Cloud Computing architecture, defined as a collection of services, applications, platforms, infrastructures and servers that allow data virtualization and data storage, are presented in Figure 10.16.

The patient's biomedical information, captured from sensors, is a view containing data with personal character, which require anonymization for future complex analysis. Cloud Computing architecture (Figure 10.17) was developed without compromising the privacy.

Cloud Computing architecture for healthcare systems must offer service life cycle management, security, responsiveness, intelligent service deployment, portability, environmental sustainability, service reliability, service availability and quality assurance (Aileni et al., 2015a).

10.5 CONCLUSIONS

Wearable sensors for healthcare monitoring should consider the:

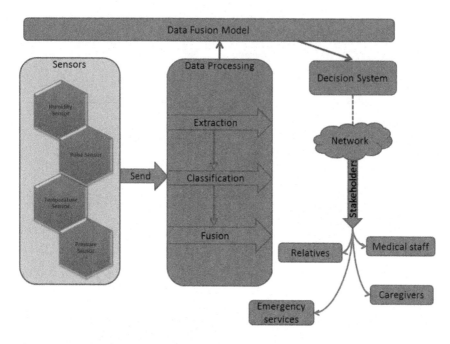

Figure 10.14 Data fusion model (Aileni and Dinca, 2015).

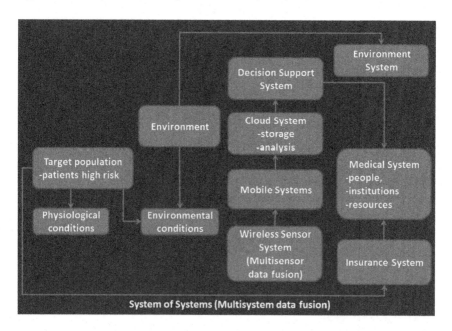

Figure 10.15 Multisystem data fusion (Aileni and Dinca, 2015).

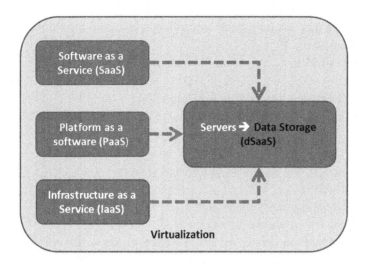

Figure 10.16 Cloud Computing architecture (Aileni et al., 2015a).

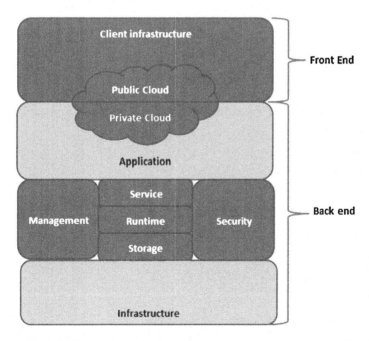

Figure 10.17 Cloud Computing remodeling by using privacy requirements (Aileni et al., 2015a).

- Fault tolerance control implementation
- Multisensory data fusion (model control by predictive algorithms for optimal decisions)
- Sensor data processing algorithm for reducing the noise and data sampling
- Data correlation for sensors values measured

Wearable electronics integrated in textile, experience data loss and low accuracy signals due to the textile structure properties.

The advantages of using Cloud Computing for healthcare are:

- Storage space for big data from biomedical sensors
- Data availability
- Data security and privacy by design
- Data analysis and predictive modeling

The work of medical personnel and caregivers is supported with decisions proposed thanks to the collected and analyzed data. Furthermore, the overall cost for the taking care for elderly is expected to decrease.

10.6 FUTURE WORK

For developing the monitoring system it will be required to analyze, collect and storage big data. The parameters from patients will be developed to support decision system analyses.

We will try to use FPGA model based on wireless body network sensors for real time monitoring of the elderly patients with diabetes and cardiovascular diseases.

ACKNOWLEDGEMENTS

Scientific research for this book chapter was developed with the financial support obtained from Wallonie-Bruxelles International (WBI) in 2015.

This book chapter was published with support of the COST Action IC 1303 Algorithms, Architectures and Platforms for Enhanced Living Environments (AAPELE) in 2016.

REFERENCES

http://medical-dictionary.thefreedictionary.com/monitoring.
http://whatsthebigdata.com/2014/10/16/home-monitoring-mhealth-wearable-devices-2013-2019/.
http://www.businesswire.com/news/home/20151222005421/en/Research-Markets-Global-Biomedical-Sensors-Market.

http://www.memsjournal.com/2014/10/wearable-sensor-market-to-expand-sevenfold-in-five-years.html.

Aileni, R.M., Dinca, L., 2015. Electroconductive Materials with High Potential for Wearable Electronic Devices Integration. IEEE, ECAI, Bucharest.

Aileni, R.M., Iftene, A., 2014. Software application for health state monitoring and data modeling for health risk factor evaluation, based on wearable sensors and observer patterns for sensors data tracking. In: ICVL. Bucharest, Romania.

Aileni, R.M., Pasca, S., Valderrama, C., 2015a. Biomedical sensors data fusion algorithm for enhancing the efficiency of fault-tolerant systems in case of wearable electronics device. In: ROLCG. IEEE.

Aileni, R.M., Pasca, S., Valderrama, C., 2015b. Cloud computing for big data from biomedical sensors monitoring storage and analyze. In: ROLCG. IEEE.

Augustyniak, P., Smolen, M., Nikrut, Z., Kantoch, E., 2014. Seamless tracing of human behavior using complementary wearable and house-embedded sensors. Sensors 14 (5), 7831–7856.

Camarinha-Matos, L.M., Rosas, J., Oliveira, A.I., Ferrada, F., 2015. Care services ecosystem for ambient assisted living. Enterp. Inf. Syst. 9 (5–6), 607–633.

Dasios, A., Gavalas, D., Pantziou, G., Konstantopoulos, C., 2015. Hands-on experiences in deploying cost-effective ambient-assisted living systems. Sensors 15 (6), 14487–14512.

Fontecha, J., Hervas, R., Mondejar, T., Gonzales, I., 2015. Towards context-aware and user-centered analysis in assistive environments. J. Med. Syst. 39 (10).

Forkan, A., Khalil, I., Tari, Z., 2014. CoCaMAAL: A cloud-oriented context-aware middleware in ambient assisted living. Future Gener. Comput. Syst. 35, 114–127. Special Section: Integration of Cloud Computing and Body Sensor Networks.

Kang, T.H., Merritt, C., Karaguzel, B., Wilson, J., Franzon, P., Pourdeyhimi, B., Grant, E., Nagle, T., 2006. Sensors on textile substrates for home-based healthcare monitoring. In: Proceedings of the 1st Distributed Diagnosis and Home Healthcare (D2H2) Conference.

Moore, J., Zouridakis, G., 2004. Biomedical Technology and Devices Handbook. CRC Press, US.

Siciliano, P., Leone, A., Diraco, G., Distance, C., Malfatti, M., Gonzo, L., Grassi, M., Lombardi, A., Rescio, G., Malcovati, P., 2009. A networked multisensor system for ambient assisted living application. In: 3rd International Workshop on Advances in Sensors and Interfaces, pp. 132–136.

Terrose, M., Freitas, R., Simoes, R., Gabriel, J., Marques, A.T., 2013. Active assistance for senior healthcare: a wearable system for fall detection. In: 8th Iberian Conference on Information Systems and Technologies (CISTI).

Xiaoming, T., 2015. Handbook of Smart Textiles. Springer, Hong Kong.

CHAPTER 11

Cloud-Oriented Domain for AAL

Vasos Hadjioannou*, Constandinos X. Mavromoustakis*,
George Mastorakis[†], Ciprian Dobre[‡], Rossitza I. Goleva[§],
Nuno M. Garcia[¶]

*Department of Computer Science, University of Nicosia, 46 Makedonitissa Avenue, 1700 Nicosia, Cyprus
[†]Department of Business Administration, Technological Educational Institute of Crete, Heraklion, Crete, Greece
[‡]University of Politehnica of Bucharest, Bucharest, Romania
[§]Technical University of Sofia, Bulgaria
[¶]University of Beira Interior, Portugal

11.1 IOT AND CLOUD COMPUTING IN AAL

The emergence of the upcoming 5G will provide the possibility of realizing the Internet of Things (IoT); a network of interconnected objects which, similarly to 5G, uses an MP2P (Mobile Peer-to-Peer) way of communication between the participant network nodes. The ubiquitous nature of IoT makes it possible to be effectively utilized almost everywhere, revolutionizing this way, not only a plethora of fields (transportation, healthcare, environmental monitoring, etc.), but the everyday lives of people around the globe as well, since any kind of device will be able to stay connected to anyone at any given anytime (Mastorakis et al., 2014).

While IoT can prove to be a great technological advancement, it is not the only player in the game, and in order to win the round it will require teamwork and coordination with other compatible participants, such as cloud computing.

Cloud computing can be considered as a gathering of resources that can be utilized remotely by those who have access to the cloud (Borges et al., 2012). It can provide its users with various services through a network, in the form of Software as a Service (SaaS), which makes it possible for users to use an application through the cloud without having the need to install it on their machine, Infrastructure as a Service (IaaS), that enables the utilization of hardware without having to purchase it, and Platform as a Service (PaaS), which provides the proper environment for a user to develop a desired software. The various services provided by cloud computing can be observed from Figure 11.1. Cloud computing is an already existing, and

Ambient Assisted Living and Enhanced Living Environments.
DOI: http://dx.doi.org/10.1016/B978-0-12-805195-5.00011-9

271

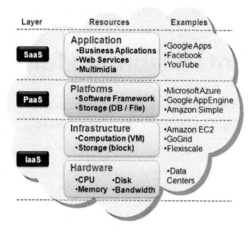

Figure 11.1 Cloud computing services.

currently used technology which is capable of working side-by-side with IoT for reaching greater heights.

The network nodes that comprise IoT can be any everyday object that can sense, process (not all), and transmit data. For this reason, IoT can be efficiently utilized in the field of Ambient Assisted Living (AAL), since the devices that take part in IoT will be able to behave as data gathering actors and collect any sort of information about a designated individual (such as location, habits, physiological condition, etc.). AAL's goal is to utilize the current technological advances in order to deliver healthcare services to those in need, in their preferred environment. Either it is used for monitoring the condition of a patient in a remote manner, or simply for improving the lifestyle of an individual, AAL services will need to have a "control center" that is in charge of managing any activities, as well as store and process the generated data. Due to the large number of devices that will comprise the Internet of Things, a huge amount of data will be created on a daily basis which will require ample processing, storage, and management. Cloud computing will be able to effectively provide such high demand of resources due to its reliability, scalability and cost-effectiveness. The cloud will be able to receive the sensed/generated data from the devices of IoT, process the information, and act accordingly, in order to provide a better and more secure living environment for patients that are receiving remote treatment, people whose physiological condition is in need of constant surveillance (elderly, chronically ill), or even perfectly healthy people that simply wish to have a better quality of life.

11.2 WHAT IOT CLOUD SYSTEMS HAVE TO OFFER

11.2.1 Features of IoT

The Internet of Things is considered to be the evolution of the current Internet, where every device will be able to connect and exchange information with one another using an MP2P way of communication. Every object will have the opportunity to be part in this network, such as smart-phones, smart-shoes, tablets, etc., as long as they are able to gather (sense) and transmit data. These smart devices are not the only participants though since other objects are able to interact with the physical world as well, gather raw data from their surroundings and transmit information, such as sensors and RFID (Radio-Frequency Identification) tags. Additionally, every node in an IoT network will communicate through a wireless medium, but due to the enormous number of connected devices, the vast address space of IPv6 will be needed (Pop et al., in press; Ciocan et al., 2014).

The omnipresent nature of IoT is the main reason why it is able to be utilized almost everywhere. Since any every day item will be able to sense and transmit data, it makes the perfect technology for AAL since a big part of home healthcare is to gather information about the designated person. Through the sensing capabilities of IoT, any kind of information can be gathered, such as location, physiological signals (blood pressure, heart rate, etc.), which can be provided to the personal doctor of a patient, alarm relatives in case of emergency, and in general enable the system in charge to decide whether any action needs to be taken after analyzing the information (Kryftis et al., 2016; Mavromoustakis et al., in press-a).

11.2.2 Features of Cloud Computing

The utilization of Information and Communication Technologies (ICT) in, not only AAL but in, e-Healthcare in general, brings about a high demand of computational resources to be dedicated for the provided healthcare services, as seen in Figure 11.2. One of the main reasons why cloud computing is considered to be appealing in AAL, is due to the sheer quantity of computational resources that it has to offer. As mentioned before, the amount

Figure 11.2 Cloud-based telemedicine applications.

of data that is generated by AAL applications is quite large, and it needs the storage and processing capabilities of the cloud in order to manage it. Additionally, the resources of a cloud are assigned via virtualization, making it possible to dynamically allocate the required resources based on the demand at a given time. This way, even devices with poor capabilities, are able to use more resource-expensive applications by simply offloading all the heavy computations to the cloud, and receiving only the result of the processing (this is especially useful when it comes to wearable technology) (Shen et al., 2013; Kryftis et al., 2015, 2014).

It is important to note that the reason cloud computing is suited for AAL is not only because of the resources it provides (Wang et al., 2014). The scalable nature of it, that allows the smooth expansion of the range of its applications, as well as its cost-effectiveness, which allows a more economic deployment, are also noticeable features that make cloud computing an even more attractive solution. The ease of access to the cloud is also an important factor; since the available services are accessed through the web, it becomes this way easy to use them anywhere at any time, making the cloud location independent, which is a crucial feature to provide in the case of AAL since its objective is to deliver a suitable living environment, through ICT, which is able to provide the required healthcare services (Memon et al., 2014; Mavromoustakis et al., 2014b).

There are two types of clouds that can be utilized in e-healthcare and both of them carry with them their own benefits and issues; the public and private clouds. Public clouds are operated by third-party providers and consist of several services that can be delivered to the general public in a pay-per-use fashion. These services can be in the form of hardware or software and can be accessed through the Internet by any paying user. Such clouds are considered appealing to be used for providing healthcare services, due to the enormous amount of resources they have to offer, but they also suffer from data transmission delay caused by geographical distances between the user and the cloud, as well as network congestion. On the other hand, private clouds are operated and used by an organization and can be considered as an extension of their datacenter, configured to deliver some extra services similar to the ones provided by the public clouds (like computation and storage capabilities). Private clouds, unlike public ones, are accessed through a private network and are dedicated to serve only the purposes of the organization. Whilst private clouds are considered to be more protected and secure, which is of crucial importance to healthcare service provision, they don't possess as much resources as public clouds do. Therefore, not only the organization will have to purchase and maintain the cloud hardware, but the provided services may also deteriorate during burst periods (Shen et al., 2013).

11.2.3 Remote Monitoring (Telemonitoring)

One of the most important services of today's healthcare systems is the monitoring of a patient's condition by measuring and analyzing the vital signs of the human body, in order to determine whether there is the need of immediate action or not. However, for people that suffer from chronic diseases, the elderly, and in general people whose health is slowly deteriorating, it can be a tiresome, expensive, and time consuming procedure to visit a health center or a hospital regularly in order to have their health status monitored.

IoT cloud systems are able to provide a solution to this problem by the means of remote monitoring (or telemonitoring). Telemonitoring enables a patient to have his physiological signals measured and analyzed while at home or outdoors, by simply allowing various sensing devices to collect and transmit his body's data to the cloud, through the MP2P architecture, where the information will be processed. All the information that is regularly collected can be stored in the cloud, allowing, this way, statistical results about someone's health to be generated, which can be later examined by the individual himself, or by his personal doctor.

11.2.3.1 Remote Sensing

There are various ways to sense information about the human body and its surroundings, especially when IoT is involved. One of the simplest ways to achieve this is with the deployment of sensors in a designated area. While such sensors will be able to obtain information like the location of a patient, along with his current activities, they will have limited capabilities in sensing the more sensitive physiological signals of a human being. A solution to this issue is the utilization of wearable technology. Such devices are considered to be specialized sensing equipment capable of gathering information about the current health condition of the user, and can be worn by anyone without interrupting the flow of their daily routine (More on wearable technology and Wireless Body Area Networks (WBAN) later on) (Batalla et al., 2015).

Another practical way for sensing information is by utilizing the various sensors embedded in smart phones through designated sensing applications. State of the art mobile phones, nowadays, are equipped with a variety of sensors, such as gyroscope, GPS, proximity sensor, etc., which can be efficiently used in providing ample data about the status of the monitored person. For example, the authors in Fahim et al. (2014) have used the accelerometer of a smart phone to detect the body movements of the user, and through a server located in a cloud, they were able to determine whether the user was walking, running, cycling or hopping.

Additionally, many applications have taken advantage of the mobile's camera and flashlight in order to measure the heart rate of the user, as seen in Figure 11.3. This is achieved by detecting the photoplethysmographic (PPG) signal, which is measured using the fingertips of a human person. By illuminating the surface of the fingertip with the phone's white LED flashlight, the camera is able to capture the extremely small color changes on the skin, caused by the expansion of the capillary vessels, which occurs with every heart beat. The calculation of the heart rate signal is accomplished by detecting the moments when the pulse is at its peak. The drawback of this method lies in its inability to continuously measure the heart rate of the user, since it is a procedure where a person is required to remain still for a few seconds with his finger in front of the camera and flashlight (Bolkhovsky et al., 2012; Mertz, 2012).

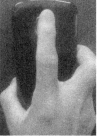

Figure 11.3 Smartphone heart rate measurement.

As stated in Sheng et al. (2012), there are two methods that can be used for mobile phone sensing; the participatory sensing, and opportunistic sensing. In the case of participatory sensing, the mobile users are responsible for the kind of sensing that takes place, and determine themselves what to sense and how to go about it. A good example of participatory sensing is the previously mentioned heart rate measuring, where the users decide what to sense and accomplish the task on their own volition. On the other hand, opportunistic sensing happens automatically without having the need to involve the user during the process; all the sensing happens in the background, without intervening with any of the user's activities.

Unfortunately, smartphone sensing does not come without its drawbacks. Firstly, the limited battery capacity of a smartphone poses the potential danger that sensing applications will consume most of the phone's energy. Since gathering data from the outside world consumes a relatively significant amount of power, it is rather unfortunate, but very much plausible, that the phone's battery will run out after a few sensing jobs. Additionally, computing resources are also drained while a phone's sensors are active, making smartphone sensing an even more challenging task (Sheng et al., 2012; Mavromoustakis et al., 2015, in press-b).

11.2.3.2 Electronic Health Records

The digital storage of, previously printed-out, health records is without a doubt advantageous in a plethora of ways. Electronic Health Records (EHRs), not only provide a more time-conserving and cost-efficient solution, they also emphasize in maintaining the privacy of patients, since access to such records is limited only to authorized individuals. Additionally, having EHRs enables the easier management of numerous records, along with the assurance of their safety (backups) and ease of access.

The implementation of EHRs was generally realized by using the client-server architecture, where a dedicated server was deployed in order to store any receiving data coming from the client side, as well as to reply to the client's requests. However, the costly equipment along with the difficulty of its implementation and maintenance are the two of the major drawbacks of this method, which prevented it from being widely utilized. On the other hand, cloud computing is able to tackle these issues head-on due to its resourcefulness and virtualized nature. It is reasonably a much better solution which, not only it offers its users with reliable, efficient, and agile services (storage, computation power, etc.), but does it while liberating them from the need of owning and maintaining their own hardware.

Additionally, when talking about health records, it doesn't necessarily mean textual data, since the usage of image-generating medical instruments (like ultrasound scanners) is common nowadays. It is noticeable that the size of such graphical data by far exceeds the size of text, making the use of cloud computing even more attractive due to its large storage capacities. The generated images can be saved and transferred to the cloud using DICOM (Digital Imaging and Communications in Medicine), a standard used for managing, storing, and transmitting medical illustrations, which defines a file format for saving the images, along with a communications protocol for transmitting them.

While taking advantage of cloud computing provides a much faster and reliable solution in storing health records, it must be taken into consideration that EHRs contain valuable and sensitive data about patients, such as medical history, personal information, habits, etc., which need to remain private at all costs. Therefore it is crucial that the services provided by the cloud, along with any information stored in it, are protected and cannot be tampered with, or accessed, by unauthorized parties. Furthermore, even if the cloud is safe from malicious attacks, it is also important that cloud providers are able to maintain the reliability of their services no matter what the case, by keeping backups or having a disaster prevention/recovery plan, averting this way from any unfortunate situations caused by outages, faulty hardware or accidents (Jin and Chen, 2015; Ciobanu et al., 2014).

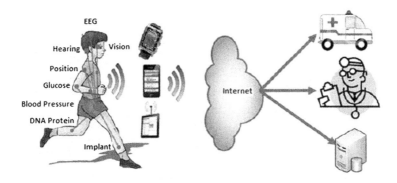

Figure 11.4 Setup of WBAN.

11.3 WEARABLE TECHNOLOGY IN AAL

11.3.1 Wireless Body Area Network

WBAN is a wireless network which is comprised of wearable devices that have the ability to measure and gather data regarding the physiological condition of the user. It is commonly utilized in healthcare due to the fact that it enables an easier, and much more cost-effective, method for remotely monitoring the condition of a patient. The information is gathered with the helped of various sensors embedded in the wearable devices, and forwarded to the servers of the corresponding medical institution, or physician, as shown in Figure 11.4. Since the sensors are placed on the bodies of human beings, it can be considered as an on-the-move WSN (Wireless Sensor Network), configured in a star topology, where the sensors worn by the users are regarded to be the endpoints of the star, and the datacenter, where all the information is gathered, to be the central node (Cheng and Huang, 2013).

Similarly, to any other technology, WBANs come with a few issues and challenges regarding their design and deployment (Uddin, 2015). The participant devices are designed to be worn and it is therefore sensible that they need to be comfortable, or in the worst case bearable to put on, constraining this way the device's weight and size, and thus limiting its computing capabilities. They also suffer from lack of scalability since they cannot support an increased number of users without having to undergo major alterations in the system's hardware or software. Additionally, as aforementioned, WBANs utilize the star topology, which means they've also inherited the star's flaw of having a single point of failure, and since the central node is responsible for communicating with every other node in the network, if it fails, the entire network collapses as a result.

11.3.2 IoT Cloud Systems in WBANs

IoT cloud systems can help with improving WBANs in a number of ways, by making them more versatile, easier to manage, and more efficient and reliable. Due to its MP2P way of connectivity, the Internet of Things is able facilitate a much better, and more efficient, means of communication, not only between the endpoints and the cloud, but between the endpoints themselves, allowing this way more fluent data transfers. Additionally, since almost every object can be a potential participant in the network, IoT automatically expands the range of possibly connected devices which minimizes, but not eradicates, the need of specialized equipment, and provides the opportunity of delivering more value added services to the users using everyday objects.

Furthermore, WBANs benefit also from the cloud computing aspect of an IoT cloud system since it helps with dealing with some of the major drawbacks of Wireless Body Area Networks. For starters, the huge amount of resources the cloud has to offer makes it possible for wearable devices to offload heavy computation to the cloud, which in turn enables vendors to give less emphasis to the capabilities of the device, thus developing more compact and easier to wear products. A cloud will also be able to allocate its resources based on the demand of the devices, giving this way more computational power to those that require it (Mavromoustakis et al., 2014c, 2014a).

By using machine learning algorithms, the cloud can also be educated and be made capable of diagnosing and predicting illnesses beforehand by analyzing the collected information of the numerous users. This way, it will be able to identify conditions as soon as they present themselves, as well as raise alarms in case of emergencies or symptoms of potential ailments. Furthermore, cloud systems are designed to be scalable therefore they can support any number of users without having the need of making considerable hardware/software changes (Wang et al., 2014).

In general, the cooperation between wearable devices and IoT cloud systems encourages the development of new medical services and equipment, due to the overflowing resource pool the cloud has to offer as well as the communication method of IoT, along with the improvement in quality of current technology.

11.3.3 Influence of Wearable Technology

Whilst wearable technology comes with a great deal of benefits and can become a valuable asset in AAL, as well as in healthcare in general, it is important to hear both sides of the story since it can possibly bring forth negative results as well. In fact, the utilization of wearable devices could potentially cause serious effects on the psychology of the user, since having such devices constantly monitoring a patient's physiological state might eventually convince him that he is currently in poor health, which may lead to even more serious mental illnesses, such as depression, stress, low self confidence and low self esteem.

Moreover, the effects of wearable devices on the emotional state of the user should be considered as well. The automatic data gathering provided by such technology, although it allows easier, faster, and more efficient monitoring of a patient's status, it is also affectionless, and lacks the human interaction and reassurance most people wish for. Especially in the case of the elderly, who are not used to, and don't fully understand the technology that is around them, such method of monitoring might cause the deterioration of their mental, physical, and emotional state due to feeling lonely and possibly forgotten (Chen et al., 2015). Of course, not every person will feel this way but it is important to state that the opposite of such behavior could also be problematic. The feeling of reassurance some people might get, that they are constantly monitored and cared for, might result in negligence of their own health due to the belief that wearable technology is looking after them (Papanikolaou et al., 2014).

11.4 CHALLENGES AND ISSUES OF IOT CLOUD SYSTEMS IN AAL

Unfortunately, before widely deploying the combined efforts of IoT and cloud computing in the field of Ambient Assisted Living, as a whole, there are a few things that need to be considered, as well as problems that need to be dealt with. Since the utilization of IoT cloud systems is a new, state of the art, technology, it is important that every aspect and possibility of what might go wrong is considered, especially when dealing with services that target the well being of the users, and when healthcare applications heavily depend on it.

11.4.1 Security and Privacy

One of the most important challenges that needs to be sorted out is the delivery of secure services that are able to, not only maintain their integrity and reliability, but to also ensure the privacy of the users and service providers. In order for secure systems to be developed, it is vital that the security of these services is something that is considered while designing them, and not something that is implemented afterwards.

Such systems should be able to block, detect, identify and neutralize any potential threat, either that is a passive attack, or active an active one. Passive attacks are easy to handle (e.g. using encryption) but difficult to detect since they are in the form of *release of message contents* and *traffic analysis*, where the attacker's purpose is to obtain information during transmissions, without altering the data in any way. On the other hand, the easy to detect, but difficult to counter, active attacks are launched in order to intentionally disrupt the proper operation of a system, and are usually used to modify the stored datasets, or prevent the proper delivery of the offered services (e.g. Denial of Service attacks) (Stallings, 2010).

Even if the system (the cloud in this case) is secure and is able to fend off any intruders that try to infiltrate it, there is still danger of compromisation by attacks that target the communication channels and data transmissions. For example, a passive attack in the form of *eavesdropping* is able to sniff information that is transferred through the wireless medium. One might say that an effective way to defend against such an attack is by using encryption, and while that might true in most cases, encryption doesn't necessarily guarantee protection from passive attacks in an IoT cloud system. Since any kind of object that is able to gather and transmit information can be part of the Internet of Things, it is only natural that there will be some devices (such as RFIDs or sensors) that do not have the necessary processing capabilities to support any encryption/decryption algorithms, thus rendering them unable to ensure the confidentiality of the transferred information (Lee et al., 2012).

Additionally, FDI (False Data Injection) attacks, where the attacker's goal is to modify or insert fake data in a system, compromising this way its integrity, can become a serious problem in the case of AAL, especially when it comes to WBANs. In IoT cloud systems, the objective of wearable devices is to gather data about the designated user and forward that information to the cloud in order for it to be processed and stored. What happens though in the case where a specific device is used by another individual? Whatever the intention behind such action is, either malicious or not, the device will keep sensing and transmitting data thus, producing this way, a less intellectual, but equally dangerous FDI attack, since the cloud will falsely suppose that the data received by the device represents the health condition of the actual patient. Therefore, it is important that the cloud can somehow recognize and authenticate the user of each device before allowing them to use it.

11.4.2 Large Datasets and Device Interoperability

Even the scalable and resourceful cloud might find it challenging in handling the data generated by IoT. As aforementioned, a huge number of interconnected objects will be a part of the Internet of Things, in which they will be constantly gathering and transmitting information from their surroundings, whereas the cloud will need to be capable of storing, processing, and managing this massive amount of data. Additionally, when it comes to monitoring the physiological condition of a patient, the data that is rapidly generated by the various sensors will require a swift analysis, as well as real-time response (Wang and Ranjan, 2015; Skourletopoulos et al., in press).

Finally, it should be noted that IoT devices are developed from a number of different vendors, some of them using different standards, operations, and communication protocols, but they will be connected to the same cloud, which is once again developed by someone else. This means that the services provided by the cloud, as well as the data exchange between cloud and devices, should be agreed on and be supported by all the participants of the network, enabling this way interoperability between the differently designed devices that will take part in an IoT cloud system (Truong and Dustdar, 2015).

REFERENCES

Batalla, J.M., Kantor, M., Mavromoustakis, C.X., Skourletopoulos, G., Mastorakis, G., 2015. A novel methodology for efficient throughput evaluation in virtualized routers. In: Proc. of IEEE International Conference on Communications 2015 (IEEE ICC 2015). London, UK, 08–12 June 2015, pp. 6899–6905.

Bolkhovsky, Jeffrey B., Scully, Christopher G., Chon, Ki H., 2012. Statistical analysis of heart rate and heart rate variability monitoring through the use of smart phone cameras. In: Engineering in Medicine and Biology (EMBC), 2012 Annual International Conference of the IEEE, pp. 1610–1613.

Borges, Helder Pereira, de Souza, Jose Neuman, Schulze, Bruno, et al., 2012. Automatic generation of platforms in cloud computing. In: Network Operations and Management Symposium (NOMS). IEEE, pp. 1311–1318.

Chen, Min, Zhang, Yin, Li, Yong, et al., 2015. AIWAC: affective interaction through wearable computing and cloud technology. IEEE Wirel. Commun.

Cheng, Shih Heng, Huang, Ching Yao, 2013. Coloring-based inter-WBAN scheduling for mobile wireless body area networks. IEEE Trans. Parallel Distrib. Syst. 24 (2).

Ciobanu, Nicolae-Valentin, Comaneci, Dragos-George, Dobre, Ciprian, Mavromoustakis, Constandinos X., Mastorakis, George, 2014. OpenMobs: mobile broadband Internet connection sharing. In: Proceedings of the 6th International Conference on Mobile Networks and Management (MONAMI 2014). September 22–24, 2014, Wuerzburg, Germany.

Ciocan, Mihai, Dobre, Ciprian, Mavromoustakis, Constandinos X., Mastorakis, George, 2014. Analysis of vehicular storage and dissemination services based on floating content. In: Proc. of International Workshop on Enhanced Living EnvironMENTs (ELEMENT 2014), 6th International Conference on Mobile Networks and Management (MON-AMI 2014). Wuerzburg, Germany, September 2014.

Fahim, Muhammad, Lee, Sungyoung, Yoon, Yongik, 2014. SUPAR: smartphone as a ubiquitous physical activity recognizer for u-healthcare services. In: Engineering in Medicine and Biology Society (EMBC), 2014 36th Annual International Conference of the IEEE, pp. 3666–3669.

Jin, Zhanpeng, Chen, Yu, 2015. Telemedicine in the Cloud Era: Prospects and Challenges. IEEE CS.

Kryftis, Yiannos, Mavromoustakis, Constandinos X., Batalla, Jordi Mongay, Mastorakis, George, Pallis, Evangelos, Skourletopoulos, George, 2014. Resource usage prediction for optimal and balanced provision of multimedia services. In: Proceedings of the 19th IEEE International Workshop on Computer-Aided Modeling Analysis and Design of Communication Links and Networks (IEEE CAMAD 2014). Athens, Greece.

Kryftis, Yiannos, Mavromoustakis, Constandinos X., Mastorakis, George, Batalla, Jordi Mongay, Chatzimisios, Periklis, 2015. Epidemic models using resource prediction mechanism for optimal provision of multimedia services. In: 2015 IEEE 20th International Workshop on Computer Aided Modeling and Design of Communication Links and Networks (CAMAD) — (IEEE CAMAD 2015). 7–9 September 2015. University of Surrey, Guildford, UK/General Track, pp. 91–96.

Kryftis, Yiannos, Mastorakis, George, Mavromoustakis, Constandinos X., Batalla, Jordi Mongay, Pallis, Evangelos, Kormentzas, Georgios, 2016. Efficient entertainment services provision over a novel network architecture. IEEE Wirel. Commun 1, 14–21.

Lee, Eun-Kyu, Gerla, Mario, Oh, Soon Y., 2012. Physical layer security in wireless smart grid. IEEE Commun. Mag.

Mastorakis, George, Trihas, Nikolaos, Mavromoustakis, Constandinos X., Perakakis, Emmanouil, Kopanakis, Ioannis, 2014. A cloud computing model for efficient marketing planning in tourism. Int. J. Online Mark. 4 (3), 14–30.

Mavromoustakis, Constandinos, Andreou, Andreas, Mastorakis, George, Bourdena, Athina, Batalla, Jordi Mongay, Dobre, Ciprian, 2014a. On the performance evaluation of a novel offloading-based energy conservation mechanism for wireless devices. In: Proceedings of the 6th International Conference on Mobile Networks and Management (MONAMI 2014). September 22–24, 2014, Wuerzburg, Germany.

Mavromoustakis, Constandinos X., Bourdena, Athina, Mastorakis, George, Pallis, Evangelos, Kormentzas, Georgios, Dimitriou, Christos D., 2015. An energy-aware scheme for efficient spectrum utilization in a 5G mobile cognitive radio network architecture. In: Special Issue on Energy Efficient 5G Wireless Technologies. Telecommun. Syst 59 (1), 63–75.

Mavromoustakis, Constandinos X., Mastorakis, George, Bourdena, Athina, Pallis, Evangelos, Kormentzas, Georgios, Dimitriou, Christos, in press-a. Joint energy and delay-aware scheme for 5G mobile cognitive radio networks. In: Proceedings GlobeCom 2014, Track GlobeCom 2014 — Symposium on Selected Areas in Communications: GC14 SAC Green Communication Systems and Networks — GC14 SAC Green Communication Systems and Networks. Austin, TX, USA, 8–12 December 2014, pp. 2665–2671.

Mavromoustakis, Constandinos X., Mastorakis, George, Bourdena, Athina, Pallis, Evangelos, 2014b. Energy efficient resource sharing using a traffic-oriented routing scheme for cognitive radio networks. IET Netw. 3 (1), 54–63 (IEEE DL).

Mavromoustakis, Constandinos X., Mastorakis, George, Bourdena, Athina, Pallis, Evangelos, Kormentzas, Georgios, Rodrigues, Joel J.P.C., in press-b. Context-oriented opportunistic cloud offload processing for energy conservation in wireless devices. In: Globecom 2014 — Cloud Computing Systems, Networks, and Applications — Globecom 2014 Workshop — The Second International Workshop on Cloud Computing Systems, Networks, and Applications (CCSNA). GC14, Austin, TX, USA, 8–12 December 2014, pp. 24–30.

Mavromoustakis, Constandinos X., Mastorakis, George, Papadakis, Stelios, Andreou, Andreas, Bourdena, Athina, Stratakis, Dimitris, 2014c. Energy consumption optimization through pre-scheduled opportunistic offloading in wireless devices. In: The Sixth International Conference on Emerging Network Intelligence (EMERGING 2014). August 24–28, 2014, Rome, Italy, pp. 22–28.

Memon, Mukhtiar, Rahr, Stefan, Pedersen, Christian Fischer, et al., 2014. Ambient assisted living healthcare frameworks, platforms, standards, and quality attributes. Sensors 2014, 4312–4341.

Mertz, Leslie, 2012. Ultrasound? Fetal monitoring? Spectrometer? There's an app for that! IEEE Pulse 3 (2).

Papanikolaou, Katerina, Mavromoustakis, Constandinos X., Mastorakis, George, Bourdena, Athina, Dobre, Ciprian, 2014. Energy consumption optimization using social interaction in the mobile cloud. In: Proc. of International Workshop on Enhanced Living EnvironMENTs (ELEMENT 2014), 6th International Conference on Mobile Networks and Management (MONAMI 2014). Wuerzburg, Germany, September 2014.

Pop, Cristian, Ciobanu, Radu, Marin, Radu C., Dobre, Ciprian, Mavromoustakis, Constandinos X., Mastorakis, George, Rodrigues, Joel J.P.C., in press. Data dissemination in vehicular networks using context spaces. In: IEEE GLOBECOM 2015, Fourth International Workshop on Cloud Computing Systems, Networks, and Applications (CCSNA).

Shen, Qinghua, Liang, Xiaohui, Shen, Xuemin (Sherman), et al., 2013. RECCE: a reliable and efficient cloud cooperation scheme in E-healthcare. In: Global Communications Conference (GLOBECOM). IEEE, pp. 2736–2741.

Sheng, Xiang, Xiao, Xuejie, Tang, Jian, et al., 2012. Sensing as a service: a cloud computing system for mobile phone sensing. In: IEEE Sensors.

Skourletopoulos, Georgios, Mavromoustakis, Constandinos X., Mastorakis, George, Rodrigues, Joel J.P.C., Chatzimisios, Periklis, Batalla, Jordi Mongay, in press. A fluctuation-based modelling approach to quantification of the technical debt on mobile cloud-based service level. In: IEEE GLOBECOM 2015, Fourth International Workshop on Cloud Computing Systems, Networks, and Applications (CCSNA). December 6–10.

Stallings, William, 2010. Cryptography and Network Security: Principles and Practice, 5th edition. Prentice Hall Press, pp. 19–22.

Truong, Hong-Linh, Dustdar, Schahram, 2015. Principles for engineering IoT cloud systems. IEEE Cloud Comput.

Uddin, Ammad, 2015. A Survey of Challenges and Applications of Wireless Body Area Network (WBAN) and Role of a Virtual Doctor Server in Existing Architecture. IEEE Computer Society.

Wang, Lizhe, Ranjan, Rajiv, 2015. Processing distributed Internet of things data in clouds. IEEE Cloud Comput.

Wang, Xiaoliang, Gui, Qiong, Liu, Bingwei, et al., 2014. Enabling smart personalized healthcare: a hybrid mobile-cloud approach for ECG telemonitoring. IEEE J. Biomed. Health Inform. 18 (3), 739–745.

CHAPTER 12

Adaptive Workspace Interface for Facilitating the Knowledge Transfer from Retired Elders to Start-up Companies

Tudor Cioara*, Ionut Anghel*, Dan Valea*, Ioan Salomie*,
Victor Sanchez Martin†, Alejandro García Marchena‡, Elisa Jimeno‡,
Martijn Vastenburg§
* *Technical University of Cluj-Napoca, Romania*
† *Eindhoven University of Technology, Netherlands*
‡ *Ingeniería Y Soluciones Informáticas del Sur S.L., Spain*
§ *ConnectedCare R&D Office Arnhem, Netherlands*

12.1 INTRODUCTION

Recent studies have shown that the lack of knowledge is the leading cause of small business failures, 46% of new businesses failing due to lack of marketing and managerial knowledge (Kelly, 2013). At the same time EU labor force has 62.2 million persons aged over 50 in 2010 and over 40% of them are decision makers in their companies having extremely valuable knowledge that after their retirement will be lost (Borg et al., 2011). Europeans are retiring early on average at 61 years due to reasons such as health (a sedentary job can contribute to chronic diseases), caring for family, freedom to pursue dreams and goals that cannot be satisfied while having a full time job, difficulty to learn or adapt to new technologies or the fact that their experience is not properly recognized by the youngers (Vendramin et al., 2012). Thus their experience and knowledge is lost, not providing any economic value after their retirement. For many older adults the retirement is a difficult decision because their jobs represent a way of feeling useful for themselves and the society, keeping them active and motivated. Moreover, 41% of people aged over 55 tend to be keener on working beyond the age at which they are entitled to a pension (Active Ageing, 2012). Older adults' job performance decreases around the age of 50, this being more visible in tasks involving problem solving, learning, and speed (Neves and Amaro,

Ambient Assisted Living and Enhanced Living Environments.
DOI: http://dx.doi.org/10.1016/B978-0-12-805195-5.00012-0

2012), while in tasks where experience is essential, they maintain a high productivity level. In our vision the recent advancements in the ICT (Information and Communications Technology) sector may help older adults to stay active beyond their retirement age, while enjoying the benefits and flexibility of the work schedule given by retirement, thus helping them to remain economically active and motivating them to continue to be involved in the workforce thus promoting their lifelong workability.

Our approach is to benefit from elders experience and try to transfer this experience to younger and more inexperienced people by means of developing a web-based interface which gives them the flexibility they need to work anytime and anywhere (at home, in the park, on the road, etc.). The main problem with this approach is that usually the elders remain excluded from technology, since they consider traditional web based interfaces as overly technical and difficult to use. For example the decline of memory capacity with age causes a slowdown in processing speed, and reduction of spatial abilities (Eckert, 2011) this is why the elders experience disorientation when navigating through menu structures of web applications.

Adaptive user interfaces involve changing the UI based on a certain profile or characteristics, thus is a promising approach to facilitate different skill levels and limitations of computer users. Two different types of user interface adaptations are listed in the state of art literature: *adaptable interfaces* and *adaptive interfaces*.

Adaptable interface allows the user to modify the interface settings and customize them to his/her preferences. Due to difficulties imposed by ageing the classical strategy of rendering the application interface as simple as possible and provide a lower common denominator for all users doesn't work for older adults. They are not keen to customize manually the interface to their needs and compared to the average computer users, older people are considerably different regarding their motor skills.

Adaptive interfaces aim at automatically accommodating and personalizing the presentation of the user interface (UI) to the characteristics of its user, in different ways so that the user can interact with it in a better and comfortable manner than he/she would do it by default. Usually, the user characteristics considered for adaptation are the user's profile and preferences, capabilities and experience and are acquired by means of monitoring. Nevertheless, for certain environments (Ambient Assisted Living, Smart/Intelligent/Pervasive Environments, etc.) context characteristics and device's capabilities are highly considered. There are two main perspectives user interface adaptation: changing the interface's layout and features and changing the content displayed by the interface. The first perspective on user interface adaptation is based on changing the interface features or its default configuration. For example considering the user previous interaction and experience, the interface may dynamically display the buttons the user interacts with mostly. Also in a very bright environment smartphones' displays tend to increase the brightness of their screens automatically to avoid problems with sunlight. In the first example the platform is aware of the user experience with the presented controls and it makes the corresponding adaptation based on these interactions. In the second one, a certain previously met context situation triggers the adaptation. The second perspective on user interface adaptation deals with the interface content adaptation. Taking into account social characteristics of the user, age, religious considerations, etc. may result in filtered content for him/her.

In this context our solution is to bring the valuable experience of elderly to start-ups and small companies, addressing intergenerational knowledge transfer of skills and competencies based on experience. One of the main challenges in reaching this goal is to develop an adaptive workspace interface customized to the elders' specific needs which enables them to provide counseling and support. The interaction of the elders with the workspace will be monitored to:

- learn the user model from traces of interaction with workspace interface;
- adapt the interface and content of the workspace to the conditions of the user by applying the necessary changes to provide them an easy interaction and usage.

The implementation and usage of an adaptive interface for elders specific needs will:

- engage and motivate the elderly in optimal collaboration in the work team for a prolonged period of time;
- protect the elderly from failing into apathy and frailty after their retirement by providing them the means to keep their minds active and to transfer their valuable knowledge.

The rest of the paper is organized as follows. Section 12.2 discusses related work. The proposed adaptive workspace architecture is detailed in Section 12.3. Section 12.4 presents the adaptation features, Section 12.5 focuses on how the data collected from sensors is modeled, Section 12.6 discusses the adaptation decision making process while Section 12.7 shows a usage scenario and results. Finally, Section 12.8 concludes the paper.

12.2 RELATED WORK

Adaptive interfaces are designed and developed by considering different aspects for achieving adaptation, out of which the most important are (See, 2015): human behavior and user preferences. Both aspects involve acquiring information about individual users which can either be implicitly or explicitly obtained (Jason et al., 2010). After data is gathered, an inference engine can use it to identify possible adaptation ways or actions. The adaptation can take place on different levels (Jameson, 2012): information, presentation, user interface and functionality. Regarding the benefits of adaptive user interfaces few studies exists in the state of the art literature. In Lavie and Meyer (2010) the authors show that in routine tasks, which is the case for elders, the fully adaptive systems are beneficial and their usage has as result shorter performance time and reduced lane deviations.

The most tailored approach for adaptation in the research literature implies defining a user model to represent the gathered data in a computer interpretable manner. Model-based approaches have been used in the past to reduce the complexity of systems design and are a promising solution for building adaptive user interfaces. Three types of user models are identified in the literature (Mejía et al., 2012): static user model (user interaction with the system is not monitored (Hothi and Hall, 1998)), dynamic user model (the way the user interacts with the system is monitored and the model is updated (Johnson and Taatgen, 2005)) and user models based on statistical information (Chen and Magoulas, 2005). At the same time adaptation can be enabled at design-time or at run-time.

Design-time adaptation approaches rely on generating different adapted user interfaces based on predefined versions with different levels of abstraction (templates). Different methodologies define possible variations of the user interface at design time and the possibility to select one of them at run-time given a specific condition (Limbourg et al., 2004; Mori et al., 2004; Sadiqa and Pirhonen, 2011). A static user model representing user's characteristics from medicine and sociology domains that can be used by an adaptive software interface is discussed in Mejía et al. (2012). The authors classify the user characteristics represented in the model as: physical, demographic, experience, cognitive and psychological. The model is expressed as UML diagrams and used as basis for developing a CRM application. Biswas et al. (2005) propose a static user model for developing software for impaired users. The model is used for clustering users according to their physical disability and cognitive level. Context-aware adaptive modeling is addressed in Feng and Liu (2015) for mobile services considering user knowledge, physiology and inclination. Based on the proposed model the authors guide the interface adaptation process by using design time defined scenarios. In Reinecke and Bernstein (2011) the authors propose adaptive systems which automatically generate personalized interfaces by considering cultural preferences of users. The approach builds a user model based on cultural particularities and various cultural influences which is used to adapt the interface of a to-do list online tool. Actions for adaptation include information density, navigation, page structure, colors, guidance and support. A method for integrating adaptive interface behavior in enterprise applications is the scope of Akiki et al. (2014). The method is based on CEDAR (Akiki et al., 2012), a model-driven service-oriented, architecture for developing adaptive enterprise applications UIs. The adaptation actions consist from removing features that are not required by users and by adapting different widget properties such as: size, location, type, etc.

Run-time adaptation approaches update and use the models for adapting the UI dynamically. A dynamic model for adaptive interfaces applied to the domain of contact centers is proposed in Jason et al. (2010). The authors represent in the user model features derived from motion characteristics (mouse velocity and acceleration) and features related to the interaction technique (average dwell time, nr of items visited, selection time, etc.). They also define task models to represent tasks that the user can perform with the system at a particular time. The usage of multiple related models at runtime to reflect different aspects of the user interaction and create multimodal and context-sensitive user interfaces for adaptive applications is proposed in Blumendorf et al. (2010). The authors consider models for user profile, environment and user tasks representation. In Gajos et al. (2010) a system for generating UIs that are automatically adapted to each user's motor abilities is proposed. The approach considers simple UI dialogs for adaptation and supports user feedback for the adaptation rules. Clerckx et al. (2005) describes a design process and architecture for developing context-aware interfaces for a tourist guiding mobile application. In Nivethika et al. (2013) the authors propose a conceptual prototype framework for mobile applications to make the user interfaces adaptive to the user by learning his/her history of interactions with applications. Historical data regarding user interaction was used as training sets for the inference engine. Similarly in Jain et al. (2013) a framework to adapt the user interface of smartphones or tablets, based on the context or scenario in which user is present, and incorporating learning from past user actions is detailed. The adaptation actions focus on application icons, menus, buttons window positioning or layout and color scheme. The user profiles are inputs to a machine learning algorithm which predicts the best possible screen configuration with respect to the user context. A multimodal interface architecture that is capable of automatic adaptation to the user by learning from the user's interaction patterns is the subject of Garzon and Cebulla (2010). Adaptations are defined as modifications of models which are specified during design time and at runtime. The learning process is based on collecting user logs, analyzing them to determine the user behavior and the adaptation rules through an AprioriAll algorithm. In Bruninx et al. (2007) the authors combine user and device models with a high-level User Interface Description Language (UIML) for dynamically adapting user interfaces. To model the user and device profiles the authors use ontologies as input for a reasoning process which generates adaptation actions such as color modification, content presentation or font scaling. Shakshuki et al. (2015) proposes a multi-agent system for monitoring patients health data. Based on the users' behavior, agents within the system use a reinforcement learning technique to dynamically generate an adaptive user interface.

12.3 ADAPTIVE WORKSPACE ARCHITECTURE

Figure 12.1 presents the architecture of the adaptive workspace system for elders' to carry out tasks for companies and to transfer their experience and knowledge during this process.

The web user interface of the system is the **Interface Collaborative Adaptive Workspace (ICAW)** which will support counseling processes and facilitate the online interaction between elderly and small companies. To remove barriers for seniors to use the proposed workspace, the developed web interface will automatically assess the limitations that come with ageing in a personalized manner. The senior will be able to activate automatic adaptation, but can also select a semi-automatic adaptation mode in which the senior has to confirm all proposed changes in the interface. The interaction of the elder with the system is monitored and data is collected by means of sensors or self-reporting questionnaires (**Elders Interface Interaction Monitoring**). The acquired data is stored in a relational database, the **UI Interaction DB**. Based on the monitored data various potential senior's limitations are determined and used then to take adaptation decisions to personalize the **ICAW** web page and to improve the interface usability

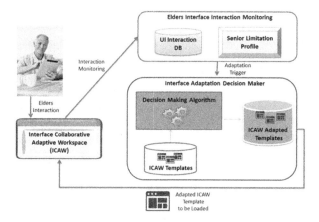

Figure 12.1 High-level architecture design for the Adaptation Decision Maker.

(**Interface Adaptation Decision Maker**). The Decision Making Algorithm implements a genetic evolutionary algorithm for finding the optimal combination of features for the **ICAW** matching the elder's impairment profile and monitored data. The decision making starts from predefined **ICAW** templates and using the operators of mutation and cross–over, optimal **Adapted ICAW Templates** are generated. During the next login, the elder is asked if he/she wants to load the new adapted interface template inferred by the decision making algorithm.

12.4 WORKSPACE ADAPTATION FEATURES

The Interface Collaborative Adaptive Workspace is the main web page with which the senior interacts to complete the jobs agreed with a company or to pass the relevant knowledge/experience. A number of adaptation features have been selected by a team of design experts, and are considered during the adaptation process to increase the web interface usability by personalizing it to elders' limitations.

The first adaptation feature considered is the workspace web page *overall layout*. The workspace web page is based on a predefined set of layout templates that ensures the appropriate responsiveness of the user interface to the elders' limitations inferred out of his/her interaction. Figure 12.2 presents the set of elders' layout defined and used in adaptation.

When selecting among the various layout temples defined the adaptation decision making algorithm needs to take into account the following constraints:

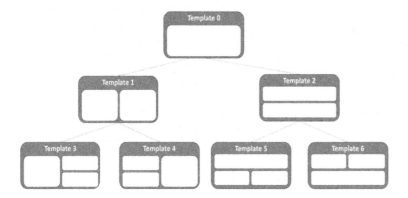

Figure 12.2 Workspace layout templates.

- Maintain focus on the primary content in all templates. The elders that use the workspace want to interact with the content they feel is relevant for the job they are currently carrying out.
- The most relevant part of the web page is the upside left part so there the relevant content needs to be always displayed.
- The new content brought by changing the layout must help older adults in better using and understanding the primary content.

Another adaptation feature considered is the *font of the web page content* which size and style can be changed. The following constraints are considered:

- When vision problems are detected larger fonts should be used (preferable 16 point and above).
- Italics and capitalized letters should be avoided, as well as decorative fonts.
- Use of colors for headings and emphasis are not recommended for people with low vision.
- The contrast is a fundamental factor for increasing the web page content readability. High contrast levels are recommended.

Older adults are usually dealing with web page navigation problems. Thus the *web interface menu and navigation among pages* are considered in adaptation decisions. The menu should be placed in the right side of the screen (since the important content is placed in the left side) while the most accessed navigation buttons should be dynamically promoted towards the top of the screen. Also the visual size of menu items is used to show their importance on the screen. The elders tend to get lost in navigation due to age related cognitive problems, thus the recent pressed buttons, the back or home buttons need to be emphasized.

An important adaptation feature is to carefully *choose between text and icon display* based on the elders' cognitive limitation and his/her profile. One important adaptation principle is that icons should be used as much as possible due to the fact that relevant research shows that they usage improve the productivity and reliability of work. At the same time the icons can bring a certain level of ambiguity and in this case the label text may be used to disambiguate mainly during the first phases of learning to use the workspace. Regarding elders cognitive limitations the icons have proven to be efficient for short term memory problems (Sanchez, 2012) and cognitive confusion.

Regarding content of the workspace interface we have defined five main content panels: Jobs Description, Chat Messages, Appointments, File Sharing, and Working Area. The appropriate content will be dynamically selected and mapped on the workspace layout template considering the elder's profile, preferences, selected job and device capabilities.

12.5 SENSOR DATA COLLECTION AND LIMITATIONS PROFILE

Our innovative approach is based on monitoring the elders' interaction with the workspace (Sensor Data Collection module) to adapt the interface layout and content to their cognitive conditions by applying the necessary changes to provide them an easier interaction and increased usability as well as to engage and motivate the elderly in optimal collaboration for a prolonged period of time. A set of sensors (e.g. touchscreen, webcam, microphone and others) will be used to gather information about the way elders interact with the workspace aiming at inferring information about the cognitive abilities and engagement level such as potential impairments that decrease the interaction quality, problems to focus on the screen, difficulties when reading what it is shown in the screen, problems to manage the keyboard or the items on the screen, etc.

Analyzing the state of the art approaches we have identified potential age related problems regarding vision, hearing, memory, attention, coordination and locomotors.

There are several **vision related problems** that are common for older people and may impact the interaction with our system such as: reduced peripheral vision, difficulty to read letters and color perception difficulty.

The *peripheral vision* of a person can be defined as the skill to localize, recognize and respond to the information in different areas of the visual range around the main object in which the attention is centered. To identify such problems the Tangent Screen test is conducted (Martel, 2016). During this test the elder sits three feet away from the screen with a target in the center and he/she is asked to press a button when an object moving into the peripheral view is seen. The *limitations regarding letters lecture* are assessed based on calculating the distance between the device and the user. To achieve this, a proximity sensor and square placed on the user's face are used. The closer to the device the user is, the bigger the square will get and the shorter the distance will be. This will make it possible to identify the existent distance. This distance measurement has to be computed according to the device used, computer or mobile phone. The luminosity level plays an important role in this process, so it will be collected with a specific sensor. The *elders' correct perception of color* is evaluated using the Ishihara test which consists in a number of colored plates containing circle of dots appearing randomized in color and size (Color Blindness). Within the pattern dots form a number or shape clearly visible to those with normal color vision and invisible or difficult to see for those with color vision problems.

Hearing loss in ageing may be present in various forms but the most common ones considered by us are: difficulty to hear high-pitched sounds, difficulty to hear specific frequencies and general hearing loss. The evaluation is conducted using a technique called pure tone audiometry (Kutz, 2015). An audiometer plays a series of tones through headphones which vary in pitch (frequency in hertz) and loudness (intensity in decibels). The evaluation of the possible difficulty to differentiate high-pitched sounds is made by using questions that form the specific hearing questionnaire. These questions have to do with high-pitched sounds, for instance a baby crying, a recorder melody or the twit of birds. Regarding *ambient noise*, diverse noise grades can be detected. Each of these grades contains a range of decibels. For example, it can be observed that sounds from a library, soft rain or a conversation have acceptable values. On the other hand, sounds like the ones from trucks, airplanes taking off or drills lead to a high hearing risk, as they are very often present in daily life. Because of this reason, many people suffer nowadays from a high hearing deficit in situations where the ambient noise is relatively high. Then, a set of questions regarding the noise level are used to evaluate the ambient. That makes it necessary to have a microphone in the device, rather PC or smartphone, to track the ambient noise during the time. Other problem to analyze is the possible *difficulty in* *differentiating sounds in various frequencies.* To carry out a proper diagnosis, a set of questions dealing with hearing audios and/or sounds was made for the specific hearing questionnaire. Objects like for instance drums or thunders were considered deep sounds, while the fall of the water or whistling were catalogued as high-pitched sounds. At last, the evaluation of the *global hearing loss* is made by detecting the volume of the device in use. The higher volume is, the higher the hearing loss will be.

Elders' cognition decline is reflected in problems regarding attention deficit or **memory problems**. The evaluation of the patient's attention will focus on a series of quiz questions that measure if the elders is easily distracted or leaves hard tasks unfinished. With respect to the age related memory diagnosis, there are two main problems to be detected: short and long term memory problems. *Short term memory* is considered an important skill for daily life of any person, since it is indispensable in any situation to remember where the person has put any particular object or why he/she is heading to a certain location. Such capacity is determined through the realization of questions related to day-to-day deeds. *Long term memory* is evaluated through questions about things that have occurred years ago. The questions correspond to moments of the youth or questions regarding of relatives.

Another aspect to be considered is related to **elders' movement precision (coordination) and speed**. *Movement precision* of the user can be estimated through the analysis of the tremors of the hands when doing some task with the mouse or touchscreen. We have used statistical software to analyze the type of mouse movement the elders does in a determined period of time. This way it was observed that the line obtained in healthy people is straight while in people with tremors this line presented waves greater or lesser depending on the level of tremors. The elders' *movement speed* is evaluated by means of parameters like the number of buttons pressed per minute or the number of clicks considering sensors such as mouse, keyboard, etc.

Table 12.1 presents a summary of the main identified age related limitations that impact the elders in using the envisioned workspace interface and must be considered during the adaptation processes in order to automatically personalize the web interface for elder needs and as a result to increase overall system usability.

A **Senior Limitations Profile** is constructed (see Figure 12.3), represented by means of ontologies and used for semantic modeling of elders'

Table 12.1 Workspace layout templates

Type of condition	Subtype	Limitation level	Data source	Assessment technique
Perception	Hearing	Pitch sound discrimination	Audiometer, microphone, questioners	Pure tone audiometry
		Environmental noise		
		Hearing loss		
	Vision	Peripheral vision reduced	Application with object moving into peripheral view	Tangent screen test
		Letter discrimination decline	Proximity sensor	Distance between the user and device
		Color discrimination decline	Ishihara plates application	Ishihara test score
Cognition	Memory	Long memory problem	Questioners	Memory quiz
		Short memory problem		
	Attention	Attention deficit		Attention span test
Physical	Locomotors	Coordination loss	Mouse, touchscreen	Statistical analysis software
		Movement speed loss	Mouse, keyboard	No. of clicks per minute

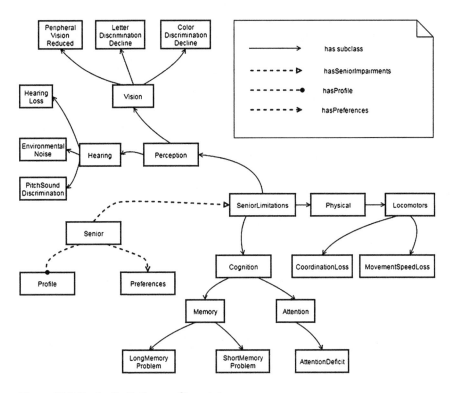

Figure 12.3 Senior limitation profile ontology.

limitations and to serve as a base for the decision making algorithm responsible for adapting the interface.

12.6 ADAPTATION DECISION MAKING

The Adaptation Decision Maker goal is to automatically decide and select the best configuration (both layout and content) for presenting the information to the elderly when using and interacting with the workspace. The *workspace layout adaptation* is referring to adjusting template features such as font size, style, window splitting, items organization on the screen, menus adjustments, page item size, etc. The *workspace content adaptation* addresses users' preferences e.g. for literacy or numeracy, reduced or more content, multi-modal output (audio or video instead of text), etc. Constructing and selecting the optimal workspace layout and content adapted template to match the elders' cognitive capabilities and limitations is not a trivial task and cannot be addressed using conventional search and deci-

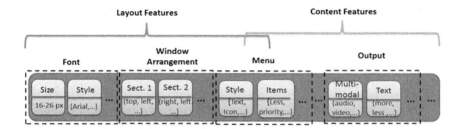

Figure 12.4 Workspace template chromosome representation.

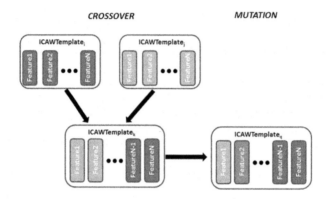

Figure 12.5 Mutation and cross over operators on templates.

sion making techniques. The optimization problem is NP-hard requiring a greedy search strategy capable of identifying the optimal or near optimal adapted template in reasonable time frame. To solve this optimization problem we have used a genetic algorithm. We have defined a chromosome as a collection layout and content features representing a workspace template and each feature represents a gene having a predefined range of potential valid values (see Figure 12.4). As a consequence, the chromosomes are modeled as a vector of templates adaptation features determining the dimension of the solution search space.

The initial population of chromosomes is randomly generated and then is evolved using single point crossover and random mutations (see Figure 12.5). A fitness function was defined and used at each algorithm step to evaluate the population chromosomes in terms of feasibility as a template and usability considering the matching degree with elders' limitations.

The best chromosomes scoring the highest values of fitness function is selected at each step, as shown in Equation (12.1):

$$F(Template) = 100 - \alpha_1 F_a(Menu) - \alpha_2 F_f(Font) - \alpha_3 F_t(Views) \tag{12.1}$$

where:

$$\alpha_1 + \alpha_2 + \alpha_3 = 1, \quad \alpha_1, \alpha_2, \alpha_3 \in [0, 1] \tag{12.2}$$

$$F_a(Menu) = |Abstraction_{recommended} - Abstraction_{generated\_menu}| \tag{12.3}$$

$$F_f(Font) = |Font_{recommended} - Font_{generated}| \tag{12.4}$$

$$F_t(Views) = \sum_{View_i \in Views} S(view_i) \tag{12.5}$$

$$S(View_i) = \frac{\sum_{\substack{section_{ij} \in View_i \\ page_j \in pages}} score(section_{ij}, page_j)}{nr\_of\_sections(View_i)} \tag{12.6}$$

$$score(section_{ij}, page_j) = \frac{\sum_{k=1}^{n} |P[k](section_{ij}) - P[k](page_j)|}{n} + \left| \frac{x^2 - t^2}{x^2 + t^2} \right| \tag{12.7}$$

In Equation (12.7) the following parameters are used:

- $P[k](e) \in \{importance, size, orientation\}$ — property k of element e
- $Pages = \{Requests, Messages, Appointments, File Sharing\}$
- n — number of properties for $section_{ij}$
- x — maximum number of elements which can be displayed in a section
- t — number of elements to be display

The pseudo code description of the adaptation decision making algorithm is presented in Figure 12.6.

12.7 USAGE SCENARIO AND RESULTS

Marta is a recently retired French woman. She worked for years in the legal department of a company. Marta is feeling bored and has nothing interesting to do despite she has plenty of time; she is falling into apathy and frailty after retirement and nothing is motivating her as in the days she was working. Marta was a hard worker and liked her work very much but unfortunately her company has already found a younger employee to take over her duties. Lately, Marta was suffering from some stress when dealing with technology and computers at work. That affected her productivity and, even she wanted to keep working at the company, she cannot compete with the efficiency of the younger when managing ICT tools, and

```
ALGORITHM: Adaptation Decision Making Algorithm
Input: SeniorProfile, ICAWTemplates, popSize
Output: ICAWAdaptedTemplates
Begin
   t = 1
   Population(t) = Randomly_Generate_Population(popSize, SeniorProfile, GUITemplates)
   Evaluate (Population(t))
   while (stopping condition not satisfied) do
   ChildPopulation = {}
   for i=1 to popSize do
      childSolution = Crossover(Population(t).get(i), Population(t).get(i+1))
      childSolution = Mutation(childSolution)
      ChildPopulation.add(childSolution)
   end foreach
   Evaluate(ChildPopulation)
   Population(t+1) = Select_Best_Solutions(Population(t), ChildPopulation)
   t = t+1
   end while
   return Get_Highest_Fitness(Population)
End
```

Figure 12.6 Adaptation decision making genetic algorithm.

she sadly accepted retirement. Marta knows that having an active life after 55 years old will improve their cognitive status. After weeks looking for activities to enroll for occupying her time, she discovered the adaptive workspace system and selects a job posted by a technological small company from Spain which aimed at commercializing its products by selling it to different European countries. The company didn't have enough knowledge on France regulations to commercialize products so Marta can provide legal support. After accepting this job Marta was invited to participate in a work team of three more people, a senior lawyer experienced in legal regulations in France, a women expert in business models and market analysis and a man who worked for years in e-commerce. Now Marta is using the Interface Collaborative Adaptive Workspace web page for passing her knowledge and communicates with other members of the team by chatting, videoconferencing or simply talking. Based on the tablet sensors she is using, the interaction with the system is tracked and Marta's age related limitations are determined. The rest of this section presents the adaptation decision process results (Figures 12.7–12.12) and the inferred Marta's limitations (Tables 12.2–12.5) in four different scenarios.

Table 12.2 User interface adaptation — Scenario 1

User interface adaptation action	Adaptation goal	Limitation addressed
Mix two/three pages into a single one and create a menu button for quick access to the new generated page.	To ease the elder's access to different content simultaneously in the workspace.	Coordination Loss Problem

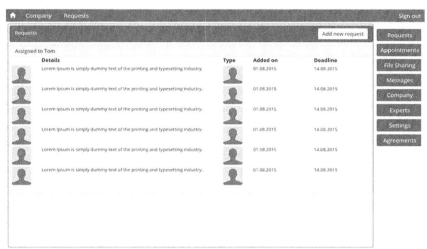

Figure 12.7 ICAW before adaptation — Scenario 1.

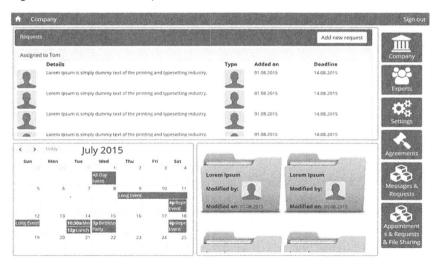

Figure 12.8 ICAW after adaptation — Scenario 1.

Table 12.3 User interface adaptation — Scenario 2

User interface adaptation action	Adaptation goal	Limitation addressed
Change user interface background and text color.	To avoid situation when text and background color are similar or generate confusion.	Color discrimination declines with loss of blue/green vision.

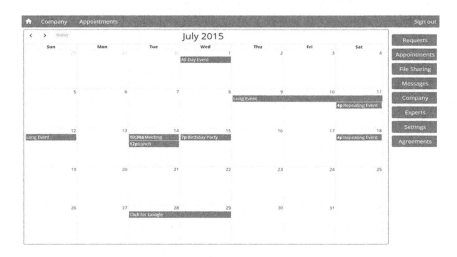

Figure 12.9 ICAW before adaptation — Scenario 2.

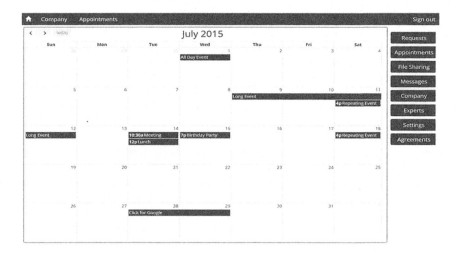

Figure 12.10 ICAW after adaptation — Scenario 2.

Table 12.4 User interface adaptation — Scenario 3

User interface adaptation action	Adaptation goal	Limitation addressed
Adapt font family and size.	Increase the text readability.	Letter discrimination declines.

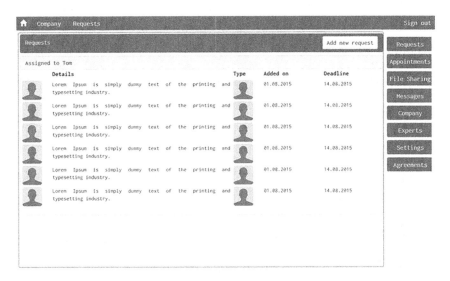

Figure 12.11 ICAW after adaptation of Figure 12.7 version — Scenario 3.

Table 12.5 User interface adaptation – Scenario 4

User interface adaptation action	Adaptation goal	Limitation addressed
Add/replace images with text.	Use images to help elder to recall navigation steps.	Short term memory problem.

12.8 CONCLUSIONS

In this paper we have proposed a decision making and adaptation technique to enact the automatic generation of a personalized elders workspace according to their cognitive profile and impairments. This will increase the engagement and motivation of elders while protecting them from fail into apathy and frailty after their retirement by providing the means to keep their minds active and to transfer their valuable knowledge to small companies. Our innovative approach is based on monitoring the elders' interaction with the workspace to adapt the interface layout and content

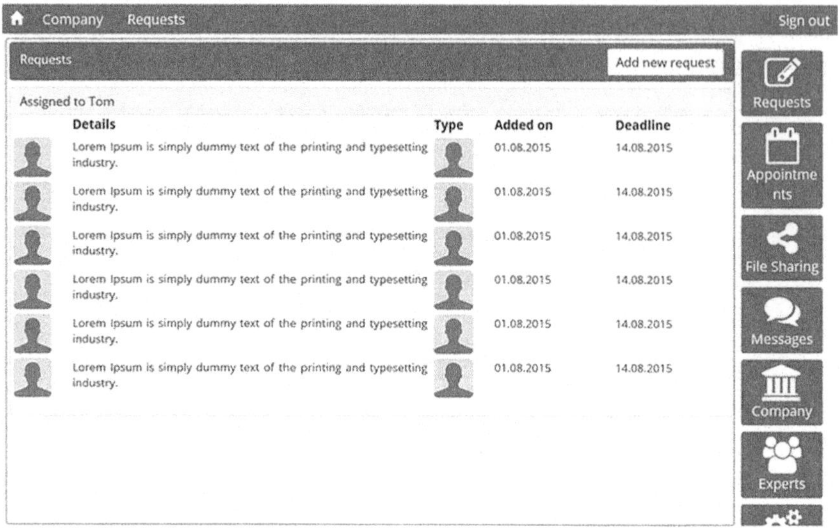

Figure 12.12 ICAW after adaptation of Figure 12.7 version — Scenario 4.

to their cognitive conditions by applying the necessary changes. A set of sensors are used to gather information about the way elders interact with the workspace aiming at inferring information about the cognitive abilities and engagement level. A Senior Limitations Profile is constructed, represented by means of ontologies and used for semantic modeling of elders' limitations and to serve as a base for the decision making algorithm responsible for adapting the interface. The Adaptation Decision Maker is based on a genetic algorithm to automatically decide and select the best configuration (both layout and content) for presenting the information to the elderly when using and interacting with the workspace. Results have been presented for different adaptation scenarios showing how the implemented architecture behaves under different elder interactions with the web interface. As a next step, a pilot study will be conducted in which the Adaptation Decision Maker will be evaluated by senior users. As part of the pilot study, seniors will be asked to compare the automatic adaptations to manual adaptations for different scenarios, thereby providing a better understanding of when automatic adaptations are preferred as opposed to manual settings.

ACKNOWLEDGEMENTS

This work has been carried out in the context of the Ambient Assisted Living Joint Programme project Elders-Up! (Elders UP!) and was supported by a grant of the Romanian National Authority for Scientific Research, CCCDI – UEFISCDI, project number AAL26/2014.

REFERENCES

Active Ageing — Special Eurobarometer 378/Wave EB76.2. Directorate-General for Communication TNS opinion & social. January 2012. Page 14.

Akiki, P.A., Bandara, A.K., Yu, Y., 2012. Using interpreted runtime models for devising adaptive user interfaces of enterprise applications. In: Proceedings of the 14th International Conference on Enterprise Information Systems. 28 June–1 July 2012, Wroclaw, Poland.

Akiki, P.A., Arosha, B., Yijun, Y., 2014. Integrating adaptive user interface capabilities in enterprise applications. In: ICSE '14 36th International Conference on Software Engineering Hyderabad. India, May 31–June 07, 2014. ACM.

Biswas, P., Bhattacharya, S., Samanta, D., 2005. User model to design adaptable interfaces for motor-impaired users. In: TENCON 2005 IEEE Region 10. 21–24 Nov. 2005, pp. 1–6.

Blumendorf, M., Lehmann, G., Albayrak, S., 2010. Bridging models and systems at runtime to build adaptive user interfaces. In: EICS '10 ACM SIGCHI Symposium on Engineering Interactive Computing Systems. Berlin, Germany, June 19–23, 2010. ACM.

Borg, P.P., et al. (Eds.), 2011. Active Ageing and Solidarity Between Generations — A Statistical Portrait of the European Union 2012. Publications Office of the European Union, Luxembourg.

Bruninx, C., Raymaekers, C., Luyten, K., Coninx, K., 2007. Runtime personalization of multi-device user interfaces: enhanced accessibility for media consumption in heterogeneous environments by user interface adaptation. In: Semantic Media Adaptation and Personalization, Second International Workshop. 17–18 Dec. 2007, pp. 62–67.

Chen, Y.S., Magoulas, G.D., 2005. Adaptable and Adaptive Hypermedia Systems. IRM Press.

Clerckx, T., Luyten, K., Coninx, K., 2005. DynaMo-AID: a design process and a runtime architecture for dynamic model-based user interface development. In: Bastide, R., Palanque, P.A., Roth, J. (Eds.), Engineering Human Computer Interaction and Interactive Systems. Springer, pp. 77–95.

Color Blindness. Ishihara Color Test. http://www.colour-blindness.com/colour-blindness-tests/ishihara-colour-test-plates/.

Eckert, M.A., 2011. Slowing down: age-related neurobiological predictors of processing speed. Front. Neurosci. 5, 25.

Elders UP! EU AAL project. http://www.eldersup-aal.eu.

Feng, J., Liu, Y., 2015. Intelligent context-aware and adaptive interface for mobile LBS. Comput. Intell. Neurosci. 2015. Article ID 489793.

Gajos, K.Z., Weld, D.S., Wobbrock, J.O., 2010. Automatically generating personalized user interfaces with supple. Artif. Intell. 174 (12–13), 910–950.

Garzon, S.R., Cebulla, M., 2010. Model-based personalization within an adaptable human-machine interface environment that is capable of learning from user interactions. In: IEEE International Conference on Advances in Computer-Human Interaction, pp. 191–198.

Hothi, J., Hall, W., 1998. An evaluation of adapted hypermedia techniques using static user modelling. In: Proceedings of the 2nd Workshop on Adaptive Hypertext and Hypermedia, HYPERTEXT'98. Pittsburgh, USA, June 20–24.

Jain, R., Bose, J., Arif, T., 2013. Contextual adaptive user interface for Android devices. In: Annual IEEE India Conference (INDICON). 13–15 Dec. 2013, pp. 1–5.

Jameson, A., 2012. Adaptive interfaces and agents. In: The Human–Computer Interaction Handbook: Fundamentals, Evolving Technologies and Emerging Applications. Lawrence Erlbaum Associates, pp. 305–330.

Jason, B., Calitz, A., Greyling, J., 2010. The evaluation of an adaptive user interface model. In: Proceedings of the 2010 Annual Research Conference of the South African Institute of Computer Scientists and Information Technologists. Bela Bela, South Africa — October 11–13. ACM, pp. 132–143.

Johnson, A., Taatgen, N., 2005. User Modeling. Handbook of Human Factors in Web Design. Lawrence Erlbaum Associates.

Kelly, K., 2013. Lack of marketing knowledge leading cause of start-up failures according to new infographic from the national association of small business professionals. PRWEB, New York. May 13 2013.

Kutz, J.W., 2015. Audiology pure-tone testing. http://emedicine.medscape.com/article/1822962-overview.

Lavie, T., Meyer, J., 2010. Benefits and costs of adaptive user interfaces. Int. J. Hum.-Comput. Stud. 68 (8), 508–524.

Limbourg, Q., Vanderdonckt, J., Michotte, B., Bouillon, L., Jaquero, V.L., 2004. Usixml: A Language Supporting Multi-Path Development of User Interfaces. Lecture Notes in Computer Science, vol. 3425. Springer.

Martel, J., 2016. Visual field exam, Healthline, January 6. http://www.healthline.com/health/visual-field#Overview1.

Mejía, A., Juárez-Ramírez, R., Inzunza, S., Valenzuela, R., 2012. Implementing adaptive interfaces: a user model for the development of usability in interactive systems. In: Proceedings of the CUBE International Information Technology Conference. Pune, India, September 03–06. ACM, pp. 598–604.

Mori, J., Paternò, F., Santoro, C., 2004. Design and development of multidevice user interfaces through multiple logical descriptions. IEEE Trans. Softw. Eng. 30 (8), 507–520.

Neves, B.B., Amaro, F., 2012. Too old for technology? How the elderly of Lisbon use and perceive ICT. J. Commun. Inform. 8 (1).

Nivethika, M., Vithiya, I., Anntharshika, S., Deegalla, S., 2013. Personalized and adaptive user interface framework for mobile application. In: Advances in Computing, International Conference on Communications and Informatics (ICACCI). 22–25 Aug. 2013, pp. 1913–1918.

Reinecke, K., Bernstein, A., 2011. Improving performance, perceived usability, and aesthetics with culturally adaptive user interfaces. ACM Trans. Comput.-Hum. Interact. 18, 2.

Sadiqa, M., Pirhonen, A., 2011. Design time, run time, and artificial intelligence techniques for mobility of user interface. Proc. Comput. Sci. 3, 1120–1125.

Sanchez, E., 2012. Icons vs labels vs both. http://edwardsanchez.me/blog/13589712.

See, S.L., 2015. Big data applications: adaptive user interfaces to enhance managerial decision making. In: Proceedings of the 17th International Conference on Electronic Commerce 2015 (ICEC '15). Seoul, Korea, August. ACM.

Shakshuki, E.M., Reid, M., Sheltami, T.R., 2015. An adaptive user interface in healthcare. Proc. Comput. Sci. (ISSN 1877-0509) 56, 49–58.

Vendramin, P., Valenduc, G., Molinié, A.-F., Volkoff, S., Ajzen, M., Léonard, E., 2012. Sustainable work and the ageing workforce. Report 09 December 2012, Reference No: EF1266, ISBN 978-92-897-1100-5.

CHAPTER 13

Telemonitoring as a Core Component to Enforce Remote Biofeedback Control Systems

Sérgio Guerreiro[*,†]
*Universidade da Beira Interior, Covilhã, Portugal
†Lusófona University, Lisbon, Portugal

13.1 INTRODUCTION

Remote monitoring is a crucial, and in most situations, an exclusive capability to provide medical assistance to the populations. Non-accessible or limited access environments are the result of manifold contexts: crisis situations (*e.g.*, floods, earthquakes, Humanitarian assistance, nuclear power plants, *etc.*), space exploration, offshore navigation and exploration, populations living in the desert, *etc.* Moreover, the long-term diseases (*e.g.*, *Diabetes Mellitus*, Parkinson disease, *etc.*) demand continuous medical assistance, which by its turn augments the population's economical pressure.

Therefore, in the last decades, the telemedicine area is researching, developing, and testing, new solutions that are able to integrate the tremendous technological advances with the current medical practices, in order to offer a cheaper, a more efficient and a more effective health care service.

In this respect, Norris (2002) states that telemedicine uses information and telecommunications technology to transfer medical information for diagnosis, therapy and education. Besides, World Health Organization (WHO) defines that telemedicine is an area including *(i)* the remote clinical services related with the health care and *(ii)* the education between practitioner and patient to optimize the process's efficiency.

Regarding a taxonomic classification, telemedicine encompasses the following sub disciplines: tele consultation, tele intervention, tele monitoring and tele education. To fulfill these extensive disciplines, telemedicine is located in the intersection of knowledge areas such as medicine, information systems, computer science, computer communication, social issues, *etc.* Nevertheless, the concerns related with the social and ethical aspects of

Ambient Assisted Living and Enhanced Living Environments.
DOI: http://dx.doi.org/10.1016/B978-0-12-805195-5.00013-2

telemedicine usage are raising and therefore are demanding a scientific and industrial response.

Recalling the remote monitoring problem, we locate this research in the scope of conceptualization using core concepts from control systems and telemonitoring.

On the one hand, we consider the research in the scope of the dynamic systems control (DSC) theory, as presented by the references of Franklin et al. (1991), Bertalanffy (1969). In DSC it is commonly agreed that the following components are demanded to enforce a full control system: the process to be controlled, the observation, the set of control actions, a reference to be followed and a controller. The general purpose of a control system is to react whenever the disturbance affects the behavior of the system or whenever a new input is established. By other words, when the system is not producing the desired output for the imposed input, as defined by Franklin et al. (1991). A feedback control pattern calculates the system input accordingly with the actual misalignment obtained between the output and input. In this pattern, the control actuation calculation takes into consideration the disturbance and the system dynamics. Because the system output depends on the disturbance imposed in the system and on the system dynamics itself. Moreover, to produce results, all control systems require the capabilities of observation and actuation.

On the other hand, instantiating a DSC to the specific context of telemonitoring is referred by Spelmezan et al. (2009) as a biofeedback control solution. For instance, if a person is warned about incorrect movements, at run-time, while walking or running, by a smart phone message (sound, graphic visual display, or text), to correct the gait pattern, so that he/she is less likely to experience an injury. Consequently, a preventive behavior change would occur instead of a reactive, corrective one. Cheung and Davis (2011), Crowell and Davis (2011), Noehren et al. (2010) presented examples of the feasibility of using feedback to change runner's gait.

To this extent, we identify the need to develop new approaches to enforce biofeedback control systems where the patients are remotely located from the practitioners and the health care personnel. This paper starts by state-of-the-art review encompassing both the theoretical approaches and the real applications that are available in the market. This state-of-the-art review has been accomplished by a team group, and disparate applications were chosen in order to include both endogenous and exogenous Human factors, e.g., *Diabetes mellitus* and space exploration. Afterwards the limitations, and opportunities, of the previous state-of-the-art are identified.

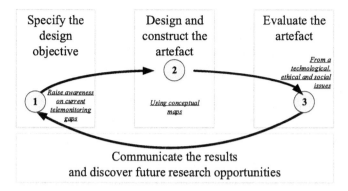

Figure 13.1 Essential design science research (DSR) phases.

Next, an ontology is proposed to bound the core concepts and their relationships regarding the remote biofeedback control systems. This ontology is further used to map the actual *(i)* technological constraints and *(ii)* the involved ethical & social aspects with the regard to the previous identified concepts. Therefore, on the one hand, we are able to identify the ontological parts that are already covered by today's technology along with the ethical & social perspective, and on the other hand, the ontological parts that are still under research. Finally, the paper is concluded and future work is identified.

Regarding the methodological approach, this paper applies the principles of the design science research (DSR), as in von Alan et al. (2004), Winter (2008), as a way to identify the constraints and the boundaries for the problem at hand. Figure 13.1 exhibits the essential DSR phases. Firstly, the design objective of raising awareness on current telemonitoring gaps is motivated. Afterwards, the second DSR phase, involves a conceptual mapping design, using conceptual maps (CMaps); and thirdly, an evaluation about the technological, and the ethical & social issues. Designing and then evaluation the design is a research initiative that aims at discovering the future research opportunities. Furthermore, when decoupling the conceptual layer of the solution from the specific technological implementation details, it allows a future technological independent development. The usage of CMaps has also the benefit of sharing a common understanding of the concepts between the researchers with different backgrounds, as explained and exemplified in Grandry et al. (2013), Gaaloul et al. (2014). Only one representation with properties of conciseness, comprehensiveness, consistency and coherence of the solution to the problem is to be designed.

The remaining of this paper is organized as follows. Firstly, in section 13.2, the background concepts related with dynamic control systems and conceptual maps are introduced. Afterwards, section 13.3, presents and discusses the state of the art applications related with telemonitoring. Then, in section 13.4, a conceptual map design conceptualizes the domain of application related with remote biofeedback control systems using the previous state-of-the-art. The conceptual map is discussed in the section 13.5. Finally, in the end, section 13.6 concludes the paper and points to future work.

13.2 BACKGROUND

This section presents the background realms that are used through the rest of the paper. Firstly, the general concepts about the dynamic systems control are refereed: evolution, main scientific approaches and main concepts. Secondly, the CMaps are briefly introduced. CMaps will be used in section 13.4 to bound the core concepts and their relationships regarding the remote biofeedback control systems.

13.2.1 Dynamic Systems Control

The aim of controlling dynamic systems is to gain efficiency in their operation, thus achieving the same results but with less resources usage. From a historical point of view, the first significant work in automatic control was James Watt's centrifugal governor for the speed control of a steam engine in the eighteenth century (Ogata, 1997). Since then, major advances have been made in the research of the systems control theory; many of them were driven by the advent of the digital computers. The James Watt's speed control system is one of the oldest examples of a control system. Here, the controlled system is the engine and the controlled variable is the speed of the engine. The speed of engine is in this example observed by the centrifugal force that is read by the sensor. The difference between the desired speed and the actual speed is the error signal. The control signal is in this example the amount of fuel that is applied to the engine. The command of the power cylinder that opens or closes the valve contains the relation between the error signal and the control signal that is delivered.

Control systems are mainly developed by the automation field, raised by the need to implement automatic supervision since the industrial revolution. A control approach is composed of two main parts: the controller and the controlled process. The controller is responsible to control the process, where a feedback control system is the system that maintains a prescribed relationship between the output and the reference input by comparing them and using the difference as a means of control (Franklin et al., 1991). The difference arises because of two reasons: *(i)* the interaction of the controlled process with the exogenous[1] factors (for instance, a disturbance) and *(ii)* the endogenous[2] factors (for instance, a breakdown). The main scientific approaches for DSC are:

Analytical, are the mathematical formalization efforts to the process model and controller model definition. It requires the complete process description in order to define the correspondingly controller. Examples of these efforts are: Brockett (1993) or Recht and D'Andrea (2004).

Cybernetic, are the automatic controllers, which are designed and implemented, in machine dependent form. A terminology is derived from the 70's. Their execution can only be done using machines. Three different classes of cybernetic controllers are found, *(i)* classic control theory (Ribeiro, 2002; Ogata, 1997), *(ii)* expert systems (Uraikul et al., 2007) and *(iii)* agent based (Castro and Oliveira, 2011).

Black-box, are considered the integration mechanisms between Human and machines in order to optimize the Enterprise performance, using a black-box perspective or also named as functional oriented perspective. Three different classes of black-box controllers are identified: *(i)* knowledge management (Matos, 2006; Magalhães, 2005), *(ii)* homeostatic (Jaeger and Baliga, 1985) and *(iii)* data driven (Force USA, 2001). Hofstede (1978) argues that only exist two forms of control systems: the cybernetic and the homeostatic. However, the knowledge management area (Magalhães, 2005), the data driven (Force USA, 2001) and the analytical perspectives (Recht and D'Andrea, 2004) are also considered major forms of control systems by their significance therefore they were included in this research as disparate approaches.

By definition, the purpose of a control system is to react whenever the disturbance affects the behavior of the system or whenever a new input is established, as in Franklin et al. (1991). By other words, when the system is

---

[1] Originating from outside.
[2] Proceeding from within.

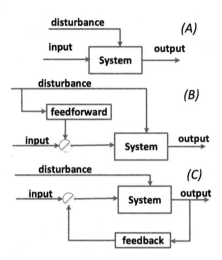

Figure 13.2 Design pattern of control systems. (A) Without control, (B) feedforward control and (C) feedback control. *Adapted from Guerreiro (2014).*

not producing the desired output for the imposed input. Control act in the input at the same time as the disturbance is affecting the system.

In this sense, some distinct configurations exist for a control system. Figure 13.2 depicts the set of classical design patterns for a control system as discussed in Guerreiro (2014). In the top, Figure 13.2(A), shows a system that is not being controlled. The disturbance always affects the output delivered by the system. In this pattern, it is not possible to guarantee the behavior of the system output. For example, considering an indoor air conditioning system, without control represents setting a static fan speed. Therefore, any temperature fluctuations are not taking into account. In the middle, Figure 13.2(B), a feedforward pattern shows that the system input changes accordingly with the actual disturbance. Therefore, the system dynamics is not included in the control actuation. In the indoor air conditioning example represents a fan speed depending on the fluctuations perceived from the outside air temperature. Figure 13.2(C) depicts a feedback control pattern that calculates the system input accordingly with the actual misalignment obtained between the output and input. In this pattern, the control actuation calculation takes into consideration the disturbance and the system dynamics, because the system output depends on the disturbance imposed in the system and on the system dynamics itself. Returning to the example, the air conditioning fan speed depends on the actual measurement from the inside air temperature. Each pair <*Observation, Control actuation*> constitute a Control pattern.

13.2.2 Conceptual Maps

CMaps are used to render a set of concepts and relationships related any knowledge domain, as detailed in Cañas et al. (2005). The main advantages are *(i)* knowledge model creation within new domains or non-explored domains, *(ii)* the capability to create knowledge visualizations, *(iii)* to complement previous existing knowledge models and *(iv)* to facilitate the stakeholders' discussion when they have diverse interpretation about a certain domain.

One promising application for the conceptual maps is the support for mining solutions that are appearing in the information systems field as the result of complex, large scale, and distributed, sets of data available from the business transactions environments. For instance, the techniques of: data mining as in Witten and Frank (2005), big data as in Laudon and Laudon (2016) and process mining as in Van Der Aalst (2011). As referred in Dudok et al. (2015) these solutions require precise domain conceptualization to offer algorithmic capabilities aligned with the business transactions environments. The supporting data operations of, *e.g.*, extraction, transformation, discover, load, assessment, *etc.*, are only possible if the data is contextualized, at a higher abstraction layer. This capability is offered by CMaps.

More complex maps are offered by Topic maps (Topic maps, 2015), which separates the occurrence in the world from the concepts. Topic maps are actually an ISO/IEC 13250-6:2010 standard.

13.3 TELEMONITORING STATE-OF-THE-ART

This section presents and discusses the state-of-the-art related with telemonitoring. Telemonitoring is encompassed as a class within the taxonomic classification of telemedicine (Norris, 2002). The other classes are: teleconsultation, teleintervention, and teleeducation. In this section, firstly, the telemonitoring state-of-the-art domain is presented and some applications examples are presented. Secondly, the limitations and opportunities are identified.

13.3.1 Telemonitoring Applications

Telemonitoring is defined as the use of information and communication technologies in order to monitor patients remotely. This variant of telemedicine has an endless number of applications, for instance, hospitalized patients, at home, at space, at remote places, at conflict situations, *e.g.* As indicated by Meystre (2005), a large set of parameters are already refereed in the scope of telemonitoring: *Cardiovascular parameters:* heart rate, fetal heart rate, blood pressure, electrocardiogram, stethoscope; *Haematological parameters:* Coagulation; *Respiratory parameters:* oximetry, spirometry, respiratory rate, CO_2 production, CO_2 consumption; *Neurological parameters:* electrocardiogram, electromyographic, intra cranial pressure; *Metabolic parameters:* Body weight, basal metabolism, glucose, lactate and ethanol in the blood temperature; *Urological parameters:* intra vesicular pressure; *Gynaecological parameters:* intrauterine pressure; *Other parameters:* geographic location, etc.

13.3.1.1 Parkinson Disease Telemonitoring

Ageing population worldwide is increasing over the centuries, with a fast rising, as presented in Bustamante et al. (2010). A direct consequence is the increase in diseases. One of the main diseases groups that affect this generation are neurodegenerative diseases such as the Parkinson's disease (PD). Accordingly with Bustamante et al. (2010), it is expected a doubling of cases by 2020. The appropriate medical care of these patients has become increasingly complex and expensive. The long length of stay in hospitals for monitoring and treatment of patients and the problems associated with this stay, contribute to increased costs and mortality. The best way to increase the quality and length of life of people affected by neurodegenerative diseases, is to improve forms of treatment. And before this step, is to improve the diagnosis and monitoring of these diseases.

Therefore, one of the actual challenges is to develop easy using devices for the patient to perform periodic tests without the need to go to hospital to monitor their health state. Human friendly devices are an incentive for patients to use on a regular basis, leading to a larger amount of data available to physicians who follow these patients.

In LeMoyne et al. (2009), PD is characterized as a chronic disorder that affects movement. It was initially characterized by James Parkinson in 1817. According to statistics from the United States, the moving disorders affect about one million people, and its incidence is proportional to age, usually occurring around age of 55. Predominant symptoms of the disease are the result of an increased rigidity of muscle tone, gait "dragged" decreased balance, and tremors at rest state. However, during voluntary movement, the tremors may diminish or even disappear. The characteristic resting tremors are quantified about 4–5 per second. Currently, medical practice for the evaluation of motor disability of PD is based on a neurological examination to the patient during his visit to a clinic. However, these tests do not reveal some important information to the neurologist, since part of the review is restricted to rely on the patient's memory and perception of their own symptoms. In addition, most patients may not be aware of some symptoms or not be able to identify early and properly, may unknowingly exaggerate or reduce the severity of symptoms. Tzallas et al. (2014) propose a solution to these problems with objective assessments, named as UPDRS (Unified Parkinson's Disease Rating Scale) that is the traditional method for evaluating the disease. However, this scale presents certain difficulties in the sense that their evaluation requires highly specialized medical facilities and inevitably limited, restricting bearing the quality of this classification. Also another problem associated with this scale holds to the fact that evaluation of the disease being affected at a given time on a given day. Thus, mobile devices have been designed to overcome these problems, the assessment of the disease can be carried out in a continuous manner and with relative independence of medical resources, as refereed by Surangsrirat and Thanawattano (2012).

Three alternatives for treating the disease are: *(i)* therapy using drugs, *(ii)* pallidotomy, and *(iii)* deep brain stimulation. The pharmacological therapy of this disease has been developed by the administration of L-dopa which can improve patient movement disorders. As *dopamine* can not pass across the blood-brain barrier, therefore *levodopa*[3] (*L-dopa*) is administered. This compound is capable of passing through the barrier. The effective use of drug therapy eventually decreases, and produces side effects such as *dyskinesia* induced by drugs. Another alternative, for medically intractable scenarios is the pallidotomy, which involves surgery to the internal segment of the *globus pallidus*, as in LeMoyne et al. (2009). The third alternative using deep brain stimulation is suitable for patients with resistance or adverse reactions to treatment through medication. The general components of deep brain stimulation device, such as the Medtronic system are electrodes, extensive cables, and an internal pulse generator that controls the electrode.

---

[3] A precursor of *dopamine*.

LeMoyne et al. (2009) shows a clinical task stimulator device sync for optimum effect that can be applied on four parameters: electrode polarity, amplitude, pulse width and frequency.

Additional methods, such as presented by Surangsrirat and Thanawattano (2012), are desirable to help in the diagnosis process and to improve the detection, and its severity, of early PD cases. This study have reported that analysis of a spiral may be useful in the diagnosis of motor dysfunction in PD patients. The methods based on an open spiral design allows the quantitative assessment of the hand movement in PD patients, for instance in Wang et al. (2010). The spiral design, based on the implementation of a mobile application, is safe, inexpensive and easy to use for the detection of movement disorders.

An alternative strategy for the evaluation of DP involves the implementation of an application for the iPhone based accelerometers in a wireless system can be found in LeMoyne et al. (2010). This application allows the characteristic tremor of PD, which transmits a signal through the accelerometer system, can be recorded at any time and transmitted by email to a remote location for post-processing.

Moreover, the need to develop a quantitative evaluation system, low cost, with small size, and providing a large amount of information, pressured the appearance of measurement systems with accelerometers based in the touch of fingers. This movement can be easily used in a clinical or home environment. The characteristics of the fingers of the touch movement as well as the range of the touch fingers, amplitude and speed are considered important for the diagnosis of PD, as in Okuno et al. (2007), Yokoe et al. (2009). Surangsrirat and Thanawattano (2012) found through the study that the patients took almost twice as long in carrying out spirals when compared to healthy individuals.

LeMoyne et al. (2010) implemented an application for the iPhone, which functions as an accelerometer system, thereby providing an easier and effective monitoring of the disease through the typical tremor. The results show that the healthy individuals were able to hold his hand in a relatively static position, resulting in acceleration graph showing minimal fluctuation with respect to 1 g of gravity. The coefficient of variation of this individual was 0.075. For the individuals with DP disease a strongly fluctuating around the value of 1 g of gravity is found, where the coefficient of variation for these individuals was 0.307, which is a significantly higher value when compared with the healthy individuals.

Bustamante et al. (2010) have developed two new testing devices for patients with PD. One test is based on the finger tap (FTT) and the other in the manual force measurement. In the FTT test was asked patients to touch with two fingers of the same hand as quickly as possible into each other, the main parameter to be measured by doctors is the frequency of the ring. In the manual force measuring test, the parameter to be measured is the force that the patient has to grasp an object. The data provided show accurate and precise values for the parameters of interest to physicians. The authors argue that the system's ease of use makes it possible to perform more frequent tests, so that the monitoring of changes in motor functions of patients is more accurate.

In another study by Okuno et al. (2007), is developed a contact force measuring system for touching the finger PD patients to show the effectiveness of the system in quantitative diagnosis. The system comprised two tri-axial accelerometers of piezoelectric elements, a touch sensor made of thin stainless steel plates, an analogic to digital converter and a PC. It was determined a transfer function representing the relationship between the contact force and the output from the accelerometer during the contact phase of the fingers. The FTT test was administered to 27 healthy subjects and 16 PD subjects. The test result for each parameter was evaluated by a neurologist. The subjects performed the continuous movement during 60 seconds and the contact force is estimated. It was shown that the contact force was lower for the parameter with highest score the UPDRS for finger touch test.

In the study of Yokoe et al. (2009), a system consisting of a touch sensor and an accelerometer to meet objective parameters of the finger touch test (FT) related to the speed was developed. The amplitude, rate and number of FTs in PD patients are measured. The authors consider the accuracy and reproducibility in the FT test performance analysis essential to assess the severity of PD.

13.3.1.2 Battle Field Telemonitoring

One example of a battlefield device is the Advanced Care and Portable telemedical Alert Monitor (AMON) (Anliker et al., 2004). This portable telemonitoring system warns about problems in the respiratory and cardiac human systems.

Grishin et al. (2012) specify the following requirements for a telemonitoring solution to be used with soldiers: (i) the individual to be monitored is a healthy person in high-activity situations with the possibility of being injured (either fatal or not); (ii) the parameters to be monitored are ECG, respiration rate, body temperature, physical activity, sleep quality, amount of haemoglobin saturated with oxygen in the blood, EEG and location; (iii) the objective of the monitoring is to automatically, and autonomously, assess, in real time, the status of a soldier (if they are alive or dead, awake or unconscious, injured or moving); (iv) the collected information is accessed through automatic algorithms that determine functional states of a soldier and allow itself to monitor its own state; (v) duration of monitoring in hours and days; (vi) high accuracy for the information; and (vii) high need for an independent assessment.

Other author, Freund (2008) presents a list of physical considerations to implement a soldier telemonitoring system: (i) low weight, volume and power consumption; (ii) simple computer architecture to facilitate updates; (iii) process development using a block or spiral; (iv) simple and small; (v) minimize logistics; and (vi) cheap to produce. Accordingly with this author, the system would be able to determine the following events: (i) is the hardware connected? (ii) which is the battery status? and is the power adequate? (iii) why is the device disconnected? (iv) where is the soldier? (v) communicate and attack; and (vi) run with equipment.

Based in the opinions of military doctors with combat experience, Tharion and Kaushik (2006) suggest a Graphical User Interface (GUI) with the following properties: (i) a differentiation system color which distinguishes subjects in need of immediate relief (red) those in need of non-priority relief (yellow) those without aid (green), further having an indicator for those whose state is unknown because to disconnect sensor (blue); (ii) incorporates information on heart and respiratory rate, body position and movement, skin temperature and deep organs (liver, heart ...) and sleep time; (iii) incorporates data location (map distance), clock and alarm, BIDS (system that detects the impact of ammunition) button to activate a distress signal, fluid consumption registration, evacuation status and patient registration; and (iv) provides information of the cognitive, thermal state, hydration and signs of life.

Another current research item in this application is data transmission. Most telemedicine systems today transmits its data via mobile networks (Global System Mobile, GSM) or the Internet, as explained in Cermack (2006). The missions in remote locations are heavily relied on satellite links, the duration of transmission and signal reception (the so-called lag) is usually relatively large. There are some basic parameters in choosing the system through which the information should be sent: (i) coverage Needs: local, regional or global; (ii) availability: temporary or permanent; (iii) mode and urgency of the transfer of information: real-time or store and forward; (iv) minimum and Maximum Bandwidth (upload and download); (v) type of transmission protocols; (vi) compatibility with multiple transmitters and receivers; (vii) data encryption and security; (viii) cost per amount of information conveyed; and (ix) system average costs considering the purchase and cost of data traffic.

In recent years there have been huge advances in the development of wireless sensor network technologies (WSNs). Generally they consist of a large number of small sensor nodes, inexpensive and with low expenditure of energy. This devices have a micro controller, a memory and a radio transceiver. The Body Sensor Network (BSN) is an especialization of the family of these WSN. As in Lim et al. (2010), it consists of various physiological sensor nodes that are placed near or in the human body for monitoring parameters such as temperature, heart rate, EEG, ECG, *etc.* Associated with BSN, Cho et al. (2008) explains that information exchange system by peer-to-peer (P2P) allows the exchange of information between each device. However, it is still necessary to transmit the information to a command center. To this end, the Unmanned Aerial Vehicles (UAV) perform the monitoring of the soldier, in real time, through the wireless vital signs acquisition, leading to a reduction of the low rescue the actuation time. UAVs can also locate the soldiers and make short communications between soldier and command center (Ba and Wang, 2012).

13.3.1.3 Crisis Response Telemonitoring

The crisis response management is triggered when a disaster situations occurs. Accordingly with Abchir et al. (2003), a disaster is consider when a serious perturbation of the functioning of a community or a society involving human casualties, material, economic or an environmental large extent, which impact exceeds the capacity of the community or society affected to cope with their own resources. Disasters can be classified as human and natural. Human or man-made disasters are those resulting from actions and or Human missions and are related to any activities performed by the same. Some examples are the wars, plane crashes (in some cases), traffic, fires, contamination of rivers or dams breaking. Natural disasters can be unpredictable and are the inevitable result of a natural phenomena of great intensity. Most of the time, these events are generated by internal and external dynamics of the Earth. Those arising from the internal dynamics are earthquakes, tsunamis, volcanic activity and tsunami while external dynamic involving storms, tornadoes, floods, landslides, among others. Some recent examples are, the hurricane Katrina in Northern America in 2005 (Kost et al., 2006; Klein and Nagel, 2007; Blackwell and Bosse, 2007), the hurricane Ike in Greater Antilles and Northern America in 2008 (Vo et al., 2010), earthquake in Haiti in 2010 (Adelakun, 2014), the tsunami in Japan in 2011 or Eyjafjallajökull volcano eruptions in 2010.

The first documented use of telemonitoring in disaster situations are referred by Garshnek and Burkle (1999): 1985 Mexico City earthquake and 1988 Armenian earthquake. By this time, telemonitoring allowed the audiovisual exchange with the U.S. to minimize the negative consequences of these unpredictable natural phenomenons.

Ajami and Lamoochi (2014) state that telemonitoring in remote places has the key objectives *(i)* to obtain information about injured people in these devastated areas and *(ii)* to transfer the information to research centers (or other rescue organizations) through the use of mobile devices. Satellite communications are pointed as one of the core technologies to solve the dependency of terrestrial communication lines disruptions, which could be affected during a disaster situation. Other issues are pointed, concerning the public education to ensure the previous necessary training of specialized personnel and preparing prior training plans to acquire knowledge about remote places.

In regard to the broad scope of telemedicine, a disaster situation results in the creation of remote places, with few communications capabilities and few available infrastructures. Therefore a myriad of telemedicine practices, as referred in the literature, are required: tele radiology (Scheinfeld et al., 2003), tele psychiatry (Doze et al., 1999), tele education (Binks and Benger, 2007) and virtual hospital (Gorini et al., 2011).

13.3.1.4 Diabetes Mellitus Telemonitoring

Diabetes mellitus is a group of diseases associated with various metabolic disorders characterized by hyperglycemia resulting from defects in insulin secretion, insulin action or both. Its pathogenesis involves genetic and environmental factors. The long-term persistence of metabolic disorders can cause susceptibility to specific complications and also promote atherosclerosis. Chronic hyperglycemia of diabetes is associated with long-term damage (Gavin et al., 1997).

Accordingly to Stone et al. (2012), diabetes mellitus (DM) affects approximately 8.3% of the US population and is associated with high medical costs. Intensive glycemic control has been used to delay or prevent the development of micro vascular complications of diabetes. However, an estimated 43.2% to 55.6% of adults with diabetes do not meet the goal of the American Diabetes Association for glycemic control. Along with the rising incidence of type-2 DM, the impact of disease is also increasing. Morbidity and mortality in patients with DM remain high despite the availability and use of various anti diabetic therapies (Pressman et al., 2014).

This disease is associated with a wide variety of clinical presentations. These range from autoimmune destruction of pancreatic β-cells and the consequent insulin deficiency and abnormalities that result in resistance to insulin. As in Kuzuya et al. (2002), the basis for abnormalities in carbohydrate, fat and protein metabolism in diabetes is impaired insulin action in target tissues. Impaired insulin action results from inadequate secretion of the same and/or insufficient tissue responses to insulin at one or more points in the complex pathways of hormone action. The impaired secretion and defects in insulin action frequently coexist in the same patient, and it is often unclear which abnormality is the primary cause of hyperglycemia. On the one hand, symptoms of hyperglycemia include polyuria, polydipsia, weight loss, sometimes with polyphagia and blurred vision. The growth deficiency and susceptibility to certain infections may also accompany chronic hyperglycemia. The most serious consequences of diabetes are hyperglycemia with ketoacidosis or hyperosmolar non-ketotic syndrome. On the other hand, the long-term complications of diabetes include retinopathy with potential loss of vision; nephropathy leading to renal failure; peripheral neuropathy with risk of foot ulcers, amputation, and neuropathic arthropathy, causing symptoms gastrointestinal, genitourinary and cardiovascular and sexual dysfunction (Gavin et al., 1997).

Generically, DM telemonitoring includes glucose records transmission to healthcare providers and results in an improved glycemic control (Pressman et al., 2014). These authors state that several monitoring studies based on mobile devices showed improvements in the patients with diabetes glycosylated hemoglobin (HbA1c), blood pressure (BP) and low density lipoprotein (LDL). The objective of this study was to evaluate a research telemetry device at home to improve glucose and blood pressure control over six months for patients with diabetes type 2. The device is used to transmit blood glucose weekly, weight, and blood pressure readings for a doctor who specializes in diabetes care.

The DiaTel system (Stone et al., 2012) is a telemonitoring solution that includes the daily transmission of glucose data and medication adjustments within 24–72 hours (also named as active care management: ACM) resulted in statistically significant reductions in HbA1c (glycated haemoglobin) at 3 months (1.7% vs 0.7%) and 6 months (1.7% vs 0.8%; $p < 0.001$ for each) compared with a phone call to coordinate the monthly care (CC), offering self-management education diabetes and referral for medication adjustment of the primary care provider (PCP). The extent of DiaTel objective was to assess whether these initial improvements could be sustained with the operations of the same or lesser degree in the participating DiaTel, re-enrolled in a 6-month study extension. In particular, assessed the impact of transmission of continuous data, initial ACM, and continued with monthly phone calls.

Moreover, the advent of mobile devices offer a huge variety of applications, *e.g.*, Glooko, OnTrack, dbees.com, Glucool Diabetes, *etc*. These applications have usual capabilities of: sharing data with practitioners, reminders for data acquisition, diet and exercise control, visualize historical data, *etc.*

13.3.1.5 Dermatological Telemonitoring

Teledermatology is a subcategory of telemedicine used by the dermatologists to evaluate videos or images from the skin of patients, accompanied by relevant information provided by them. The fundamental purpose of Teledermatology is offer get access to cutaneous specialized medicine to remote places. As a secondary objective, the increase in diagnostic efficiency Hospitals decreasing the number of visits by the patient (as in Wurm et al. (2008)).

According to Heffner et al. (2009) is revealed that dermatologists were more successful in terms of diagnosis and treatment when using a two-stage evaluation: *(i)* evaluating images of skin problems in real time through video conferencing systems and later *(ii)* analyzing the pictures. This study also allowed the assessment of disease at an early stage. In situations, where is not possible of having a face to face consultation with a dermatologist doctor, a teledermatology system used by a general practitioner (not specialized in dermatology) increases the reliability of diagnostic.

Wurm et al. (2008) refer that Norway was the first European country to introduce teledermatology systems. In 1989, a teledermatology service was established with a real-time linking between the University Hospital of Tromsø and the primary care center in Kirkenes situated about 800 km away (corresponding to a travel of approximately 12 hours).

Finland also played an important role in the development of teledermatology. The patients and their general practitioners participated in consultations of the Primary Health Care Center in Ikaalinen with a dermatologist to 55 km in Hospital university of Tampere (TAUH). Consultations were carried out using standard commercial equipment videoconferencing, a modified camera and a dermatoscope. During an eight-month study, 25 patients participated in these teledermatology consultations. After teleconsultation, in 19 cases (76%) the treatment of patients have been changed, the diagnoses were changed in 13 cases (52%) and 18 patients (72%) did not need to go to the University Hospital of Tampere (as in Wurm et al. (2008)).

From a mass market perspective, many mobile applications are available for teledermatology practice, *e.g.*, the *Teledermatology Pilot* where the patients, students and dermatologists can ask questions and exchange information; the *DigitalDermDr* allowing to sends skin photos to dermatologists, this service commits with 48 hours reply; the *Dermutopia* explains acne treatments and basic dermatological treatments and the *eDermoscopy* that allows to take, high quality polarized and non-polarized, images with an increase quality of up to 20*x*. In addition, it serves as a platform for the second professional opinion within a team of international dermatology experts. The doctor can send any of the images that consider suspicious directly from the electronic system to the team of experts and then get a second opinion.

13.3.1.6 Cardiac Diseases Telemonitoring

The coronary artery disease are quite common and are related to the accumulation of cholesterol in the inner layer of the arterial wall, acting as a barrier to the passage of blood stream (National Heart, Lung, and Blood Institute, 2015); heart attacks occur when blood flow to the heart is interrupted for a period of time sufficient to damage the myocardium contractile heart muscle (American Heart Association, 2015); unstable angina occur due to coronary artery obstruction or spasm, causing chest pain due to decreased oxygen supply to the heart muscle (National Heart, Lung, and Blood Institute, 2015); heart murmurs are indicative of a problem caused by the failure of a heart valve or a hole in the wall between heart chambers (American Heart Association, 2015) and finally, atrial fibrillation occur when the heart rate is rapid and irregular due to contractions uncoordinated ear (National Heart, Lung, and Blood Institute, 2015). Other conditions could be referred to equally serious implications for the health of patients.

Cleland et al. (2005) state that telemonitoring as telemedicine coverage area, through monitoring and constant individual monitoring enables the health professional with the ability to predict and/or prevent possible urgent complications, acting to prolong and improve the patient's health.

13.3.1.7 Space Exploration Telemonitoring

Nicogossian et al. (2001) explain that remote monitoring of *(i)* the crew health, *(ii)* the aerospace vehicles and *(iii)* the surrounding environment is considered essential in the operations of the space exploration programs. The astronaut safety and the mission success encompasses a large number of challenges. Therefore, the continuous development of new technologies with greater control and performance for remote monitoring is required. With increasing length and complexity of space missions, the telemedicine dependency tends to augment, while at the same time, provides novel solutions for terrestrial problematic, *e.g.*, by adapting the technology to enable improve medical care in remote areas. Today, joint efforts of engineers, technicians and practitioners, offer integrated systems that are able to monitor not only the health, but the functions of the ship and the environmental parameters. These are the three factors that convey a positive feedback about the mission.

In regard to the physiological crew health response, Assad and de Weck (2015) study the available data since 1961 and compare the adaptation in short- and long-term missions. For short-term missions, the adaptation is faster and the symptoms disappear more quickly. On the other hand, for long-duration missions it can take years and require physiotherapy to recover. Nowadays, new equipments are being developed to minimize the presented impacts in the Human health.

Moreover, it was reported by Aponte et al. (2006) that spaceflights induce different immune responses and many potentially harmful, some occurring immediately after arrival from space, while others develop throughout the length of the mission. Some factors that cause these changes are microgravity, stress, decrease in bone density and radiation. Until present date, some effects are identified on the immune system having a reduction in T cell counts, a decrease in the concentration of NK cells and their functionality, decreased immunity mediated by cytosines and some modified immunoglobulins, increasing the susceptibility to infections. This concern is more relevant when considering interplanetary missions (outside the Earth orbit). The Earth's surface is protected from cosmic radiation and solar radiation by the atmosphere and the Earth's magnetic field, which does not occur in other planets. The Moon, our natural satellite, has not both, making different scenario. However, the risk is reduced due to proximity, making it possible to treat the lunar base with the help of telemedicine in near real time and the evacuation is relatively fast. On the other hand, planets like Mars, with very poor atmosphere, radiation, no magnetic field and reduction of immune response entails very significant risks. The distance between Mars and Earth is extensive and variable, changing with the translational movement of both planets and making the cancellation of the mission a difficult process that take several months to return to Earth. Moreover, bidirectional signal transmission can take between until 46 minutes, making it impossible for procedures in real time, such as conference calls or tele intervention. Only store and forward systems can be considered. Medical intervention will have to become more autonomous and less dependent on land controls, requiring more resources and equipment such as increased workload of medical training of the crew.

In light of these requirements, there are many research projects available in the literature related with telemonitoring applications for space exploration, namely, Space sock, by Fei et al. (2010), which is a system composed of 4 non-evasive sensors; LifeGuard, from Mundt et al. (2005), which aims at real-time locating the individuals during operations in remote environments; Constellation Program, by Hill (2011), presenting a programmable suit accordingly with the mission; V2Suit, from Duda et al. (2015), which compensate the gravity absence using a viscous drag; and Biosuit, available from Canina et al. (2006), which is a project developed by MIT, aiming the integration of biosensors and maintaining the comfort and movements freedom.

13.3.2 Limitations and Opportunities

Table 13.1 summarizes the limitations and opportunities raised by each telemonitoring application. In addition, for each application a textual description is presented.

Parkinson Disease Telemonitoring

The main disadvantages of new methods involving technologies is the inevitable resistance to them, mostly by people over age. In addition, the devices designed for monitoring PD often fall into disuse, ending in a

Table 13.1 Limitations and opportunities for telemonitoring applications.

Application	Limitations	Opportunities
Parkinson disease telemonitoring	- Resistance, mostly by people over age - Fall into disuse, ending in a partial functional prototype	- Accessing health conditions in patients' comfort zone - Improve issues of speed, memory for data storage, battery and physical dimensions
Battle field telemonitoring	- Energy consumption - Confidentiality	- Reliable data transfer system - Fast transmission of vital signs using light-weighted equipment
Crisis response telemonitoring	- Geographical barriers - Access to communications and reliable information	- Minimize barriers between health care providers and patients - Study of economic benefits and cost-effectiveness
Diabetes mellitus telemonitoring	- Virtual medical treatment has the risk of error due to reduced human interaction - Privacy of electronic health information	- Data encryption, storage medium protection and access control mechanisms
Dermatological telemonitoring	- Not integrated with the national health care strategies	- Equipments with augmented capabilities (better mobile devices image quality)
Cardiac diseases telemonitoring	- Few studies to the role of telemonitoring. Major part is related with telesurgery	- Aiding telemonitoring for telesurgery has a synergistic effect for pre and post-operative - Predict and prevent cardiac disease
Space exploration telemonitoring	- There is no condition to perform complex medical procedures or available resources to carry out	- Develop information systems for medical decision support, data collection and storage - Develop new materials - Develop nano-technologies and biotechnology techniques

partial functional prototype. Although the results of studies for different monitoring technologies claim to be positive, these will never replace a neurologist or any health care provider.

The spirals evaluation studies are very preliminary, and serves as a basis for other future studies to improve this technique. Also, the population used for the study is restricted, which severely limits the applicability of results to other individuals. It could also be noted that when the data obtained is sent to a doctor that accompanies the patient' case report, has the advantage of accessing their health condition without the patients having to move out from their comfort zone.

Regarding the studies related with the mobile applications using accelerometer systems, are also very recent and in an early stage. The issues of speed, memory for data storage, battery, reduction in physical dimensions, lifetime, *etc.*, are still to be addressed. The data provided by this device should also be further used for choosing the most appropriate treatment to use. More studies with larger population should also be performed, instead of just using two subjects, healthy and another patient. In the case of sensor devices for the touch of finger movement, in future work should focus on the PC application, so as to improve the user interface and offer new features for data analysis. One of the main features that are being planned to be implemented consists in calculating the FTT finger Strike sign. This will provide physicians and clinicians more precise information on the parameter associated to this signal as well as the test itself.

Battle Field Telemonitoring

The recession for this application are limited due to few references existing in the literature. We noticed this issue and relate it primarily with a confidentiality issue. Nevertheless, a major concern for the battle field systems is the energy consumption. A system must be available and working properly for the battle time frame. Bluetooth is refereed by some authors as a promising technology to be considered in the future. Moreover, reliable data transfer systems between soldiers acting in the battle field, and the command center is cornerstone to enforce decision making. One practical solution is presented by the UAV systems that collects information from soldiers and the command center. UAV is able to solve the problem of loss of one of the information nodes, replacing it with another. Finally, the transmission of vital signs, using light-weighted equipment, with fast data transmission is a promising research field.

Crisis Response Telemonitoring

The destruction in affected areas is often so massive that becomes totally inaccessible, making it difficult for immediate medical assistance. The main goal of telemedicine in such cases is to try to save as many lives as possible. Telemedicine holds the promise of improving access to health care, especially in areas where there are geographical barriers and need to reduce costs. However, the main problem in telemedicine is not the lack of technology, but is how to apply it (Simmons et al., 2008). The ability to survive a traumatic accident depends on the received medical attention immediately, and the success or failure of a disaster response is often determined by immediate access to communications and reliable information, as in Qiantori et al. (2012). In particular, telemedicine can help communities in remote or rural areas with few health and personal services, because it overcomes distance and time barriers between health care providers and patients.

On the one hand, the critical success factors identified in the literature include: establishing clear program goals; winning the governmental and institutional support; adapting existing friendly interfaces; determination of accessibility and connectivity constraints; implementing standards and protocols; and dissemination of evaluation results. On the other hand, the lack of studies documenting the economic benefits and cost-effectiveness of the application of telemedicine is also a limitation.

Diabetes Mellitus Telemonitoring

The factors that can contribute to a defective glycemic control level include: *(i)* inadequate monitoring of home glucose monitoring, *(ii)* incorrect medication or lifestyle changes, *(iii)* patient education about the disease and *(iv)* access limited to providers for the diabetes control. In the absence of accurate data and adequate blood glucose values at home, suppliers can be hesitant to prescribe oral hypoglycemic agents or insulin regimes aggressively due to fear of hypoglycemia. Moreover, virtual medical treatment also entails reduced human interaction among professionals and medical patients, and therefore increases the risk of error when medical services are delivered in the absence of a registered professional. From the literature, it is identified an increased concern of compromising the privacy of electronic health information. Therefore, data encryption, storage medium protection and access controlled digital environments are demanded. In addition, there may be poor record of blood glucose values of the patient, and a quality assurance risk which may be compromising.

Dermatological Telemonitoring

From the literature we found that understanding teledermatology as a specialization of telemonitoring and then embedding it in the national health care strategy is a key recommendation for the development of this technology. Moreover, teledermatology demands new materials (*e.g.* the image quality obtained by mobile devices) to be easily accepted by all the health care actors.

Cardiac Diseases Telemonitoring

Even with the assumption that cardiac diseases could be more predicted and/or prevented (as in Cleland et al. (2005)) few solutions are encountered in the literature. A large effort is related with telesurgery techniques, however, few studies are given to the role of aiding telemonitoring for telesurgery. A joint effort between both results in a synergistic effect and with clear benefits, both pre and post-operative side.

Space Exploration Telemonitoring

The strategy for medical emergencies and therapy is similar to the remote sites strategy: stabilization and evacuation (if necessary). There is no condition to performed complex medical procedures or available resources to carry out. The support for basic and advanced life is limited and would absorb a lot of human resources. On the other hand, all the astronauts receive 40 to 60 hours of medical training, however, even with the support of telemedicine, medical intervention conditions are not similar to those found on Earth. Thus, since the survival of the crew is the priority, in some conditions, the evacuation may become a necessary option. Three key points are refereed, in the literature, to develop space exploration telemonitoring: nano-technologies, biotechnology and information systems. It is expected that autonomous nano devices will be key players in telemonitoring to act as biosensors (embedded in textiles), processing elements and signal analysis. Moreover, it is expected new materials for the equipment design. In the scope of information systems, the medical decision support and data collection and storage, allowing astronauts to better control and management of information and medical events are also expected.

13.4 REMOTE BIOFEEDBACK CONTROL SYSTEMS: ONTOLOGICAL DESIGN

This section presents the author research using a conceptual map design regarding the telemonitoring related work using the *(i)* synthesis of the knowledge extracted from section 13.3 and the *(ii)* dynamic control systems from subsection 13.2.1. The following concepts and relationships are also defined using a textual description and aim at presenting an intellectual solution which is independent from ICT implementations. In the end, the designed ontology characterizes the goal of this paper: what are the core components of a remote biofeedback control system?

13.4.1 Concepts and Relationships

Seven concepts are the genesis of the designed conceptual mapping (as identified by the shaded box in Figure 13.3): cardiac, skin, Parkinson and diabetes mellitus diseases; space exploration; crisis response; and battle field. These initial concepts represent the applications that have been researched by the state-of-the-art section. From here, each bibliographic reference has been reviewed and its concepts were extracted to a list.[4] In addition, the relationships between those concepts were also designed. Whenever exists a textual explanation in any reference that connect to two or more concepts, an arrow is used. This approach allowed the creation of a conceptual map as depicted in Figure 13.3.[5] The following concepts were extracted from section 13.3: Accurate, Aerospace vehicles, Affordable, Communication capabilities, Decision making, Distributed environment, Emergency teams, Health Care Actor, Human health, Injured people, Life support systems, Measurement, Military Actor, Mobile device, New form of treatment, Power, Remote places, Secured connections, Sensor, Surrounding environment, and Tele Monitoring.

From the relationships, we identify the following. First similarity is identified. On the one hand, from the literature, it is agreed that diabetes mellitus, Parkinson, skin and cardiac diseases demand new form of treatments, which, by its turn, demands Tele Monitoring. On the other hand, crisis response, battle field and space exploration demand life support systems (pressured by remote places locations), which, by its turn also demands

---

[4] To easy recognition the concepts are presented by a Sans Serif font.
[5] The conceptual map was produced using *IHMC CMap Tools* public available at http://cmap.ihmc.us/.

Tele Monitoring. On both set of applications Tele Monitoring is identified as a solution for their existing problems.

With similar importance, Tele Monitoring is also referred in the literature as possible solution to enforce the prediction of Parkinson, skin and cardiac disease situations.

It is stated that health care, military and emergency teams actors are responsible for decision making. However, in the literature is recurrent the requirement from battle field and crisis response applications for new forms of decision making.

From a capability perspective, Tele Monitoring enables the measurement of Human health, injured people, aerospace vehicles and the surrounding environment using Mobile devices. Conversely, those measurements provide new sources of data for new form of treatment and decision making.

Finally, to obtain measurement, the mobile devices are pointed as large amount sources of data. These are affordable and have sensors, which are accurate, but, are heavily dependent on communication capabilities using secured connections and power. Moreover, nowadays, the mobile devices are embedded in distributed environments.

13.4.2 Conceptual Mapping

Figure 13.3 exhibits the conceptual mapping of remote biofeedback control systems. The major benefits of this mapping are: *(i)* to capture the tacit knowledge from the literature, *(ii)* to facilitate the understanding of a large quantity of written information in an understandable form for the researchers, and *(iii)* to find gaps in the knowledge structure to define new research opportunities. Furthermore, following the rigor imposed by DSR methodology, the conceptual map is analyzed using the four principles of *(i)* abstraction, *(ii)* originality, *(iii)* justification and *(iv)* benefit as proposed by Österle et al. (2011). In light of this, the CMap of Figure 13.3 delivers the following results:

1. *Abstraction* (the CMap must be applicable to a class of problems) — the presented CMap could be used to evaluate other telemonitoring applications, considering the fact that it is based in the synthesis of seven different applications and is supported by 49 bibliographic references. The applications are divided in a first set of medical context: cardiac disease, skin disease, Parkinson disease and diabetes mellitus disease; and a second set of major referred societal context: Space exploration, Crisis response and Battle field. This taxonomy is based in the most

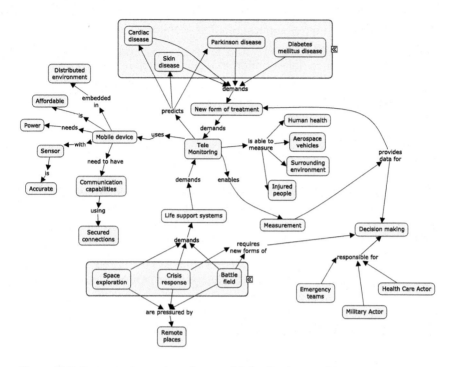

Figure 13.3 Conceptual mapping of remote biofeedback control systems.

relevant contexts that were obtained from the used scientific databases: IEEExplore, ACM library and PubMed.

2. *Originality* (the CMap must substantially contribute to the advancement of the body of knowledge) — by gathering an extensive state-of-the-art combined with Dynamic Systems Control principles, the proposed CMap exhibits the currently most demanded solutions for telemonitoring by both scientific and industrial efforts.

3. *Justification* (the CMap must be justified in a comprehensible manner and must allow validation) — the presented CMap identifies the Dynamic Systems Control enforcement principles, as abstracted in section 13.2.1, namely, observation is supported by **measurement** and control actuation is supported by **new form of treatment** and **decision making**. From a control pattern point of view (*cf.* Figure 13.2), the used bibliographic references recall to the feedforward pattern where the Human is a system to be controlled and the measurements are used to produce the control actuation. This consideration complies with the formal requirement for a biofeedback control systems: using vital signals

to decide upon the best actuation to be taken. Moreover, this CMap identifies an assessable network of concepts that drive to their enforcement: need to use **mobile device**, what to **measure** and who are the stakeholders involved.

4. *Benefit* (the CMap must deliver benefit, immediately or in the future for the respective stakeholder groups) — the proposed CMap benefits researchers in biofeedback telemonitoring control systems field making clear the limitations, opportunities, and gaps, that are demanding new solutions. A similar approach could be consulted in Guerreiro and Tribolet (2013). In this sense, the conceptualization presented in this paper is useful to serve as a guide to subsequent research in this domain.

13.5 DISCUSSION

Following the multidimensional approach used by Laudon and Laudon (2016), this section discusses the technological, ethical and social aspects that are raised during the course of state-of-the-art review.

13.5.1 Technological Constraints

Accordingly to PD telemonitoring literature, the partial functional prototypes sometimes fall into disuse. Furthermore, in technological terms, the following aspects are identified: *(i)* the great amount of data collected by the mobile devices sensors, is nowadays identified as a difficult issue for the identification of most appropriate data treatment to apply, *(ii)* the increasingly amount of power consumed, *(iii)* the inability to acquire any kind of signals, *e.g.*, human emotions, *(iv)* the telecommunications dependency for reliable data transfers and *(v)* high cost involved in the technologies development.

The vital signs acquisition, combined with light-weighted equipment (embedded in textiles) and devices, seams a promising research field.

13.5.2 Ethical & Social Aspects

Privacy of electronic health information appears to be a core concern in many works. This issue will demand new solutions in the years to come. Data encryption, storage medium protection and quality assurance risk are identified as technological solution. However, full access control models are demanded with a design integrated with all the subsystems involved in the telemonitoring solutions. This requirement is imposing research challenges to many scientific areas: computer science, information systems, telemedicine, *etc.*

On the one hand, concerns with the intellectual property is identified with major intensity in the mobile devices applications landscape. On the other hand, in the specific case of battle field telemonitoring, the military confidentiality minimizes the available information.

The adequacy of technological solutions with social surrounding environment is pointed as a critical success factor. To exemplify, the authors refer a need to *(i)* establish clear program goals, *(ii)* involve governments and institutions and *(iii)* disseminate of evaluation results. Moreover, it is important to remark that some countries already integrated the telemonitoring in their national health care strategy, *e.g.*, Finland.

Resistance to mobile technology is also referred in the telemonitoring literature. This resistance increases by ageing people. In addition, literature identifies that sending data directly to the health care actors, without moving from home, is recognized as a comfort enhancement.

Moreover, telemonitoring do not replace the health care actors. But rather, it acts as an initial assessment, supporting the professionals with more data, aiding the decision making processes.

13.6 CONCLUSIONS AND FUTURE WORK

Telemedicine area is researching, developing, and testing, new solutions that are able to integrate the tremendous technological advances with the current medical practices, in order to offer a cheaper, a more efficient and a more effective health care service. In this sense, telemonitoring is located as a sub discipline of telemedicine and aiming the use of information and communication technologies in order to monitor patients remotely.

The aim of this paper is to raise awareness on the current gaps and opportunities of the telemonitoring body of knowledge. Specifically, this paper analyzes the state-of-the art in the domain of seven remote telemonitoring control systems solutions that support Human contexts such as Parkinson disease, battle field, crisis responses, diabetes mellitus disease, dermatology, cardiac diseases and space exploration.

This recession supports the design of an ontology, using conceptual maps, that bound the core concepts and their relationships regarding the telemonitoring control systems solutions. This research result has the benefit of sharing a common understanding of the concepts between different researchers with distinct backgrounds. To obtain this result, we analyzed the contemporary state-of-the-art contained in 49 bibliographic references, and the common concepts and relationships are synthesized.

To that end, a design science research approach is followed, where *(i)* the design objective is to raise awareness on current telemonitoring gaps, *(ii)* the constructed artefact is a CMap and finally *(iii)* the artefact is evaluated by an argumentative discussion in regard to technological constraints, ethical and social aspects.

Future work will involve *(i)* a broader bibliographic recension, *(ii)* the addition of other evaluation strategy based in semi-structured interviews to cross check the ontology and *(iii)* use more refined ontology models, *e.g.*, WOSL or DEMO.

ACKNOWLEDGEMENTS

The author of this paper would like to acknowledge the 2014/2015 students of Telemedicine course at Universidade da Beira Interior, Covilhã, Portugal, for their contributions to this project. Moreover, the author expresses his gratitude by the helpful comments and suggestions presented by the reviewers.

REFERENCES

Abchir, M.B., Basabe, P., et al., 2003. United Nations international strategy for disaster reduction. In: Living with Risk, pp. 1–412.

Adelakun, O., 2014. The need for telemedicine and e-health in Haiti. In: Informatics in Africa Conference HELINA14, p. 58.

Ajami, S., Lamoochi, P., 2014. Use of telemedicine in disaster and remote places. J. Educat. Health Promot. 3.

American Heart Association, 2015. Building healthier lives, free of cardiovascular diseases and stroke. Accessed in September, http://www.heart.org/HEARTORG/.

Anliker, U., et al., 2004. Amon: a wearable multiparameter medical monitoring and alert system. IEEE Trans. Inf. Technol. Biomed. 8 (4), 415–427.

Aponte, V., Finch, D., Klaus, D., 2006. Considerations for non-invasive in-flight monitoring of astronaut immune status with potential use of MEMS and NEMS devices. Life Sci. 79 (14), 1317–1333.

Assad, A., de Weck, O.L., 2015. Model of medical supply and astronaut health for long-duration human space flight. Acta Astronaut. 106, 47–62.

Ba, X., Wang, P., 2012. Design of soldier status monitoring and command and control system based on Beidou system. In: 2012 2nd International Conference on Computer Science and Network Technology (ICCSNT). IEEE, pp. 1362–1366.

Bertalanffy, L.v., 1969. General Systems Theory. George Braziller, New York.

Binks, S., Benger, J., 2007. Tele-education in emergency care. J. Emerg. Med. 24 (11), 782–784.

Blackwell, T., Bosse, M., 2007. Use of an innovative design mobile hospital in the medical response to Hurricane Katrina. Ann. Emerg. Med. 49 (5), 580–588.

Brockett, R., 1993. Hybrid models for motion control systems. In: Essays on Control: Perspectives in the Theory and Its Applications, pp. 29–53.

Bustamante, P., Grandez, K., Solas, G., Arrizabalaga, S., 2010. A low-cost platform for testing activities in Parkinson and ALS patients. In: 2010 12th IEEE International Conference on e-Health Networking Applications and Services (Healthcom). IEEE, pp. 302–307.

Cañas, A.J., Carff, R., Hill, G., Carvalho, M., Arguedas, M., Eskridge, T.C., Lott, J., Carvajal, R., 2005. Concept maps: integrating knowledge and information visualization. In: Knowledge and Information Visualization. Springer, pp. 205–219.

Canina, M., Newman, D.J., Trotti, G.L., 2006. Preliminary considerations for wearable sensors for astronauts in exploration scenarios. In: 3rd IEEE/EMBS International Summer School on Medical Devices and Biosensors, 2006. IEEE, pp. 16–19.

Castro, A.J.M., Oliveira, E., 2011. A new concept for disruption management in airline operations control. Proc. Inst. Mech. Eng., G J. Aerosp. Eng. 225 (3), 269–290. http://dx.doi.org/10.1243/09544100JAERO864. http://pig.sagepub.com/content/225/3/269.abstract. http://pig.sagepub.com/content/225/3/269.full.pdf+html.

Cermack, M., 2006. Monitoring and telemedicine support in remote environments and in human space flight. Br. J. Anaesth. 97 (1), 107–114.

Cheung, R.T., Davis, I.S., 2011. Landing pattern modification to improve patellofemoral pain in runners: a case series. J. Orthop. Sports Phys. Ther. 41 (12), 914–919.

Cho, D.-K., Chang, C.-W., Tsai, M.-H., Gerla, M., 2008. Networked medical monitoring in the battlefield. In: MILCOM 2008, IEEE Military Communications Conference 2008. IEEE, pp. 1–7.

Cleland, J.G., Louis, A.A., Rigby, A.S., Janssens, U., Balk, A.H., 2005. Noninvasive home telemonitoring for patients with heart failure at high risk of recurrent admission and death: the Trans-European Network-Home-Care Management System (TEN-HMS) study. J. Am. Coll. Cardiol. 45 (10), 1654–1664.

Crowell, H.P., Davis, I.S., 2011. Gait retraining to reduce lower extremity loading in runners. Clin. Biomech. 26 (1), 78–83.

Doze, S., Simpson, J., Hailey, D., Jacobs, P., 1999. Evaluation of a telepsychiatry pilot project. J. Telemed. Telecare 5 (1), 38–46.

Duda, K.R., Vasquez, R.A., Middleton, A.J., Hansberry, M.L., Newman, D.J., Jacobs, S.E., West, J.J., 2015. The variable vector countermeasure suit (v2suit) for space habitation and exploration. Front. Syst. Neurosci. 9.

Dudok, E., Guerreiro, S., Babkin, E., Pergl, R., van Kervel, S., 2015. Enterprise operational analysis using demo and the enterprise operating system. In: Aveiro, D., Pergl, R., Valenta, M. (Eds.), Advances in Enterprise Engineering IX. In: Lecture Notes in Business Information Processing, vol. 211. Springer International Publishing.

Fei, D.-Y., Zhao, X., Boanca, C., Hughes, E., Bai, O., Merrell, R., Rafiq, A., 2010. A biomedical sensor system for real-time monitoring of astronauts physiological parameters during extra-vehicular activities. Comput. Biol. Med. 40 (7), 635–642.

Force USA, 2001. Command and control. Tech. rep., Air Force Doctrine Document 2-8.

Franklin, G., Powell, J., Emami-Naeini, A., 1991. Feedback Control of Dynamic Systems, 2d ed. Addison-Wesley Publishing Company.

Freund, B., 2008. Real-time monitoring of our warfighters health state: the good, the bad, and the ugly. Tech. rep., DTIC Document.

Gaaloul, K., Guerreiro, S., Proper, H.A., 2014. Modeling access control transactions in enterprise architecture. In: IEEE 16th Conference on Business Informatics, vol. 1. CBI 2014, Geneva, Switzerland, July 14–17, 2014, pp. 127–134.

Garshnek, V., Burkle, F.M., 1999. Applications of telemedicine and telecommunications to disaster medicine. J. Am. Med. Inform. Assoc. 6 (1), 26–37.

Gavin III, J.R., Alberti, K., Davidson, M.B., DeFronzo, R.A., et al., 1997. Report of the expert committee on the diagnosis and classification of diabetes mellitus. Diabetes Care 20 (7), 1183.

Gorini, A., Capideville, C.S., De Leo, G., Mantovani, F., Riva, G., 2011. The role of immersion and narrative in mediated presence: the virtual hospital experience. Cyberpsychol. Behav. Soc. Netw. 14 (3), 99–105.

Grandry, E., Feltus, C., Dubois, E., 2013. Conceptual integration of enterprise architecture management and security risk management. In: The Fifth Workshop on Service oriented Enterprise Architecture for Enterprise Engineering (SoEA4EE 2013), an International Workshop of the 17th IEEE International EDOC Conference. Vancouver, BC, Canada. IEEE.

Grishin, O., Grishin, V., Bryzgalov, A., Smirnov, S., 2012. The modern concept of war fighter physiological status monitoring: literature review. World Appl. Sci. J. 19 (8), 1149–1156.

Guerreiro, S., 2014. Towards multi-level organizational control framework to manage the business transaction workarounds. In: 16th International Conference on Enterprise Information Systems. INSTICC, pp. 288–294.

Guerreiro, S., Tribolet, J., 2013. Conceptualizing enterprise dynamic systems control for run-time business transactions. In: 21st European Conference on Information Systems, p. 5.

Heffner, V.A., Lyon, V.B., Brousseau, D.C., Holland, K.E., Yen, K., 2009. Store-and-forward teledermatology versus in-person visits: a comparison in pediatric teledermatology clinic. J. Am. Acad. Dermatol. 60 (6), 956–961.

Hill, T.R., 2011. Exploration space suit architecture and destination environmental-based technology development. In: 2011 IEEE Aerospace Conference. IEEE, pp. 1–14.

Hofstede, G., 1978. The poverty of management control philosophy. Acad. Manag. Rev. 3 (3), 450–461.

Jaeger, A., Baliga, B., 1985. Control systems and strategic adaptation: lessons from the Japanese experience. McGill University, Montreal, Canada Faculty of Management and Texas Tech. University, Lubbock, Texas, U.S.A. College of Business Administration Strateg. Manag. J. 6, 115–134.

Klein, K.R., Nagel, N.E., 2007. Mass medical evacuation: Hurricane Katrina and nursing experiences at the New Orleans airport. Disaster Manag. Response 5 (2), 56–61.

Kost, G.J., Tran, N.K., Tuntideelert, M., Kulrattanamaneeporn, S., Peungposop, N., 2006. Katrina, the tsunami, and point-of-care testing optimizing rapid response diagnosis in disasters. Am. J. Clin. Pathol. 126 (4), 513–520.

Kuzuya, T., et al., 2002. Report of the committee on the classification and diagnostic criteria of diabetes mellitus. Diabetes Res. Clin. Pract. 55 (1), 65–85.

Laudon, K., Laudon, J., 2016. Management Information Systems, 14th ed. Prentice Hall.

LeMoyne, R., Coroian, C., Mastroianni, T., 2009. Quantification of Parkinson's disease characteristics using wireless accelerometers. In: ICME International Conference on Complex Medical Engineering, 2009, CME. IEEE, pp. 1–5.

LeMoyne, R., Mastroianni, T., Cozza, M., Coroian, C., Grundfest, W., 2010. Implementation of an iPhone for characterizing Parkinson's disease tremor through a wireless accelerometer application. In: 2010 Annual International Conference of the IEEE, Engineering in Medicine and Biology Society (EMBC). IEEE, pp. 4954–4958.

Lim, H.B., Ma, D., Wang, B., Kalbarczyk, Z., Iyer, R.K., Watkin, K.L., 2010. A soldier health monitoring system for military applications. In: 2010 International Conference on Body Sensor Networks (BSN). IEEE, pp. 246–249.

Magalhães, R., 2005. Fundamentos da Gestão do Conhecimento Organizacional, 1st ed. Edições Sílabo, Lda.

Matos, M.G.d., 2006. Organizational engineering, an overview of current perspectives. M.S. thesis. Universidade Técnica de Lisboa, Instituto Superior Técnico.

Meystre, S., 2005. The current state of telemonitoring: a comment on the literature. Telemed. J. e-Health 11 (1), 63–69.

Mundt, C.W., et al., 2005. A multiparameter wearable physiologic monitoring system for space and terrestrial applications. IEEE Trans. Inf. Technol. Biomed. 9 (3), 382–391.

National Heart, Lung, and Blood Institute, 2015. Accessed in September, http://www.nhlbi.nih.gov/.

Nicogossian, A.E., Pober, D.F., Roy, S.A., 2001. Evolution of telemedicine in the space program and earth applications. Telemed. J. e-Health 7 (1), 1–15.

Noehren, B., Scholz, J., Davis, I., 2010. The effect of real-time gait retraining on hip kinematics, pain and function in subjects with patellofemoral pain syndrome. Br. J. Sports Med. 45 (9), 691–696.

Norris, A.C., 2002. Essentials of Telemedicine and Telecare. Wiley.

Ogata, K., 1997. Modern Control Engineering. Prentice-Hall, Inc.

Okuno, R., Yokoe, M., Fukawa, K., Sakoda, S., Akazawa, K., 2007. Measurement system of finger-tapping contact force for quantitative diagnosis of Parkinson's disease. In: 29th Annual International Conference of the IEEE, Engineering in Medicine and Biology Society, 2007. EMBS 2007. IEEE, pp. 1354–1357.

Österle, H., et al., 2011. Memorandum on design-oriented information systems research. Eur. J. Inf. Syst. 20 (1), 7–10.

Pressman, A.R., Kinoshita, L., Kirk, S., Barbosa, G.M., Chou, C., Minkoff, J., 2014. A novel telemonitoring device for improving diabetes control: protocol and results from a randomized clinical trial. Telemed. e-Health 20 (2), 109–114.

Qiantori, A., Sutiono, A.B., Hariyanto, H., Suwa, H., Ohta, T., 2012. An emergency medical communications system by low altitude platform at the early stages of a natural disaster in Indonesia. J. Med. Syst. 36 (1), 41–52.

Recht, B., D'Andrea, R., 2004. Distributed control of systems over discrete groups. IEEE Trans. Autom. Control 49 (9), 1446–1452.

Ribeiro, M.I., 2002. Análise de sistemas lineares, vols. 1 e 2. IST Press, ISBN 972-8469-13-6.

Scheinfeld, N., Kurz, J., Teplitz, E., 2003. A comparison of the concordance of digital images, live examinations, and skin biopsies for the diagnosis of hospitalized dermatology consultation patients. SKINmed: Dermat. Clinic. 2 (1), 14–19.

Simmons, S., Alverson, D., Poropatich, R., DIorio, J., DeVany, M., Doarn, C.R., 2008. Applying telehealth in natural and anthropogenic disasters. Telemed. e-Health 14 (9), 968–971.

Spelmezan, D., Schanowski, A., Borchers, J., 2009. Wearable automatic feedback devices for physical activities. In: Proceedings of the Fourth International Conference on Body Area Networks. ICST (Institute for Computer Sciences, Social-Informatics and Telecommunications Engineering), p. 1.

Stone, R.A., Sevick, M.A., Rao, R.H., Macpherson, D.S., Cheng, C., Kim, S., Hough, L.J., DeRubertis, F.R., 2012. The diabetes telemonitoring study extension: an exploratory randomized comparison of alternative interventions to maintain glycemic control after withdrawal of diabetes home telemonitoring. J. Am. Med. Inform. Assoc. 19 (6), 973–979.

Surangsrirat, D., Thanawattano, C., 2012. Android application for spiral analysis in Parkinson's disease. In: 2012 Proceedings of IEEE, Southeastcon. IEEE, pp. 1–6.

Tharion, W.J., Kaushik, S., 2006. Graphical user interface (gui) for the Warfighter Physiological Status Monitoring (WPSM) system — US army medic recommendations. Tech. rep., DTIC Document.

Topic maps, 2015. Accessed in September, http://www.iso.org/iso/catalogue_detail.htm?csnumber=43940.

Tzallas, A.T., et al., 2014. Perform: a system for monitoring, assessment and management of patients with Parkinson's disease. Sensors 14 (11), 21329–21357.

Uraikul, V., Chan, C., Tontiwachwuthikul, P., 2007. Artificial intelligence for monitoring and supervisory control of process systems. In: Special Issue on Applications of Artificial Intelligence in Process Systems Engineering. Eng. Appl. Artif. Intell. 20 (2), 115–131.

Van Der Aalst, W., 2011. Process Mining: Discovery, Conformance and Enhancement of Business Processes. Springer Science & Business Media.

Vo, A.H., Brooks, G.B., Bourdeau, M., Farr, R., Raimer, B.G., 2010. University of Texas Medical Branch telemedicine disaster response and recovery: lessons learned from Hurricane Ike. Telemed. e-Health 16 (5), 627–633.

von Alan, R.H., March, S.T., Park, J., Ram, S., 2004. Design science in information systems research. MIS Q. 28 (1), 75–105.

Wang, M., Wang, B., Zou, J., Chen, L., Shima, F., Nakamura, M., 2010. A new quantitative evaluation method of Parkinson's disease based on free spiral drawing. In: 2010 3rd International Conference on Biomedical Engineering and Informatics (BMEI), vol. 2. IEEE, pp. 694–698.

Winter, R., 2008. Design science research in Europe. Eur. J. Inf. Syst. 17 (5), 470–475.

Witten, I.H., Frank, E., 2005. Data Mining: Practical Machine Learning Tools and Techniques. Morgan Kaufmann.

Wurm, E.M., Hofmann-Wellenhof, R., Wurm, R., Soyer, H.P., 2008. Telemedicine and teledermatology: past, present and future. J. Dtsch. Dermatol. Ges. 6 (2), 106–112.

Yokoe, M., Okuno, R., Hamasaki, T., Kurachi, Y., Akazawa, K., Sakoda, S., 2009. Opening velocity, a novel parameter, for finger tapping test in patients with Parkinson's disease. Parkinsonism Relat. Disord. 15 (6), 440–444.

CHAPTER 14

The Role of Smart Homes in Intelligent Homecare and Healthcare Environments

Laura Vadillo Moreno*, María Luisa Martín Ruiz*,
Javier Malagón Hernández*, Miguel Ángel Valero Duboy*,†,
María Lindén‡

* *Telematics and Electronics Engineering Department, ETSIS Telecommunication, Politécnica de Madrid University, Spain*
† *National Reference Centre of Personal Autonomy and Technical Aids (CEAPAT), Health, Social Services and Equality Ministry, Spain*
‡ *School of Innovation, Design & Engineering, Mälardalen University, Västerås, Sweden*

14.1 INTRODUCTION

The ageing population is a widespread fact, which affects most developed countries. This ageing of the world population has important consequences, both economically and socially. An ageing population requires more care, an increase in the number of qualified people for their support, and more efficient management of health resources. This situation demands new strategies in health services to assume effectively the growing health needs of ageing people. The strategy of active ageing proposed by the World Health Organization focuses on enhancing the quality of life of ageing people, optimizing opportunities for health, participation and security. Health technologies have to rely on situations that are not "under control", unlike a medical unit, where all the environmental conditions are well known. The health actions are conducted from two important perspectives: on one hand, to reinforce and promote prevention plans to delay or prevent chronic diseases, and on the other hand, to treat timely detected diseases, providing needed care and support to minimize their consequences. These actions include the development of high quality and affordable services to provide a continuum of care that includes health promotion, disease prevention, appropriate treatment of chronic diseases and equitable access.

Smart homes offer good opportunities in these scenarios, providing a monitored environment managed by context-aware services. This infras-

Ambient Assisted Living and Enhanced Living Environments.
DOI: http://dx.doi.org/10.1016/B978-0-12-805195-5.00014-4

tructure is the basis for the creation of intelligent healthcare smart home services.

The current chapter presents an initial discussion of the smart home concept and its main scenarios. Thereafter, an introduction to the main building blocks of smart homes and their interrelationships is presented. In subsequent sections, these blocks are detailed: middleware tools, home automation technologies and sensors, the acquisition context, knowledge base, reasoning and learning. Furthermore, the chapter refers to contemporary paradigms like Big Data, cloud computing and the Internet of Things. Finally, the chapter takes into account different challenges for professionals designing smart homes.

14.2 THE SMART HOME CONCEPT

The smart home concept has been established in the last years with different, though similar, interpretations. The concept of the smart home has implied from the beginning the idea of the house as an assistant in domestic tasks. The idea of this concept was published by Joseph Deken, one of the pioneers, in his book *The Electronic Cottage* (Deken, 1982), which introduces the personal computer as a slave to carry out tasks that could improve the life of the inhabitants of the home, executing diverse tasks related to cleaning, security, temperature control, etc.

Firstly, the development of intelligent buildings originated in the late 80s and early 90s. These intelligent buildings prioritized safety systems, automation and control of elements, such as light, temperature and location. Thus, the concept of home automation was born, limited primarily to the control of housing.

Nowadays, the concept of the smart home uses different expressions with different meanings. The "domotic house" is a commonly used term centered on the automation of devices to control doors, lights, etc., while the expression "connected home" is centered more on permanent connection to the Internet to provide services to the users. A domotic house includes devices and systems that need be integrated with the home communications or external networks to provide services to the user (such as multimedia, comfort, entertainment, energy conservation or telehealth). This integration defines the concept of the digital home (Rae and Kumar, 2008). Devices used in a digital home can be just a light controller, a refrigerator or a television that knows its state, telephony, security systems, etc. All those objects will be connected to the home network to give their states or receive instructions.

However, a digital home does not properly define the smart home concept. Although some definitions of a smart home are centered on "intelligent automation" (Lutolf, 1992), akin to domotic or digital home concepts, there is no common consensus definition of this concept, and "smart" can comprise a wide range of possibilities. The characteristic of "smart" applied to the home should include the capability of reasoning over the collected data from the sensors, including previous knowledge of the context, adaptability to changing situations in the environment and their inhabitants, and learning.

This intelligence had already been taken into account in the late '90s by Kidd, who defined the concept of the "aware home", considering the use of "wearable computers" together with intelligent environments to allow learning about users' habits and behavior and carrying out intelligent automatic and adaptive tasks involving more convenient and personalized services (Kidd et al., 1999). This idea of context-aware services in the home environment is also reflected in the intervening years by Alam et al. (2012) and Anbarasi and Ishwarya (2013), the latter introducing the smart home as "an application of ubiquitous computing in which the home environment is monitored by ambient intelligence to provide context-aware services and facilitate remote home control". This definition highlights the value of observation to provide proactive services, especially valuable in supporting the independent living of people with disabilities and elder adults.

A smart home must necessarily consider the human factors dimension. The most important characteristic of a smart home should be the user or users and the satisfaction of their needs. This factor is also included in others definitions of a smart home, such as Satpathy's (2006) definition, which describes a smart home as a home which is smart enough to assist the inhabitants to live independently and comfortably with the help of technology. Valero Duboy (2015) integrates the intelligent dimension of the smart home with the dimension of human factors and defines the smart home as a digital home that:

- Understands the needs of each user (taking into account that the user needs can change depending on the context);
- Correctly solves problem situations for each person;
- Meets performance needs and expectations in the ethical, legal, security and quality aspects of life dimensions;

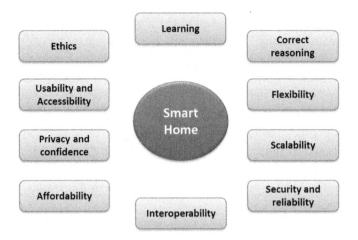

Figure 14.1 Essential aspects of smart home.

- Learns and acquires the necessary experience to incorporate responses or habits not previously known, depending on the capabilities and expectations of each person and their context of use in the residential environment.

The final definition includes the digital home concept, adding the necessary intelligence to improve the quality of life of the user, taking into account the human factors dimension. Figure 14.1 shows the key aspects of the digital home that can be defined as "smart", as identified by Valero Duboy (2015).

14.3 TELEHEALTH SCENARIO IN SMART HOME

The concept of telehealth is a broad term that can combine diverse services. It is currently used to refer to healthcare services including sanitary education, information services on health at a distance, telecare and telemedicine services.

The first ideas of healthcare in the "home of the future" revealed this integration as the essence of the smart home, highlighting its capacity for reasoning, risk resolution and learning, and its interaction with the user (Warren et al., 1999). This vision is fully aligned with the concept of ambient intelligence (AmI), which is defined by the Information Society Technologies Advisory Group (ISTAG) as a vision in which people will live in your environment, surrounded by intelligent interfaces integrated into everyday objects. These objects will be capable of recognizing, re-sponding and adapting their behavior to the needs of people. Thus, AmI environments are mainly characterized by their ubiquity, transparency and intelligence, with an emphasis on user-friendliness, more efficient services support, user empowerment, and support of human interactions (European Commission, 2001).

Digital home and building automation provides AmI with the necessary infrastructure to achieve an adequate environment for home care. Control devices allow the recording of user and environment data, analysis of it, and reaction to the environment or sending an alarm call. The situation of the user at home can be observed through the active interaction of the person and the devices or systems (i.e. putting on coffee maker or television, or using medical devices), through passive devices sited in walls or doors (i.e. presence, temperature, etc.) or through smart objects based in the new paradigm of the Internet of Things (IoT) or wearable sensors, including smart textiles.

Furthermore, the health status of the elderly is closely related to their daily behaviors (such as sleep action and routines of daily activities). Health management can be enhanced if these data can be acquired and processed in an automated way.

Thus, these devices, combined with distributed intelligent systems, can detect and respond to changing situations that are generated at home, plan-ning activities according to the particular context and predicting possible new potentially dangerous situations.

Therefore, applying the smart home concept in the healthcare scenario, we can define a smart home as the integration of digital home technologies (devices, systems and communications) with the necessary intelligence to recognize the health status and improve the person's quality of life (effi-ciently detecting and resolving situations dangerous to the person due to their health condition or their environment, or providing telemedicine ser-vices) and taking into account the needs and expectations of its inhabitants.

14.3.1 Telecare

In the healthcare scenario, the most commonly known service is the tele-care service. The telecare service is a social and sanitary care system inside the home designed to cover the needs of people that require assistance or quick help in case of emergency.

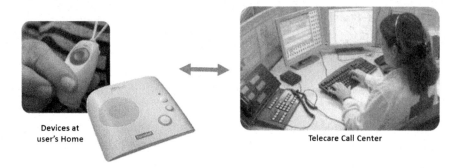

Figure 14.2 Home telecare service.

Telecare has the aim of helping elderly or disabled people stay as long as possible in their homes independently. It is an important element of security and peace of mind for the user and his family — especially for those people that live alone, those people that have fear or anxiety caused by geographic isolation, people with disability, or people that may suffer repeated risk situations, such as falls. The vision of the telecare concept is reflected in the standard definition offered by the ETSI TR 102 415 recommendation, in which we can observe a clear convergence of healthcare and social care with technology: Telecare includes the provision of social care services or health to people at home in a community, with the support based on information technology and communications (ICT) systems (ETSI, 2005).

The usual service, commonly known as home telecare, consists of a terminal unit sited in the home of the user, which includes a speaker and a microphone attached to the telephone of the user, and a handset device, which the user wears as a pendant. The service provides constant assistance: it is available 24/7. If the user needs help, he may press the button on the terminal unit or the pendant (Figure 14.2). When the user presses the button to ask for help, he is attended by a professional at the telecare center, who will provide an answer to the situation (mobilizing others human resources, such as familiars or emergency services). Telecare centers are call centers that have information about the user profile (where the person lives, known health problems, close contacts, etc.) to enable them to take better decisions when the user calls the center as a function of the profile and the situation observed during the call. This service usually includes other services such as appointment reminders or reminders to takes medication.

Despite its indisputable benefits (Celler et al., 1999), this telecare system cannot automatically detect risk situations. The user needs to press a button to obtain assistance.

A second generation of home telecare service appears as an evolution of the previous one. This generation adds other automatic alarm devices at home, such as smoke, gas and flood. The activation of any of these sensors causes a direct call to the telecare center. The alarm is also communicated to the user through an audible signal, possibly including flashing lights or vibration elements.

Fall detectors and bed and chair occupancy detectors are also included in this second-generation service, which detects when the user is seated or lying down.

Finally, a third generation of telecare systems is leading to telemonitoring services at home and during people's mobility. In these systems, user and environment variables are collected and used to help take decisions about risk situations to the person based on previous knowledge acquired and the lifestyle of the user. These data can be analyzed remotely by professionals or the relatives of the user. Also, they can be used as a marker of activities performed by the user and can provide reassurance of their well-being (Stowe and Harding, 2010).

14.3.2 Telemedicine

Telemedicine is defined as the use of ICT to provide clinical services to patients remotely. Examples of telemedicine include video consultations with specialists, medical evaluations, remote diagnostics and treatments, and digital transmission of medical images. Telemedicine is used nowadays as a valuable tool in certain specialities such as radiology and neurosurgery, in which specialists can receive and analyze digital images remotely and do not need physical contact with the patient (Hailay et al., 2002).

In this vein, medical telecare appears more adapted to medical treatment at home. The objective of medical telecare is to provide medical and specialized care (clinical diagnostics, monitoring and medical treatment) to patients in their houses in a remote way. The user at home can measure their vital signs (i.e. blood pressure, temperature and pulse) and send them to the professional. The professional can process this measures and can send a diagnosis to the user. It is also possible to add teleconference or video-conference as part of the service delivery to provide visual contact with the professional in real time. Also, these systems can include detection of any health risk to the user.

14.3.3 Monitoring in Telehealth Scenarios

Accordingly, there are three typical kinds of monitoring at home in telehealth scenarios:

- **Alert detection** as a result of monitoring based on environmental sensors (smoke detector, flood detector, fall detector, etc.). This scenario can be reactive or proactive. The intelligence includes basic cause-effect behaviors (e.g. if the smoke detector is activated, a call is sent to the telecare center). This monitoring can also include a major intelligence that appears as a result of the combination of various sensors to detect other possible risk situations (e.g. location sensors at home do not detect movement all day) (Valero et al., 2009).
- **Medical monitoring** is based on the monitoring of biomedical variables to obtain the health status and detect emergencies related to health situations, detect early diseases, or take decisions about medication prescription. It is possible also to integrate remote control with clinical decision support systems (CDSS) for medical consultation results (Mouttham et al., 2009; Villarreal et al., 2011; Islam et al., 2015; Hussain et al., 2012).
- **Activities monitoring** involves applications for activity recognition analyzing the activities that a person carries out at home. Continuous remote monitoring of patients enables the telecare system to track lifestyle changes over time and detect patterns of possible pathologies or alarm signs (Ni et al., 2015; Suryadevara et al., 2013; Ridi et al., 2015; Benmansour et al., 2015).

The need for a robust anomaly detection model is fundamental to predict facts that require immediate attention on the part of the user or specialized personnel to manage the situation.

14.4 TELEHEALTH IN SMART HOME: MAIN COMPONENTS

Models and architectures used to support healthcare services in a smart home are an area of active research. There are many designs for equipment and implements. Changes among architectures are mainly related to the kind of sensors or detectors, the context abstraction level, the communication model, the kind of reasoning system or machine learning algorithms, the security procurements, the extensibility and the reusability. However, they all apply the principles of context-aware systems, presenting very similar components and functionalities.

Context-aware systems appear within the AmI concept, proposing new ways of interaction between the person and ICT. The main characteristic of the context-aware system is the capability of adaptation to the environment. This kind of system can provide important opportunities for the development of services focused on improving the quality of life and personal autonomy of the person (Alwan et al., 2006; Solanas et al., 2014).

To provide systems adapted to the user and user environment, it is necessary to use systems that allow the efficient acquisition and organization of information and a powerful reasoning mechanism to manage this information. There are four basic elements considered in all context-aware healthcare monitoring systems in a smart home, which can be extrapolated to other scenarios (Loke, 2006; Miraoui et al., 2008; Ni et al., 2015; Hassanalieragh et al., 2015):

- The environment sensors or devices to acquire information from the physical world and actuators, which allow changes to the environment. A context-aware platform should obtain all useful information from the physical world relevant to its action domain.
- Support for context information gathering from different sources and delivery of appropriate context information to different services or functionalities.
- Storage and management of contextual information. This could include the context data representation and the registry of acquired and processed information. Furthermore, these elements should contain facilities for allowing the specification of different system behaviors in different contexts.
- Processing, which includes data interpretation for reasoning tools to process data collected by the environment in order to infer new knowledge. This part of the system is responsible for the reaction and interaction with the environment.

A typical general architecture for context-aware technology in the healthcare environments in a smart home is shown in Figure 14.3. This general architecture starts in the physical layer. The physical layer includes all physical elements, which acquire information from the physical world (environment and people). A second layer, usually designated the context adapter,

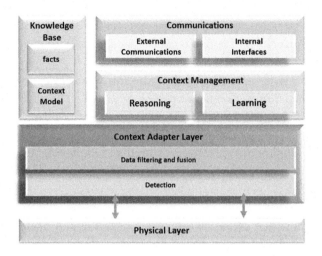

Figure 14.3 Typical general architecture of a context-aware healthcare environment in a smart home.

gathers data acquired by the physical layer and adapts these data so they can be assimilated by the rest of the system according to the context model.

The knowledge base component includes facts collected, which require storage, and the context model, which defines which data can interpret the system, the characteristics of these data, and the relationships with other data. Moreover, the context model can define the behaviors of the system as a result of the relationships between the elements of the model.

The simplest reasoning process appears as a result of including an inference engine to contrast data collected externally with the context model. Others kind of reasoning can be included, based on rules, definitions or probabilities, and can be combined as hybrid context models. Predefined reasoning behavior can be completed with machine learning algorithms to allow knowledge of the system to evolve and the adaptation process to improve with the changing conditions of the environment. Some authors have called this layer as the service layer, showing its capability of creating high-level services through aggregation and reasoning about contextual data obtained by the sensors.

Finally, these type of architecture should include a specific component to manage communications. There are two typical communications: external (e.g. to advise other people of an emergency at home) and internal (interfaces of the system with their users).

However, implementing a context-aware system requires many issues to be addressed (Satyanarayanan, 2001), among them: What are relevant data that should be taken into account? How is the context represented internally? Is historical context useful? What algorithms should be included to process the context? Under what circumstances should one part of the context be used in preference over another? Should location information be treated just like any other context information, or should it be handled differently?

14.5 MIDDLEWARE TOOLS

Most relevant middleware tools used for development context-aware systems in smart homes are based on the Open Service Gateway initiative (OSGi), the Foundation for Intelligent Physical Agents (FIPA) and web services (WS).

OSGi is an open standard for a service-oriented dynamic software based on a components model (OSGI, 2015). The specification defines modular software components and services organized into bundles. It offers utilities to create and add modules in execution time and interact with other system modules, and it simplifies tasks such as starting, finishing or updating a new component or service. However OSGi interfaces do not have the capability to provide autonomous services with advanced characteristics, remaining at the functional level (interface description, role identification, and models to provide and manage services). A typical OSGi architecture for context-aware services was described by Ricquebourg et al. (2006). He defines the PCIA model (perception-context inference-action) and presents an architecture to take into account the context of the smart home based on layers.

On the other hand, FIPA is an open standard for agent and multi-agent systems. A multi-agent system is a distributed system composed of autonomous software entities existing in a concrete environment and interacting with them. Features of AmI and context-aware environments perfectly fit with multi-agent systems characteristics (Sahli, 2008; Pech, 2013; Olaru et al., 2013).

Agents have reactive and proactive autonomous reasoning capabilities. An agent has its own beliefs, desires, intentions and goals, and can incorporate learning capability. Also, an agent can interact with other agents in the system and externally satisfy a common goal, adapting the behavior according to the situation and tasks in the system.

One of the main advantages of multi-agent systems is their modularity. Tasks can be divided into a set of basic and decoupled distributed components, making it possible for each agent to use the most appropriate paradigm to achieve its goals. This distributed functionality, together with their capability to communicate, negotiate and coordinate, influences the efficiency of the workflow, the easiness of reutilization, and the flexibility to include new components or capabilities in the system (Wooldridge, 2009).

Some platforms include OSGi and FIPA in their implementation. OSGi is used more like a service residential platform to ease the installation and execution of multiple services in an individual home gateway, while FIPA provides the necessary intelligence to offer autonomous services, which adapt their behavior to the user and environment needs.

Finally, web services are a set of practices for developing services as web applications. Web services are components that provide a concrete functionality or application to solve complex tasks over the Internet. Web services can be used by other programs or applications. They include REST, a popular tool in cloud-based APIs and services, and SOAP-based messaging protocol. Smart home context-aware applications can include some kind of external reasoning or information management provided by web services (Kang and Park, 2013; Lasierra et al., 2014).

The integration of web services (such as REST) with OSGi is common. This model implements the remote functionality using web services. OSGi is used as middleware to communicate with and between the devices, to solve communications problems among different logical devices, and to provide utilities to publishing services. In the same way, it is also possible to find the integration of agents and web services. Agents can contain and reason about the semantics of the web services, and can mediate and compose web services.

14.6 HOME AUTOMATION TECHNOLOGIES AND SENSORS

At the beginning of the chapter, a smart home is described like an extension of a home automation system (or a digital home) to which intelligence is added. This section presents some different automation technologies, leaving the following sections to discuss intelligence.

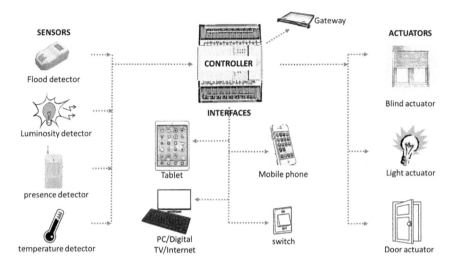

Figure 14.4 Elements of an integrated automation system.

14.6.1 System Elements

According to Kyas (2013), information in a home automation system is collected from **sensors** or **interfaces** (inputs). This information is processed by one or more **controllers** that send commands to the **actuators** (outputs). These devices are connected via an internal network, also called the home area network (HAN), in order to share and exchange information according to a **protocol**. If the network has a set of systems with different network protocols, a **residential gateway** is used for communication between these systems or for the connection of the HAN with the outside (usually the Internet). Figure 14.4 shows a representation of elements of an integrated automation system.

A summary of each of these elements follows.

- **Sensors**. These are devices that monitor the environment by detecting physical or chemical magnitudes. Sensors transform them into digital information and transmit it to the HAN. There are many types of sensor, such as environmental sensors (water, gas, smoke, temperature, wind, humidity, rain, lighting, movement, etc.), physiological sensors (measuring pulse, respiration, blood pressure, etc.) or multimedia sensors (microphones, cameras, etc.). Greater detail about sensors and their functionalities is given in section 14.6.6.

- **Interfaces**. These devices can display information about the system to the users (on a screen or smartphone) and can interact with the system. An example of an interface is a switch to command a lighting system to turn on/off; a keyboard for the opening of a door; a control for the opening of a garage door, for the opening of blinds or for the turning on or off of lights; a tablet or smartphone that runs an application to interact with the system, etc. These interfaces, also with sensors, communicate with the controller/controllers in order to decide which actuator/actuators must perform the actions.

- **Controller**. This device manages the system, either according to the information received from sensors or interfaces or according to the way, it is suggested by a stored schedule. The system may have a single controller or multiple distributed controllers.

- **Actuator**. This receives l commands from the controller to perform an action on a household component. For example, it can control an electromagnetic relay in order to turn lights off or on; an actuator may also, for example, control a motor to lower/raise the blind or to open/close a garage door.

- **Protocol**. The information transmitted by the HAN between sensors, interfaces, actuators and controllers is carried out by the transmission medium (which can be wired or wireless) according to a specific protocol. This protocol establishes the set of rules and the shape and type of message that can be sent by the HAN. Currently, there are many protocols such as KNX, X10, EnOcean, ZigBee, Z-Wave and INSTEON, among others.

- **Residential gateway**. Usually, sensors, interfaces, actuators and controllers understand a single protocol. A gateway allows the interconnection of devices that use different communication protocols to achieve a global system — for example, to interconnect a KNX system with an X10 system. Furthermore, a gateway can be used to connect the HAN with the Internet or with a local area network (LAN) at home.

The elements described above can be physically separated or integrated into one device. Depending on the technology used, a sensor can act as a controller and/or as an actuator.

14.6.2 Transmission Media

There are various technologies to transmit information between different HAN devices, such as wire or wireless. As shown below, some systems only support one type of communication, while others support multiple types.

14.6.2.1 Wired Technology

All sensors, interfaces, controllers and actuators are connected to each other via cables (Sharma and Sharma, 2014). As discussed in the network topology section below, there are two connection alternatives. In the first, sensors, interfaces and actuators have to be connected to the controllers and the devices are not connected to each other. In the second, a cable (also known as a BUS) can be used to interconnect them all.

There are two types of cable usually used: the twisted pair and the housing power line itself (PLC: power line carrier or PCS: powerline carrier system).

When a twisted pair is used, the installation must be done either when the home is being built or when doing a modification: for this reason, wired technology that uses a twisted pair is more difficult to implement. Another factor is the high cost of installation and maintenance complexity. However, this technology can be considered more robust (Reinisch et al., 2007).

The PLC technology, which uses existing electrical wires, can be deployed at any time and its expansion is simpler (Rosslin and Kim, 2010). In this technology, the devices send the information by injecting a coded signal on the electrical wire, which will be received by all the devices connected to the network. The injected signal has a message including the address of the device destination. The message is ignored by all nodes except the recipient. This technology has a big disadvantage: messages do not always reach the target. This happens due to interferences in the electrical cable which are produced by electrical or electronic devices connected to the electric network — e.g. a switched on washing machine.

14.6.2.2 Wireless Technology

The network devices are connected to each other via radio waves (Reinisch et al., 2007). This allows more flexibility when placing various devices in the desired location. This technology reduces possible electric shock problems. Systems using wireless technology are usually less expensive because they do not need wired installation. Furthermore, some sensors need have to use a wireless technology for communication, such as sensors that are placed on the human body.

Wireless technology has also some problems:

a) Use of specific protocols. Homes usually use wireless communication protocols to connect computers, laptops and smartphones. The more widespread and standardized protocols are wireless LAN (IEEE 802.11) or Bluetooth (IEEE 802.15.1). The implementation of these protocols in a device requires high processing capacity and big energy consumption. Because sensors have a small processing capacity and use batteries, this becomes a problem.

b) Energy. Most of the wireless devices, mainly sensors, need a battery. On the other hand, actuators or controllers can be powered by the electrical network. This represents a limit to the functionality and processing capacity of the sensors. That is why sensors need to be designed very carefully for efficient use of energy. There are newly emerging market devices that can be fed by the environment with piezoelectric elements, thermocouples or solar cells. Another technique used to minimize the consumption of energy is going into "sleep" mode as often as possible, minimizing the time that the sensors' state is that of "listening".

c) Interference. Wireless technology should make use of the regulated frequency bands. Only those that have the best features to minimize energy consumption and with better radio wave propagation must be chosen, even though these bands do not have a high data rate. The 900 MHz band is the most commonly used. However, these frequencies are often used by other wireless home devices, such as cordless headphones or garage door openers, and therefore they are subject to interference. Consequently, it is necessary to use robust modulation and transmission techniques to reduce the effects of narrow band interference.

d) Security. Wireless communication is an environment where security implications must be taken into consideration. An attacker could take control of an insecure system even without being in the home. Secure protocols, such as those that use encryption, can consume energy. Therefore, it is necessary to find a compromise between a robust system and the energy consumption that it requires.

14.6.3 Network Topology

There are various HAN topologies, depending on the way the devices are interconnected and the role taken by controllers in the information transmission. There are four main types of topology:

Centralized. In a centralized topology, there is a single controller device. All sensors and interfaces send their information to the controller. The controller analyzes the information and, according to its programming, sends the appropriate commands to the actuators. In this topology, the controller is the most sensitive element because, if it fails, the network ceases operation. On the other hand, it is easy to implement because only a controller has to be programmed. Also, if a wired communication system is used, the controller can be connected to a battery. In the case of interruption of the electricity supply, the system can still partially operate because the sensors are supplied with power through the cable. It is possible that some actuators are not able to carry out their function in devices that need energy.

Decentralized. In this model, there are multiple controllers, each of them connected to a set of sensors, actuators and interfaces. To provide the system with comprehensive functionality, the controllers communicate with each other. This topology is more robust than the centralized one, because the failure of a controller will leave without service only the part of the network that is connected to it.

Distributed. In this model, there are no controllers, in that the sensors and actuators have integrated the controller features and thus can communicate directly with each other. With sensors having some processing ability, they can analyze the data they receive before reporting the findings to the actuators.

This model is well suited to wireless communication systems, because the devices (sensors or actuators) can be far apart and still communicate with each other. In this case, all sensors/actuators of the network are able to forward messages that reach them, and thus ensure that communication is possible between all devices on the network, regardless of the distance that exists between them. The benefit of redundancy is also achieved, because if a device fails, another can take over its task. In this model, there would be a controller for the user to program and supervise the system.

Hybrid, mesh or mixed. This model combines centralized, decentralized and distributed topologies. It can have a central controller or several decentralized controllers. Interfaces, sensors and actuators can also be controllers and are able to forward the information coming from other devices to the network without passing through another controller.

14.6.4 Open, Proprietary or Heterogeneous System

The protocol and the software used by the smart home devices may be proprietary or may be based on an open standard.

When a system is proprietary, only the owner company can sell the devices and make improvements. Therefore, the system evolution is linked to the life and politics of that owner. This system may or may not evolve properly and may even disappear, so the smart home network would become obsolete. The only advantage that such systems have is that they can be cheaper and more robust because they do not have to follow a standard of compatibility with other manufacturers.

Open systems depend on some standards that clearly define the protocols. Therefore, any manufacturer can design the devices and develop the software. Compatibility between devices from different manufacturers, the system expansion and the system growth is therefore guaranteed. The evolution of these systems is very fast, since there are several companies that perform upgrades and offer new devices to the market. The standard can be developed by an international standard organization or can be created by alliances between companies that can admit new members or not.

Finally, there are mixed systems, where a company is responsible for the design and maintenance of the protocol, but allows any other company to manufacture devices if previously certified.

14.6.5 Popular Home Automation Technologies

In the market, there are many home automation systems that compete with each other and which are also incompatible. Table 14.1 shows the popular home automation technologies and the main features, according to the previous sections.

14.6.6 Sensors

A sensor is a physical device capable of detecting physical or chemical signals and transforming them into analogue electrical signals (Sharma and Sharma, 2014). These electrical signals are converted into digital signals

Table 14.1 Popular home automation technologies (Withanage et al., 2014; Cheng and Kunz, 2009; Alam et al., 2012; Sharma and Sharma, 2014)

	X10 [X10, 2016]	Z-Wave	ZigBee	INSTEON	EnOcean	KNX	Lonworks
Released (Year)	1975	2001	2004	2005	2008	2003	1992
Transmission media	Wired (1) AC power line	Wireless	Wireless	Both wireless and AC power line	Wireless	Wired (1) AC power line	Wired (1) AC power line
Network topology	Centralized	Hybrid	Hybrid	Hybrid	Hybrid	Hybrid	Distributed
Proprietary/open/mixed system	Open	Proprietary (2)	Open IEEE 802.15.4	Proprietary (2)	Mixed (2) ISO/IEC 14543-3-10	Open (3) ISO/IEC 14543	Open (2) ISO/IEC 14908
Advantages	Lower cost. Easy to install. It is one of the oldest: many devices are available.	Easy to install. Compatibility of devices among many different vendors. High performance.	Low-power consumption.	Easy to install. Robust PLC and RF used simultaneously. Partial compatibility with X10 devices.	Ultra-low power consumption (batteryless wireless sensors). Easy to install.	Reliable. Robust.	Reliable. Robust.

continued on next page

Table 14.1 (*continued*)

	X10 [X10, 2016]	Z-Wave	ZigBee	INSTEON	EnOcean	KNX	Lonworks
Disadvantages	Prone to noise. No encryption. No two-way communication.		Incompatibility of devices among many different vendors.	No encryption. High-power consumption.	Basic security.		Complex.
Price (4)	Low	Medium	Low	Medium	Low	High	High
Web page	www.x10.com	www.z-wave-alliance.org	www.zig-bee.org	www.in-steon.com	www.enocean-alliance.org	www.knx.org	www.eche-lon.com

(1) Although the typical installation is wired, there are transceivers to connect sensors or actuators that communicate with each other by radiofrequency.

(2) Allows:
- Other companies with licenses to manufacture devices.
- The manufacture of devices using components of the proprietary company.
- The manufacture of devices by a company or an alliance of companies.

(3) Although the protocol specification is standardized, it is necessary to pay KNX to purchase the specification and a license for software management.

(4) Price based on the cost of a light switch actuator: low 30–50€; medium 50–100€; high >100€.

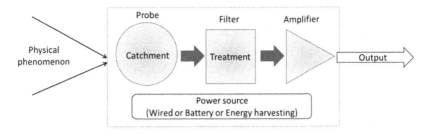

Figure 14.5 Electronic components in a sensor.

Table 14.2 Types of environmental sensor

Sensor	Measurement
Passive infrared (PIR)	Motion
Active infrared (AIR)	Motion/identification
Radio frequency identification (RFID)	Object information
Pressure	Pressure on mat, chair, etc.
Smart tiles	Pressure on floor
Magnetic switches	Door/closet opening/closing
Camera/microphone	Activity
Humidity	Water leak
Gas	Gas leak
Smoke	Fire

through an analogue to digital converter. The digital signal is processed by a microprocessor or microcontroller to produce a message that must be sent over the network to the receptor according to the protocol being used. The message is sent by a transceiver. These electronic components are powered by a battery (in wireless systems) or through cable connection (in wired systems), or they may even pick up energy from the environment (solar, piezoelectric, etc.). Figure 14.5 shows a scheme of electronic components in a sensor.

There are different types of sensor that might be considered for home automation, such as environmental and wearable. **Environmental sensors** (Ni et al., 2015; Acampora et al., 2013) are distributed around the environment — e.g. in furniture, appliances, walls and doors, etc. (See Table 14.2.)

Wearable sensors (Ni et al., 2015; Salih et al., 2013; Acampora et al., 2013) are worn by the residents. These sensors can be embedded into clothes (sensing textiles), eyeglasses, shoes and wristwatches, or positioned directly on the body. (See Table 14.3.)

Table 14.3 Types of wearable sensor

Sensor	Measurement
Accelerometer/gyroscope	Recognizing body postures (e.g. sitting, kneeling, crawling, lying, standing up, walking, running, etc.). It is also used to measure vibration or acceleration due to gravity, useful in recognizing elderly falls, for example.
Glucometer	Blood glucose
Blood pressure	Blood pressure
CO_2 gas	Oxygen concentration during human respiration
ECG	Cardiac activity
EEG	Brain activity
EMG	Muscle activity
Pulse oximetry	Blood oxygen saturation
Humidity and temperature	Body humidity and temperature

14.7 ACQUISITION CONTEXT

In typical context-aware systems architectures, there is a specific module called "acquisition of context", "acquisition of knowledge" or "context management", which is an intermediate part between the exterior (inputs of the system) and the context processing of the system. This layer can appear as a single module or more modules. Its main functionality consists of acquiring data from the devices sitting in the sensing layer and translating these data into a useful and understandable format to be processed by the system.

Context-aware architectures usually show this layer divided into two parts. The first part directly detects and collects physical data. These data are translated according to a common protocol that the rest of the system can understand in order to provide useful information to the higher layers (e.g. the acquisition context layer detects the A5 signal that the movement sensor sends when detecting movement. This signal is translated to "movement_at_bathroom_on" and stored together with a timestamp and other relevant information in structured datasets). The sequence of events, or the lack of some of them, can be relevant information in the system (e.g. detect large periods of inactivity of a person at home). A possible sequence of primitive events is shown in Augusto and Nugent (2006b) with the structure (0 at_kitchen_on, 1 cooker_on, 2 at_reception_on, etc.).

The second part can integrate this useful information provided by these primitive events registered in the first part of this layer, with the aim of providing more complete context information. The new context information should also be registered by the system.

The context detected and translated in this acquisition context layer should also be delivered to the correct services, applications or architecture modules for processing. These modules will use these sequences of events to determine normal or abnormal home behavior. Typical mechanisms of context distribution (Van Kranenburg et al., 2006) are subscription (services or modules interested in certain context information can subscribe to obtain all changes in this information and, when a change occurs, subscribed services are notified) and polling (services that use the context make inquiries to know the state of the context information that needs to be used).

14.8 KNOWLEDGE BASE

The knowledge base (KB) is one of the main parts of a health monitoring system. It includes concepts of interest to define the application domain used by the system, as well as the relationships between concepts.

A KB includes a context definition and historical events. The context definition consists of the description and representation of the "known world" of the system. It represents a context abstraction that explains what data and concepts the system understands and how they should be defined, represented and formalized in a common way by all elements and services of the system. It contains the relationships between the events, facts and concepts of the domain.

The method of context representation can improve the process of inference. Inference is used in data filtering, extracting relevant features to generate new information that may be useful for the reasoning process.

On the other hand, 'historical events' stores information collected by the acquisition context layer and results obtained via the reasoning process of the system. The historical events and decisions taken by the system can be a good source of knowledge to obtain new knowledge that feeds the system. New knowledge can include or modify behavioral rules, learning the habits of persons for improving system knowledge. Even more, evolving Big Data techniques applied to databases of different smart homes over long time periods can help to answer certain questions, such as how some illnesses can evolve over time, to help to detect some disease symptoms early.

14.8.1 Context Definition

In order to be smart, a digital home must be context-aware. The main feature of a context-aware system is its capability to obtain and manage relevant contextual information and to adapt its behavior to contextual changes in an automated way, without direct user intervention. The main objective is to provide new information or services useful to the person (Zhang et al., 2009; Esposito et al., 2010). These services can also change, and need to be able to assimilate changes and to adapt accordingly (Büscher et al., 2009).

One of the most common definitions of context was provided by Dey in 2001. He defined context as the information, which can be used to characterize the situation of an entity. Moreover, he defines entities as persons, places or objects that are considered relevant to the interaction between a user and an application, including the user and applications themselves (Dey, 2001). Similarly, Schilit et al. (1994) refer to context as location, identities of nearby people and objects, and changes to those objects. They define the aspects that a context-aware system should control during the initial analysis (known as the five Ws theory): Who (the user); What (the task that the user is doing and intends to achieve); Where (the place where it is happening); When (the moment at which it occurs); and Why (the reason for the behavior the user is displaying — this is the most important question to solve in order to know the user's needs at that moment) (Schilit et al., 1994).

Dey (2001) introduces the user's task as an important concept in context-aware systems, since task characterizes the user's situation. Moreover, he classifies elements to describe the context into five categories: individual (contains properties and attributes that describe the entity), activity (tasks done by the entity), location, time and relations (between others entities).

However, not all information obtained through sensors should be considered useful. In fact, a context-aware system should use all useful context information from the physical world to carry out the functionality for which it is designed. In this way, it is also possible that not all the necessary information can be represented in a direct way by the available sensors in the home. Pre-processing and aggregation of sensors' data can provide information that is more complex.

In context-aware healthcare monitoring scenarios, there are interesting factors that could be measured, such as:

- **User's profile(s)**. Name, gender, age, social and health situation or patient context. Furthermore, patient context can be characterized as any information that can be used to characterize a patient medical situation such as high blood pressure. According to Al-Bashayreh et al. (2013), the context information can include the patient's vital signs (body temperature, glucose level, heart rate, pulse and oxygen in blood); medical symptoms (such as dizziness); risk factors (such as obesity); prescribed medications; physical activities (e.g. sleeping, running, etc.); emotional state; and surrounding environment.
- **Environmental conditions**. Lights, temperature, humidity, flood, smoke, fire, sound, floor pressure, etc.
- **Devices and electro domestics**. Other devices installed in the home that provide information about what the user is doing — e.g. sensor TV on/off.
- **Computing context**. Network connectivity, communication bandwidth, nearby resource.
- **Location**. In order to maintain the position of the user and his/her location changes.
- **Time**. Time of day, week, month, year, season.
- **Activities related to the person's behavior**. Sleeping, eating, taking a shower, watching TV, coming back to the house, cleaning the house, ironing, etc. A common activity to supervise is taking medication in a time window and secondary effects (sleep disorders, remaining in bed, etc.). Habits can also form part of the context. The user or professional can include habits before or during the system use; also, the system can generate habits related to the repetition of activities or situations over time.
- **Possible risk situation at home**. User fall, possible intoxication by gas, inactivity, etc.

Context-aware systems that consider this kind of information can help to improve the quality of medical care. In home health applications, this information may help to improve the health status of the person, allowing professionals to take decisions about medications or treatments, allowing the identification of critical situations, or identifying possible diseases in the initial phases.

14.8.2 Context Modeling and Reasoning

Further to defining the concepts or entities, it is important also to establish the relationships between them. How to model the context information is a key issue in the process of behavioral adaptation. In order to understand and manage the context complexity and the relationships between elements such as sensors, measures, temporal information, user profiles, events and actions, a powerful mechanism is required that allows, defines and represents the context. Over the last decade, a number of context modeling techniques have been developed to deal with the treatment of data collected by sensors and devices and to facilitate an efficient adaptation process for context-aware applications. Factors to take into account when selecting a context modeling tool include (Van Bunningen et al., 2005; Bettini et al., 2009):

- Managing the heterogeneity of context information sources that differ in their features of data acquisition and presentation, update rate and semantic level.
- Taking into account the richness, expressiveness and information quality in the acquisition and treatment of data.
- Dealing with rapidly changing environments.
- Obtaining and using location and timeliness (context histories).
- Capturing the relationships and dependencies among entities included in the context.
- Dealing with imperfections in context information (inaccuracy of measures, lack of information, incomplete or ambiguous information or conflict between information sources).
- Supporting consistency, verification of the model and reasoning process.
- Ease of use by designers and non–designers, and how easily the technique relates context to world concepts.
- Capacity to exploit contextual information in making decisions.
- Capacity to incorporate human knowledge.
- Guaranteeing an efficient access to the context information by the services or components of the system, selecting relevant objects, attributes and suitable access paths to be represented in the context modeling.
- The intrinsic distributed nature of a context-aware environment makes it desirable that all parts of the system share the same interpretation of the exchanged data.

The most common approaches used to represent context information in context-aware systems are described below.

Key–value models are the simplest approach to modeling the context. The context is described with a list of simple attributes and values. It can be appropriate to represent simple attributes such as profiles (interest, name, age, contact list, etc.). However, they lack the capacity for structure and link information, and they support limited reasoning (Schilit et al., 1994).

Markup-based models use markup languages in order to model and formalize context information, typically XML. They provide a hierarchical structure to capture context information, but not to include capacities to define relationships and dependencies.

Object-oriented models provide formal models for supporting reasoning and providing inherent characteristics of object oriented paradigms (such as encapsulation, inheritance, etc.). This model implements class hierarchies and relationships between objects as attributes or sub–objects (Zhang et al., 2005).

Logic-based models are an approach widely used in context-aware systems to define logic with facts and the relationships between them. When the system includes new information, this information is contrasted with previously defined rules. In this process, new knowledge can be generated as a result. Some advantages of this model are the high degree of formality and expressiveness, the capability of reasoning, and the verification and validation of the context model.

The most common approaches used in context-aware environments are:

- First-order predicates. These provide an expressive description of context using Boolean operators and existential and universal quantifiers. The model supports operations like conjunction, disjunction, negation and quantification. It allows the definition of complex expressions that provide descriptions of properties, the structure of the context and the rules to define actions to manage it. Use of this logic allows the use of rules to infer different contexts directly (Ranganathan and Campbell, 2003; Loke, 2004).
- Fuzzy logic. This is used to handle uncertainty in context with vague, inaccurate, ambiguous, incomplete or imperfect data. It is useful in representing imprecise concepts such as tall. Fuzzy logic has the advantage to represent the fuzziness and uncertainty associated with the real world. It is able to define imprecise concepts by defining the degree of membership from 0 to 1. However, this approach per se cannot be

adequate for modeling and reasoning situations in pervasive environments. It is very relevant for a fuzzy modeling and reasoning method to be combined with a rich theoretical basis for supporting context-aware scenarios (Haghighi et al., 2008). A fuzzy logic system is composed of three main parts: fuzzy set (set of input data); rules (behavior of the system previously defined); and inference engine (to contrast the fuzzy set with the rules of the system). An example of the use of fuzzy logic in a context-aware healthcare environment is defined by Yuan and Herbert (2012). This author uses a fuzzy logic model to represent relevant variables and to build the low level and high level context. Low level context relates to the physiological context, personal context and environmental context, and high level consists of activity events and medical conditions.

- Description logic. This belongs to the first-order predicates. It is usually used together with ontologies. This kind of model defines concepts, hierarchies of concepts, properties, individuals and their relationships. This semantic provides simple reasoning capabilities over data. A DL knowledge base has usually two components: the terminology box (TBox), which includes and defines the terms of a domain; and the assertion box (ABox), which includes assertions about individuals in terms of the TBox.
- Probabilistic logic. This method associates a probability with a situation, or can differentiate if an entity is included in a specific kind (for example, a probability of 0.7 for a fall situation). In context-aware environments such as a smart home, this kind of model can be used to reason through the fusion of different sensors and contexts. When there are conflicts in the collected information, probabilities can be helpful to take decisions. However, the main disadvantage of these models is the amount of data needed to define the conditional probabilities of an event given particular evidence. The most common techniques are the Dempster-Shafe and Hidden Markov models.

Rule-based systems are useful in context-aware environments as tools to define the context and as a support for reasoning. Traditionally, these have been used to describe expert systems or robot behavior and are related to first-order predicate models. A rule-based system is composed of a fact base, knowledge base and inference engine. The knowledge base defines if <conditions> then <actions> rules, while the fact base contains the information that will be checked with the rules defined (Wang et al., 2012). Tools often used to define rule-based systems are Clips, Jess and Drools.

Bayesian networks belong to supervised learning. They are directed acyclic graphs and are appropriate for combining uncertain information from diverse sources and deducing higher-level contexts. They are composed of nodes that represent variables, which may be observable quantities, variables, hypotheses or unknown parameters; the edges represent conditional dependencies. Nodes, which are not connected, represent variables that are conditionally independent of one another. Each node has an associated probability function that takes as input a particular set of values from the parent node variables and returns the probability of the variable represented by the node.

Ontology-based models are a kind of logic based model. These models are currently the most used, valuated and recommended in order to implement AmI systems. They fit perfectly with the implicit features of context-aware environments, which require great expressivity, reasoning and adaptation capabilities. Ontologies provide a formal explicit specification of a shared conceptualization (Gruber, 1993; Borst, 1997). Ontologies based on description logics offer a formal semantic to represent concepts in a hierarchal manner, using the properties and relationships between them, and including instances and axioms or declarations that the elements must meet in a standard way for all components of the system. The linked data characteristics of ontologies and their semantic structure makes them a key tool to define the context and to simplify the inference process in an automated procedure.

Several authors have highlighted the value of ontologies in ubiquitous and context-aware systems, declaring them the best option for context modeling in comparison with other modeling methods (Chen, 2004; Strang and Linnhoff-Popien, 2004; Benyahia et al., 2012; Zhang et al., 2013; Kim and Chung, 2014). Ontologies offer many benefits:

- Higher expressivity and a formal semantic, allowing efficient reasoning of context information.
- Ability to detect redundancy and consistency and classify new instances automatically in the knowledge base, allowing the extraction of new implicit knowledge through automatic reasoning.
- Easily understood, modifiable and extendable.
- Ontologies are based on XML: as it is independent of platform, it can be used in different systems or applications.
- Reducing the difficulty of sharing information between elements of a dynamic distributed system.
- Offers facilities to integrate and evolve with other ontologies.

- A lack of ontologies implies an absence of semantics and a lack of results interpretation, reducing the performance of the system and so increasing the difficulty of sharing and evolving.
- Simple reasoning capabilities: the integration of ontologies with reasoning engines is the foundation of the decision making process.
- Mature models: there are mature standards to support ontologies, and tools to implement and manage them.

Interesting comparisons between context definition approaches, based on applying these characteristics to context-aware and ubiquitous systems, can be found in the studies carried out by Strang and Linnhoff-Popien (2004) and Nalepa and Bobek (2014). Furthermore, ontologies have been widely used and validated as a tool for context modeling in pervasive environments. Some known examples of the use of ontologies in pervasive environments are SOUPA ontology (Chen et al., 2005, 2004), which includes a set of vocabularies to represent intelligent agents with associated beliefs, desires and intentions, time, space, events, user profiles, actions and policies for security and privacy. In this line, CONON — ontology based context modeling and reasoning using OWL (Wang et al., 2009) — proposes OWL (OWL, 2013) as a common vocabulary for context representation and reasoning in pervasive environments based on agents. A complete context ontology for ambient assisted living is presented by Forkan in their CoCaMaal middleware for AAL (Forkan et al., 2014).

Ontologies have been widely included in health monitoring and care systems at home, such as the Foo Siang Fook et al. (2006) ontology, which presents a monitoring system with a context model modeled in OWL (OWL, 2013) and facilitating care-giving for Alzheimer's patients; or Paganelli's work, which describes a system based on ontologies and pervasive technologies to provide continuous care and manage home alarms (Paganelli and Giuli, 2011). The work of Ni also presents a model for human activity representation for smart home applications based on a network of ontologies classified in three categories: user ontologies, SH context ontologies and ADL ontologies (Ni et al., 2015).

Ontologies are commonly used together with intelligent agents. That is to say, ontologies allow agents to manipulate contextual information according to the model and to obtain new inferences and relationships based on the state of the environment. Chen used their SOUPA ontology to provide a shared ontology combining many vocabularies from different consensus ontologies to represent intelligent agents. They conclude that ontologies are a key requirement in pervasive computing. Similar approximations are taken by Jih to validate a context-aware service platform at home based on JADE agents and ontologies to manage the complex relationships between instances of the real world (Jih et al., 2007); or Perrot, who proposes a context-aware framework based on agents and ontologies for a distributed context in pervasive healthcare systems (Perrot et al., 2012).

14.9 REASONING

This layer receives all data obtained by the environment, pre-processed by the acquisition layer. The reasoning layer analyzes the information acquired according to its world model and the reasoning rules and algorithms previously defined in the context model. The context model and reasoning mechanism together define system behavior.

The sensor platform requires tools that allow the efficient processing and management of data according to the context in each moment. Once the context domain is defined, it is also necessary to include a definition of the behavior system through relationships between concepts and behavior rules. These rules are created from expert knowledge received from care professionals. Furthermore, as a result of accumulated experience, new rules obtained by data-mining techniques can be created and added in an automated way (or previously supervised by professionals). Other rules can also be eliminated or modified. The system can collect data from historical databases and extract new knowledge that was hidden, learning from the acquired experience.

The system behavior definition can be implemented according to the context model chosen in previously. The most common methods to define context-reasoning capabilities are rules (if... then...), fuzzy logic (to represent imprecise concepts and obtain the degree of membership at an entity), probabilistic logic (Dempster-Shafer theory and Hidden Markov models) or Bayes networks and ontologies.

Each of these techniques has its advantages and disadvantages and its special application domains. Hybrid approaches in context modeling can help to improve the reasoning over acquired data (Dargie, 2007). The use of ontologies with rule-based systems is very common in context-aware healthcare systems at home (Rakib and Ul Haque, 2014). Ontologies can define reasoning procedures through the semantic expressiveness, thanks to relationships between elements and the operations 'to', 'and', 'not', 'or', etc. New data are included as new individuals and are contrasted with definitions of the ontology. Individuals are classified in one of the categories defined in the ontology through the inference engine. Ontologies can be used together in fuzzy based systems to improve the behavior of the system and to construct personalized services (Sohn et al., 2014).

Temporal reasoning has mostly been used in combination with rule-based systems to identify dangerous situations and take measures to resolve them. The incorporation of fuzzy rules in a home system can help to automatize the environment according to observation of the daily life activities and behavior of their inhabitants. Moreover, fuzzy rules are commonly used to generate a higher-level context (e.g. medical condition, activity and accident event), allowing identification of the current state of the patient (normal, abnormal or emergency) based on the combined high-level context (Yuan and Herbert, 2012). Rules can be aggregated, modified or suppressed to allow for an effective adaptation of behavior to the environment and behavior changes in the inhabitants. This automatic rule or behavior adaptation as a result of previous experience is known as learning.

14.10 LEARNING

The system can collect data from historical databases and extract new knowledge that was hidden, identifying patterns and learning from the acquired experience. Pattern identification consists of the identification of facts or a sequence of events repeated in time. The repeatability of patterns can be detected and learned. Learning can aid prediction and decision-making. New rules can consequently be created and added in an automated way (or previously supervised by professionals), and other rules can be eliminated or modified as a result of the learning process. However, this usually requires the collection of big datasets and usually requires human effort to label data.

In home healthcare scenarios, the main application of machine learning techniques is in recognizing daily activities and routines through in-home sensors and abnormal changes in activities that could suggest a possible alarm situation (commonly named anomaly detection). Anomaly detection is used to detect and identify rare events in large datasets. The classic approach to anomaly detection uses rule-based systems; however, anomaly detection can benefit from the possibility of including statistical analysis and advanced machine-learning approaches.

There are two typical kinds of machine learning model: supervised learning models and unsupervised learning models.

Supervised models require a prior training process. The training process requires the use of many cases, and for each case the result provided by input data aims to achieve a good accuracy level. These models are used to predict. Their limitations lie in the fact that the activities that can be recognized are limited to the training data used, and this kind of system requires sufficient annotated data to train the classification models.

Typical applications of supervised learning models in healthcare domains are in classification problems. Classification consists of predicting discrete results through an input dataset to predict a category or label of the type: {"Yes", "No"}, or {"Normal", "Risk", "Alarm"}.

Several examples of algorithms include the Hidden Markov model, Bayes classifier, decision trees, neural networks and the support vector machine (Augusto and Nugent, 2006a; Chung and Liu, 2007; Fleury et al., 2010; Avic et al., 2010; Hoque and Stankovic, 2012; Babakura et al., 2014).

Unsupervised models do not require a prior training process. Input data do not have known classified results. However, the model finds similarities among data to organize and extract general rules or patterns to allow it to label new inputs. These models are used when it is not known what properties can be predicted from other different properties. They are mainly used to discover patterns or data tendencies. A typical application is clustering.

Clustering is used to find groups of instances with similar characteristics and is commonly used to model behavioral patterns (Barger et al., 2005; Nguyen et al., 2007). One of the common methods is k-means, a traditional clustering algorithm which divides data into k clusters according to similarity (Hung et al., 2013). Bayesian networks and Markov models (Yin et al., 2015) are much-repeated algorithms in these approaches.

Commonly used tools in machine learning are Weka (Weka, 2015), Apache Mahout (Apache Mahout, 2014), MATLAB (MATHLAB, 2015) and R (R Project, 2015). Weka is perhaps the leading open source project for machine learning algorithms. It is written in Java, and contains tools for data pre-processing, classification, regression, clustering, association rules and visualization.

14.11 BIG DATA, CLOUD COMPUTING AND THE INTERNET OF THINGS

Big Data is currently an interesting paradigm. It can be defined as the automated processing of great quantities of data and information provided by different and diverse data sources, with the objective of correlating and linking these data and generating new valuable knowledge.

Big Data offers new possibilities in the provision of healthcare services with added value in smart homes. The processing of large amounts of data from multiple users and homes can improve the detection of potential risk events or help to prevent them, and even obtain common symptoms of early diseases.

This new paradigm is possible nowadays due to the increasing number of devices available to collect data on the environment and users — from devices installed in diverse rooms in the home (i.e. movement sensors, contact sensors, etc.) to wearable technology or devices included in everyday objects in the user's house. The Internet is of prime importance with this paradigm: objects are connected through the Internet to transmit and receive data, and different kinds of interconnection protocols of devices in the home must be available.

The Internet of Things (IoT) is a new paradigm that is revolutionizing the capture and management of data. It includes and links smart connected devices that can provide information about people, sites, context situations, etc. (e.g. cell phones, lamps, wearable devices, sensors in the home, coffee makers, etc.). Cisco envisions over 50 billion connected devices in the world in 2020, connecting people with people, people with things, and things with things (Internet of Things, 2016).

IoT combined with AmI is very promising in the promotion of healthcare services at home (Vermesan and Friess, 2014; Lopez et al., 2013). However, IoT requires adequate infrastructure to data management, and analytics tools to provide potent reasoning and context-aware services.

Cloud computing provides the basic supporting infrastructure for the provision of Big Data services, opening powerful new ways of data processing and storing. It can provide a virtual infrastructure (hardware, development platforms, and applications or services) to integrate monitoring devices, and to store, manage and present information collected by IoT based devices in an efficient way (Gubbi et al., 2013; Mu-Hsing Kuo, 2011).

Cloud computing allows smart homes access to cloud services without need of installation — only an Internet connection is required — and it

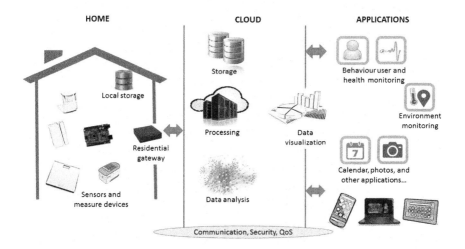

Figure 14.6 Cloud components.

allows the use of these services in any place and at any time (assuming the use of wearable technology and IoT). It allows users and care professionals to access the hardware and software of remote equipment that sustains the service. Figure 14.6 shows the main cloud components and the relationships between the smart home, sensors, IoT, cloud services and applications.

Traditional computing requires a high initial cost for the launch of the service, including hardware, adequate space (electricity, refrigeration, etc.), administration tasks and upgrade control of the systems; it is a hard mechanism to scale. Cloud computing provides numerous advantages when compared with traditional computing:

- The client user does not need special processing or storing facilities at home.
- Services can scale easily in a flexible way and cost effectively. It is easy to add more or reduce capabilities to manage changes in the demands of the clients. Organizations can pay only for their use based on the storage capacity, virtual processors and network use (Vaquero et al., 2008).
- Users can use the same service with specific conditions.
- The provider can control all the data and easily include new changes in the service without the need to introduce the new changes client by client.
- It allows sharing of the electronic history register (EHR) among different care professionals. It provides great storage space to include Big

Datasets for EHRs (patient data, clinical trials, medical images, etc.), thus saving the cost of storing data locally (Schweitzer, 2012).
- Together with Big Data, data collected by different users can be used to obtain new knowledge, which is useful for all users in near real-time, supporting provider decisions.

Cloud computing supports different kinds of service — software as a service (SaaS), platform as a service (PaaS), and infrastructure as a service (IaaS) — and different deployment models (private, public, community and hybrid cloud). Each service type offers different solutions to cloud integration with its own deployment requirements (Mu-Hsing Kuo, 2011).

Some cloud computing environments available to support healthcare applications and services include Amazon Web Services (AWS, 2015), Microsoft Azure (MA, 2015), Google App Engine (GAE, 2015), Heroku (Heroku, 2016) and Jelastic (Jelastic Cloud, 2016).

Cloud computing is commonly found together with Big Data and IoT in the creation of powerful healthcare services that can provide powerful and scalable storage and which allow the processing of context information and linking of different data sources. Multiple examples of use of these platforms can be found in literature. J. Puustjärvi and L. Puustjärvi developed a smart home ontology to include semantics for the data exchanged by the systems and devices in the smart home (Puustjärvi and Puustjärvi, 2015). Moreover, they also include the principles to integrate the smart home's data with other external data sources. Specialized sensors and wearable devices at home are used to monitor the health and general wellbeing of ageing users. The authors propose a cloud-based SaaS approach using a group of repositories to include light, refrigeration data, wellbeing, healthcare providers, family members and security. These repositories receive their data from the IoT devices located in the smart home and the users can access the repositories through specific applications. Fortino at al. also propose a system architecture based on cloud computing for the management and monitoring of body sensor networks, and develop an initial prototype as an application of the Google App Engine (a cloud computing PaaS for web applications) (Fortino et al., 2012). Fan at al. propose a sensor house with a main program running on Raspberry Pi to listen and respond to the data sensors, synchronized with a remote cloud server (Fan et al., 2015). They adopt an Ali cloud server (similar to AWS) in order to achieve remote control and observation of home appliances. Doukas at al. present a platform based on cloud computing for health management through mobile and specific wearable sensors (textile sensors to collect bio signals, such as temperature, oxygen saturation, heart rate and ECG; motion data; and contextual data like location, activity status, etc.) (Doukas and Maglogiannis, 2012). The cloud part consists of a Java EE application to manage graphical interfaces and sensor communication. They choose a PaaS provider — the Jelastic provider — to support their Java-based application, databases, load balancers, etc.

Enterprise cloud services such as Microsoft Health Vault are also available. Microsoft Health Vault is a cloud service designed to collect information about the health of the user and their medical registries (HealthVault, 2016). There are some devices (tracking steps, pulse and oximeter devices, weighing scales, accelerometers, etc.) and applications that can connect with HealthVault to aggregate data or use these data to analyze tendencies and propose recommendations, such as Fitbit, FitnessSyncer, etc. The company iControl Networks offers a connected home platform for unifying data, apps and devices in some scenarios based in the cloud (i Control Networks, 2016). In healthcare scenarios at home, this includes presence detection, pattern recognition, and voice and video communications. Samsung Smart Home SmartThings provides a cloud service to control and monitor Samsung Smart Home devices through applications using a REST API, able to connect with hundreds of smart devices (Samsung, 2016). Philips's HealthSuite platform is able to connect and manage more than seven million consumer devices, sensors and apps using the capabilities of Amazon Web Services to construct more personalized healthcare solutions (such as smart medical alert systems and baby monitoring) (HealthSuite, 2016).

14.12 CHALLENGES

The future of the smart home is promising. Big enterprises like Google, Facebook, Microsoft, Cisco, Samsung and Apple are incorporating the efforts of recent years in this field, and in the near future, new advanced services in monitoring, security, leisure, health and energy saving will be part of our lives. In this way, Table 14.4 shows some technical challenges and work lines related to technologies, reasoning and cloud.

One main challenge in a smart home derives from the difficulty in finding a common logical and physical infrastructure. There is no consensus about what protocols should be used to connect devices, and a logical infrastructure is not clear. Each work equipment implements its own solutions using its own knowledge, or has protocols already available, and many

Table 14.4 Summary of challenges

	Home automation technologies and sensors	Data acquisition and reasoning	Cloud services
Technologies	Cost-effective solutions. Easy wiring and desirable plug 'n' play smart objects. High robustness and accuracy. Low power consumption (energy management). Efficient registry of devices and resource discovery.	Mechanism to lead with inaccurate measures or lack of data. In Big Data, powerful computational resources to process a great quantity of data in real time. Virtual sensors construction (integration of measures of a set of physical devices in more complex measures). Performance and treatment of data when a specific device does not work or is defective. Mechanisms to lead with inaccurate data.	Cloud provides great capacity for storage and processing of data from different sources. It can provide a mechanism to combine these data and obtain new knowledge through Big Data algorithms. Main challenges in these solutions include scalability and flexibility with cost-effective solutions; resource exhaustion prediction and management.
Communication	Interoperability among devices with different protocols. Energy saving. Mobility. Standards for data communication with local and remote services.	Real time communication and continuous availability. Ensures the privacy and integrity of sensitive data exchange, ensuring confidentiality, integrity and authenticity.	Cloud can provide the ubiquity (anywhere and at any time) required to access and send information. However, standards for cloud communication are required to exchange sensible data in home healthcare and telecare solutions. Ensure the privacy and integrity of sensitive data exchange, ensuring confidentiality, integrity and authenticity.

(continued on next page)

Table 14.4 (continued)

	Home automation technologies and sensors	Data acquisition and reasoning	Cloud services
Information management	Generation of data in a standard way (smart gateways to data capture and pre-processing of structured data). Home healthcare services can involve a large amount of different non-structured or semi-structured data that need to be stored. Lead with low device capabilities: hardware devices in smart home usually have limited or no storage capability. Ensure the privacy and integrity of data storage and exchange.	Great capability to store data is required. Clear specifications to store information in a stable way. Integration of different sources of data. Ontology based semantic standards. Ensures the privacy and integrity of data storage and exchange of sensitive data, also ensuring confidentiality, integrity and authenticity. The information generated by the system to the user, healthcare professionals or other implicated people should be clear, concise, easy to understand and accessible. Proper access control policies for allowing authorized users to use specific services.	Cloud can solve problem of data storage. It offers unlimited virtually on-demand storage. Cloud can lead with this integration of different sources of data and services. It has the possibility of integrating medical and sensor data, ensuring availability and redundancy. An infrastructure to provide the electronic health record (EHR) and new common specifications is required to integrate EHR with new healthcare and telecare services. Standards for data sharing and information storage. Ontology based semantic standards. Ensure the privacy and integrity of data storage of sensitive data, also ensuring confidentiality, integrity and authenticity. Continuous hardware, software and human resources monitoring. Proper access control policies for allowing authorized users to use specific services.

(continued on next page)

Table 14.4 (*continued*)

	Home automation technologies and sensors	Data acquisition and reasoning	Cloud services
Processing capabilities	Smart home devices usually have limited processing resources. External capabilities of data processing. Real time interactions with cloud services and between smart objects. Local platforms to integrate and process data.	Great processing power, memory and high velocity of great data volumes can be required to process large quantity of data (Big Data). Management of data in a standard way. Real time processing. Complex reasoning over great quantity of data, maintaining coherence. Processing according to context and parametrization. Management of data representation and visualization in an efficient way.	Cloud offers unlimited processing capabilities in an on-demand model. Offers visualization at any time and anywhere. Data sharing, service discovery and composition. Real time interactions with cloud services and between smart objects and cloud.

of them tend to reinvent parts, which have already been implemented in other projects.

Most of these developed technologies use proprietary protocols. Each one of these technologies can provide solutions to concrete problems. However, idle technologies can work together to develop a more complete complementary functionality. Nevertheless, this lack of interoperability has also complicated the integration of components. Standardization of protocols plays a pivotal role in sensor monitoring solutions. To solve this problem and ameliorate the facility to include new devices (plug and play), new models should be implemented to define and standardize connectivity ways, interfaces, protocols, elements and interactions between elements and with other services. Future generations of platforms must be able to abstract the capabilities of devices and include relationships between capabilities, monitoring measures, situations and conclusions, to make agile the incorporation of devices and smart objects in the home. Hence, the architecture should be designed to be flexible and as independent as possible of the specificities of physical devices. The use of ontologies based on semantic standards could allow information mapping in order to enable information exchange among different standards, platforms and devices (Vermesan et al., 2011).

On the other hand, most applications are tested in projects and are limited to the recognition of specific situations, or are only prototypes. One of the main difficulties of managing context-aware systems is the complexity of the environment: the intelligent monitoring process has to take into account a range of factors to achieve adequately characterized people, situations, risks and reactions. Results obtained in diverse projects leave many questions unanswered. When and in what conditions does the system adequately reason? How many tests are needed with people to demonstrate the quality and reliability of the system? Is it possible to use these tests in environments as sensitive as healthcare environments? Can the system explain its behavior? It is possible to find explanations of how to manage certain situations at home, but it is more complicated to know what is happening and why. Activity recognition techniques allow us to know in an accurate way what activity is being performed by the user, but solving the "why" is clearly more difficult. That is to say, the link between the kind of activity, the behavior patterns in daily life and the personal profile of the user-patient could be explained in some situations and a possible risk situation detected or rejected (e.g. domestic activity causing an increase in blood pressure).

Furthermore, factors such as ease of use and installation, performance, reliability and security should be kept in mind when choosing devices and protocols. When designing a smart home, focus must be kept on the interaction with the user and understanding how the technology fits into the life and daily routines of the user. To be smart, efficient, secure and reliable, the smart home must necessarily be usable, accessible and affordable, and it should give the user confidence. It is essential to address the design of home healthcare services taking into account user features — factors such as age, education, experience with technology, cognitive disrepair, physical, visual, auditory or speech capabilities, known diseases, etc., or changes in their status or behavior due to secondary effects of medication or disorders related to their diseases.

Personalization of the system for the user's needs is a main challenge in healthcare monitoring systems, but in order to replicate the system it is necessary to find a common structure that allows their production. Thus the other main challenges that appear in standardization are about what devices to use and where to place them, how this structure influences the interaction with applications and, finally, how it is possible to standardize the sensor map of a house in order to obtain a map which can be used in all houses without damaging the functionality of the system. How is it possible to standardize and generalize a system in order to achieve an easy setup?

On the other hand, these context-aware healthcare environments suggest the need to obtain a great quantity of data from multiple sources in order to obtain information about the user and the current situation of the user at home. This amount of information translates into the need for a huge number of detection devices. A higher number of devices implies higher expenses, and they can have influence in the sense of rejection of the system (the user can feel controlled or invaded by devices). The progressive miniaturization of devices and current trends in the Internet of Things (IoT) and smart textiles with integrated devices in objects in the person's daily life can perhaps signify evolution in this matter.

Smart homes have an important role to play in the future of intelligent homecare and healthcare, with the potential to empower the inhabitants by making them feel safer. Personalization and customization to the individuals will allow the following of lifestyle habits and health trends, giving warnings and getting help when needed. Through this, smart homes might also be a way to encourage a healthier lifestyle, and raise the alarm when our health is deteriorating.

14.13 CONCLUSION

Future smart homes will offer diverse healthcare personalized services, which are sustained by home automation technologies and smart sensors, more efficient context-aware technologies and a telecommunications infrastructure more prepared to support adequate data exchange and store. Remote monitoring will become more and more popular in the years to come in telecare services and for patients recovering from illness. Furthermore, more advanced and efficient health services to detect symptoms of illness through recompilation of great quantities of data in houses can become a reality in a future, based on the activities of daily life and the routines of people or services.

When professionals are designing smart homes, especially within the application areas of homecare and healthcare, the starting point must be the needs of the users. The users are primarily the people living in the smart home, but can also be relatives, care staff and healthcare staff who communicate with the smart home and the people living there. There are special requirements when designing a technical system that is supposed to be used by people that do not necessarily have a technical background.

The requirements will be even greater when the task is to take responsibility for the health of people. To start with, the involvement of the users throughout the whole development process is vital. Firstly, the actions of the smart home should be desired by the people living there, and should also give them an increased feeling of safety and security. There is a risk that monitoring activities and different parameters in a home environment may seem an intrusion into the user's privacy if the advantages are not clear.

The use of technology acceptance models oriented to healthcare applications and the involvement of users (care professionals, patients or relatives) is a proven success factor for obtaining a final product (products more suitable and adapted to the user need; products which solve real problems with real and adequate solutions).

For a smart home to be a part of intelligent homecare or healthcare, all parts of the system must be reliable. The combination and integration of sensors into the system must work properly, which also includes aspects of interoperability and standards. Data communication must be safe and secure, and also supply integrity, in terms of both data information and personal integrity. Further, data aggregation and decision support are crucial, and are the part of the system that will be responsible for raising an alarm when needed and triggering data transmission. A combination of sensors can monitor human presence, health status and various activities, and the system can learn to know the normal behavior and trigger an alarm or warning when a deviation occurs.

Even though there are numerous technical solutions to set the context definition and reasoning, real applications are still difficult to find. Different health telemonitoring scenarios require different technical solutions to hardware, context and reasoning definition, as we discussed in challenges section.

In a future, available hardware technologies and intelligent context-aware services will increase the capacity to understand complex contexts, identify actions and situations, detect risk episodes and react to each situation in the best possible way.

ACKNOWLEDGEMENTS

This article is part of research conducted under the EDUCERE project (Ubiquitous Detection Ecosystem to Care and Early Stimulation for Children with Developmental Disorders; TIN2013-47803-C2-1-R), supported by the Ministry of Education and Science of Spain through the National Plan for R+D+I (research, development and innovation).

REFERENCES

Acampora, G., Cook, D.J., Rashidi, P., Vasilakos, A.V., 2013. A survey on ambient intelligence in healthcare. Proc. IEEE 101 (12), 2470–2494.

Alam, M.R., Reaz, M.B.I., Ali, M.A.M., 2012. A review of smart homes — past, present, and future. IEEE Trans. Syst. Man Cybern., Part C, Appl. Rev. 42 (6), 1190–1203.

Al-Bashayreh, M.G., Hashim, N.L., Khorma, O.T., 2013. Context-aware mobile patient monitoring frameworks: a systematic review and research agenda. J. Softw. 8 (7), 1604–1612.

Alwan, M., Dalal, S., Mack, D., Kell, S., Turner, B., Leachtenauer, J., Felder, R., 2006. Impact of monitoring technology in assisted living: outcome pilot. IEEE Trans. Inf. Technol. Biomed. 10 (1), 192–198.

Anbarasi, A., Ishwarya, M., 2013. Design and implementation of smart home using sensor network. In: Proceedings of International Conference on Optical Imaging Sensor and Security. Coimbatore, Tamil Nadu, India.

Augusto, J.C., Nugent, C.D. (Eds.), 2006a. Designing Smart Homes. Springer-Verlag, Berlin, Heidelberg, pp. 1–15.

Augusto, J.C., Nugent, C.D., 2006b. Smart homes can be smarter. In: Designing Smart Homes, pp. 1–15.

Avic, A., Bosch, S., Marin-Perianu, M., Marin-Perianu, R., Havinga, P., 2010. Activity recognition using inertial sensing for healthcare, wellbeing and sports applications: a survey. In: 23rd International Conference on Architecture of Computing Systems, pp. 1–10.

AWS, Amazon Web Services, 2015. https://aws.amazon.com/.

Babakura, A., Sulaiman, M.N., Mustapha, N., Perumal, T., 2014. HMM based decision model for smart home environment. Int. J. Smart Home 8 (1), 129–138.

Barger, T.S., Brown, D.E., Alwan, M., 2005. Health-status monitoring through analysis of behavioral patterns. IEEE Trans. Syst. Man Cybern., Part A, Syst. Hum. 35 (1), 22–27.

Benmansour, A., Bouchachia, A., Feham, M., 2015. Multioccupant activity recognition in pervasive smart home environments. ACM Comput. Surv. 48 (3), Article 34.

Benyahia, A.A., Hajjam, A., Hilaire, V., Hajjam, M., 2012. Ontological architecture for management of telemonitoring system and alerts detection. In: eHealth and Remote Monitoring.

Bettini, C., Brdiczka, O., Henricksen, K., Indulska, J., Nicklas, D., Ranganathan, A., Riboni, D., 2009. A survey of context modelling and reasoning techniques. Pervasive Mob. Comput. 6 (2), 161–180.

Borst, W.N., 1997. Construction of Engineering Ontologies for Knowledge Sharing and Reuse. Centre for Telematics and Information Technology.

Büscher, M., Coulton, P., Efstratiou, C., Gellersen, H., Hemment, D., Mehmood, R., Sangiorgi, D., 2009. Intelligent mobility systems: some socio-technical challenges and opportunities. In: Communications Infrastructure. Systems and Applications in Europe, pp. 140–152.

Celler, B., Lovell, N., Chan, D., 1999. The potential impact of home telecare on clinical practice. Med. J. Aust. 518–521.

Chen, H., 2004. An intelligent broker architecture for pervasive context-aware systems. Ph.D. Thesis. University of Maryland, Baltimore County.

Chen, H., Perich, F., Finin, T., Joshi, A., 2004. Soupa: standard ontology for ubiquitous and pervasive applications. In: The First Annual IEEE International Conference on Mobile and Ubiquitous Systems: Networking and Services, pp. 258–267.

Chen, H., Finin, T., Joshi, A., 2005. The SOUPA ontology for pervasive computing. In: Ontologies for Agents: Theory and Experiences, pp. 233–258.

Cheng, J., Kunz, T., 2009. A Survey on Smart Home Networking. Carleton University.

Chung, P.-C., Liu, C.-D., 2007. A daily behavior enabled hidden Markov model for human behavior understanding. Pattern Recognit. 41 (5), 1572–1580.

Dargie, W., 2007. The role of probabilistic schemes in multisensor context-awareness. In: Fifth Annual IEEE International Conference on Pervasive Computing and Communications Workshops, pp. 27–32.

Deken, J., 1982. Electronic Cottage: Everyday Living with Your Personal Computer in the 1980's. William Morrow & Co., Inc.

Dey, A.K., 2001. Understanding and using context. Pers. Ubiquitous Comput. 5 (1), 4–7.

Doukas, C., Maglogiannis, I., 2012. Bringing IoT and cloud computing towards pervasive healthcare. In: 2012 Sixth International Conference on Innovative Mobile and Internet Services in Ubiquitous Computing (IMIS), pp. 922–926.

European Commission, 2001. Scenarios for Ambient Intelligence in 2010.

Esposito, A., Tarricone, L., Zappatore, M., Catarinucci, L., Colella, R., 2010. A framework for context-aware home-health monitoring. Int. J. Auton. Adapt. Commun. Syst. (IJAACS) 3, 75–91.

ETSI TR 102 415 V1.1.1, 2005. Human Factors (HF) Telecare services; Issues and recommendations for user aspects.

Fan, X., Huang, H., Qi, S., Luo, X., Zeng, J., Xie, Q., Xie, C., 2015. Sensing home: a cost-effective design for smart home via heterogeneous wireless networks. Sensors 15, 30270–30292.

Fleury, A., Vacher, M., Noury, N., 2010. SVM-based multi-modal classification of activities of daily living in health smart homes: sensors, algorithms and first experimental results. IEEE Trans. Inf. Technol. Biomed. 14 (2), 274–283.

Foo Siang Fook, V., Tay, S.C., Jayachandran, M., Biswas, J., Zhang, D., 2006. An ontology-based context model in monitoring and handling agitation behavior for persons with dementia. In: Fourth Annual IEEE International Conference on Pervasive Computing and Communications Workshops, pp. 1–5.

Forkan, A., Khalil, I., Tari, B., 2014. CoCaMAAL: a cloud-oriented context-aware middleware in ambient assisted living. Future Gener. Comput. Syst. 35, 114–127.

Fortino, G., Pathan, M., Di Fatta, D., 2012. BodyCloud: integration of cloud computing and body sensor networks. In: 4th IEEE International Conference on Cloud Computing Technology and Science, pp. 851–856.

GAE, Google Cloud Platform, 2015. https://cloud.google.com/appengine/.

Gruber, T., 1993. A translation approach to portable ontologies specifications. Knowl. Acquis. 5 (2), 199–220.

Gubbi, J., Buyya, R., Marusic, S., Palaniswami, M., 2013. Internet of Things (IoT): a vision, architectural elements, and future directions. Future Gener. Comput. Syst. 29 (7), 1645–1660.

Haghighi, P.D., Krishnaswamy, S., Zaslavsky, A., Gaber, M.M., 2008. Reasoning about context in uncertain pervasive computing environments. In: Roggen, D., et al. (Eds.), EuroSSC 2008. In: LNCS, vol. 5279, pp. 112–125.

Hailay, D., Roine, R., Ohinmaa, A., 2002. Systematic review of evidence for the benefits of telemedicine. J. Telemed. Telecare, 1–7.

Hassanalieragh, M., Page, A., Soyata, T., Sharma, G., Aktas, M., Mateos, G., Kantarci, B., Andreescu, S., 2015. Health monitoring and management using Internet-of-Things (IoT) sensing with cloud-based processing: opportunities and challenges. In: IEEE International Conference on Service Computing.

HealthVault, 2016. https://www.healthvault.com/es/es.

HealthSuite, 2016. http://www.usa.philips.com/healthcare/innovation/about-health-suite.

Heroku, 2016. https://www.heroku.com/.

Hoque, E., Stankovic, J., 2012. AALO: activity recognition in smart homes using active learning in the presence of overlapped activities. In: 6th IEEE International Conference on Pervasive Computing Technologies for Healthcare, pp. 139–146.

Hung, Y-S., Chen, K-L., Yang, C-T., Deng, G-F., 2013. Web usage mining for analysing elder self-care behavior patterns. Expert Syst. Appl. 40 (2), 775–783.

Hussain, M., Khan, W.A., Afzal, M., Lee, S., 2012. Smart CDSS for smart homes. In: ICOST 2012. In: LNCS, vol. 7251, pp. 266–269.

i Control Networks, 2016. http://www.icontrol.com/developers/#partner.

Internet of Things, 2016. http://www.cisco.com/c/en/us/solutions/internet-of-things/overview.html.

Islam, S.M.R., Kwak, D., Humaun, K., Hossain, M., Kwak, K-S., 2015. The Internet of Things for health care: a comprehensive survey. IEEE Access 3, 678–708.

Jelastic Cloud, 2016. https://jelastic.com/.

Jih, W-r., Hsu, J.Y-j., Lee, T-C., Chen, L-l., 2007. A multi-agent context-aware service platform in a smart space. J. Comput. 18 (1), 45–59.

Kang, J., Park, S., 2013. Context-aware services framework based on semantic web services for automatic discovery and integration of context. Int. J. Adv. Comput. Technol. 5 (4), 439–448.

Kidd, C.D., Orr, R., Abowd, G.D., Atkeson, C.G., Essa, I.A., MacIntyre, B., Mynatt, E.D., Starner, T., Newstetter, W., 1999. The aware home: a living laboratory for ubiquitous computing research. In: Proceedings of the Second International Workshop on Cooperative Buildings, Integrating Information, Organization, and Architecture. Springer-Verlag, London (UK), pp. 191–198.

Kim, J., Chung, K-Y., 2014. Ontology-based healthcare context information model to implement ubiquitous environment. Multimed. Tools Appl. 71 (2), 873–888.

Kyas, O., 2013. How To Smart Home. A Step by Step Guide Using Internet, Z-Wave, KNX & OpenRemote. Key Concept Press.

Lasierra, N., Alesanco, A., Garcia, J., 2014. Designing an architecture for monitoring patients at home: ontologies and web services for clinical and technical management integration. IEEE J. Biomed. Health Inform. 18 (3), 896–906.

Loke, S.W., 2004. Logic programming for context-aware pervasive computing: language support, characterizing situations, and integration with the web. In: Proceedings of the IEEE/WIC/ACM International Conference on Web Intelligence, pp. 44–50.

Loke, S., 2006. Context-Aware Pervasive Systems: Architectures for a New Breed of Applications. Auerbach Pub.

Lopez, P., Fernandez, D., Jara, A.J., Skarmeta, A.F., 2013. Survey of Internet of Things technologies for clinical environments. In: 27th International Conference on Advanced Information Networking and Applications Workshops (WAINA), pp. 1349–1354.

Lutolf, R., 1992. Smart Home concept and the integration of energy meters into a home based system. In: Proc. 7th Int. Conf. Metering Apparatus Tariffs Electr., pp. 277–278.

Apache Mahout, 2014. http://mahout.apache.org/.

MATHLAB, 2015. http://es.mathworks.com/products/matlab/.

MA, Microsoft Azure, 2015. https://azure.microsoft.com/.

Miraoui, M., Tadj, C., Amar, C.B., 2008. Architectural survey of context-aware systems in pervasive computing environment. Ubiquitous Comput. Commun. J. 3 (3), 1–9.

Mouttham, A., Peyton, L., Eze, B., Saddik, A.E., 2009. Event-driven data integration for personal health monitoring. J. Emerg. Technol. Web Intell. 1, 110–118.

Mu-Hsing Kuo, A., 2011. Opportunities and challenges of cloud computing to improve health care services. J. Med. Internet Res. 13 (3).

Nalepa, G.J., Bobek, S., 2014. Rule-based solution for context-aware reasoning on mobile devices. Comput. Sci. Inf. Syst. 11 (1), 171–193.

Nguyen, A., Moore, D., McCowan, I., 2007. Unsupervised clustering of free-living human activities using ambulatory accelerometry. In: Conf. Proc. IEEE Eng. Med. Biol. Soc. 2007, pp. 4895–4898.

Ni, Q., García Hernando, A., de la Cruz, I., 2015. The elderly's independent living in smart homes: a characterization of activities and sensing infrastructure survey to facilitate services development. Sensors, 11312–11362.

Olaru, A., Florea, A., Fallah Seghrouchni, A., 2013. A context-aware multi-agent system as a middleware for ambient intelligence. Mob. Netw. Appl. 18 (3), 429–443.

OSGi Alliance Specifications, 2015. http://www.osgi.org/Specifications.

OWL: Web Ontology Language, 2013. http://www.w3.org/2001/sw/wiki/OWL.

Paganelli, F., Giuli, D., 2011. An ontology-based system for context-aware and configurable services to support home-based continuous care. IEEE Trans. Inf. Technol. Biomed. 15 (2), 324–333.

Pech, S., 2013. Software agents in industrial automation systems. IEEE Softw. 30 (3), 20–24.

Perrot, C., Finnie, G., Morrison, I., 2012. Establishing context for software agents in pervasive healthcare systems. In: 15th International Conference on Network-Based Information Systems, pp. 447–452.

Puustjärvi, J., Puustjärvi, L., 2015. The role of smart data in smart home: health monitoring case. Proc. Comput. Sci. 69, 143–151.

The R Project for Statistical Computing, 2015. https://www.r-project.org/.

Rae, S., Kumar, A., 2008. Home Networking and Digital Living Networking Alliance. Tata Consultancy Services Limited.

Rakib, A., Ul Haque, Hafiz Mahfooz, 2014. A logic for context-aware non-monotonic reasoning agents. In: Gelbukh, A., et al. (Eds.), Human-Inspired Computing and Its Applications, pp. 453–471.

Ranganathan, A., Campbell, R.H., 2003. An infrastructure for context-awareness based on first order logic. Pers. Ubiquitous Comput. 7, 353–364.

Reinisch, C., Kastner, W., Neugschwandtner, G., Granzer, W., 2007. Wireless Technologies in Home and Building Automation. In: 5th IEEE International Conference on Industrial Informatics, pp. 93–98.

Ricquebourg, V., Menga, D., Durand, D., Marhic, B., Delahoche, L., Loge, C., 2006. The Smart Home Concept: our immediate future. In: 1st IEEE International Conference on E-Learning in Industrial Electronics, pp. 23–28.

Ridi, A., Zarkadis, N., Gisler, C., Hennebert, J., 2015. Duration models for activity recognition and prediction in buildings using Hidden Markov Models. In: IEEE International Conference on Data Science and Advanced Analytics (DSAA), pp. 1–10.

Rosslin, J.R., Kim, T., 2010. Review: context aware tools for smart home development. Int. J. Smart Home 4 (1), 1–12.

Sahli, N., 2008. Survey: agent-based middlewares for context awareness. Electron. Commun. EASST 11.

Salih, A., Salih, M., Abraham, A., 2013. A review of ambient intelligence assisted healthcare monitoring. Int. J. Comput. Inf. Syst. Ind. Manag. Appl., 741–750.

Samsung, 2016. http://www.samsung.com/uk/smartthings/.

Satpathy, L., 2006. Smart housing: technology to aid aging in place. New opportunities and challenges. M.S. Thesis. Mississippi State Univ., Starkville.

Satyanarayanan, M., 2001. Pervasive computing: vision and challenges. In: IEEE Personal Communication.

Schilit, B.N., Adams, N., Want, R., 1994. Context-aware computing applications. In: IEEE Workshop on Mobile Computing Systems and Applications, pp. 85–90.

Schweitzer, E.J., 2012. Reconciliation of the cloud computing model with US federal electronic health record regulations. J. Am. Med. Inform. Assoc. 19 (2), 161–165.

Sharma, H., Sharma, S., 2014. A review of sensor networks: technologies and applications. In: Recent Advances in Engineering and Computational Sciences (RAECS), pp. 1–4.

Sohn, M., Jeong, S., Lee, H.J., 2014. Case-based context ontology construction using fuzzy set theory for personalized service in a smart home environment. Soft Comput. 18 (9), 1715–1728.

Solanas, A., Patsakis, C., Conti, M., Vlachos, I., Ramos, V., Falcone, F., Postolache, O., Perez-martinez, P., Pietro, R., Perrea, D., Martinez-Balleste, A., 2014. Smart health: a context-aware health paradigm within smart cities. IEEE Commun. Mag. 52 (8), 74–81.

Stowe, S., Harding, S., 2010. Telecare, telehealth and telemedicine. Europ. Geriatr. Med. 1 (3), 193–197.

Strang, T., Linnhoff-Popien, C., 2004. A context modeling survey. In: Workshop on Advanced Context Modelling, Reasoning and Management, UbiComp 2004 – The Sixth International Conference on Ubiquitous Computing. Nottingham/England.

Suryadevara, N.K., Mukhopadhyay, S.C., Wang, R., Rayudu, R.K., 2013. Forecasting the behavior of an elderly using wireless sensors data in a smart home. Eng. Appl. Artif. Intell. 26, 2641–2652.

Vaquero, L.M., Rodero-Merino, L., Caceres, J., Lindner, M., 2008. A break in the clouds: towards a cloud definition. Comput. Commun. Rev. 39 (1), 50–55.

Valero, M.A., Vadillo, L., Pau, I., Peñalver, Ana, 2009. An intelligent agents reasoning platform to support smart home telecare. In: IWANN 2009, Part II. In: LNCS, vol. 5518, pp. 679–686.

Valero Duboy, M.A., 2015. Hogar digital inteligente. In: Calidad de Vida y Ciudadanos, pp. 123–126.

Van Kranenburg, H., Bargh, M., Iacob, S., Peddemors, A., 2006. A context management framework for supporting context-aware distributed applications. IEEE Commun. Mag. 44, 67–74.

Van Bunningen, A.H., Feng, L., Apers, P.M.G., 2005. Context for ubiquitous data management. In: International Workshop on Ubiquitous Data Management, pp. 17–24.

Vermesan, O., Friess, P., Guillemin, P., Gusmeroli, S., Sundmaeker, H., Bassi, B., Soler Jubert, I., Mazura, M., Harrison, M., Eisenhauer, M., Doody, P., 2011. Internet of Things Strategic Research Roadmap.

Vermesan, O., Friess, P. (Eds.), 2014. Internet of Things – From Research and Innovation to Market Deployment. River Publishers.

Villarreal, V., Fontecha, J., Hervas, R., Bravo, J., 2011. An architecture to development an ambient assisted living applications: a study case in diabetes. In: Proc. 5th Int. Symp. Ubiquitous Computing and Ambient Intelligence. Riviera Maya, Mexico.

Wang, K.I.K., Abdulla, W.H., Salcic, Z., 2009. Ambient intelligence platform using multiagent system and mobile ubiquitous hardware. Pervasive Mob. Comput. 5 (5), 558–573.

Wang, H., Metha, R., Chung, L., Supakkul, S., Huang, L., 2012. Rule-based context-aware adaptation: a goal-oriented approach. Int. J. Pervasive Comput. Commun. 8 (3), 279–299.

Warren, S., et al., 1999. Designing smart health care technology into the home of the future. In: Workshops on Future Medical Devices: Home Care Technologies for the 21st Century, pp. 1–15.

Weka 3: Data Mining Software in Java, 2015. http://www.cs.waikato.ac.nz/~ml/weka/.

Withanage, C., Ashok, R., Yuen, C., Otto, K., 2014. A comparison of the popular home automation technologies. In: IEEE in Innovative Smart Grid Technologies — Asia (ISGT Asia), pp. 600–605.

Wooldridge, M., 2009. An Introduction to Multiagent SYSTEMS. John Wiley & Sons.

Yin, J., Zhang, Q., Karunanithi, M., 2015. Unsupervised daily routine and activity discovery in smart homes. In: 37th Annual International Conference of the IEEE, Engineering in Medicine and Biology Society (EMBC).

Yuan, B., Herbert, J., 2012. Fuzzy CARA — a fuzzy-based context reasoning system for pervasive healthcare. In: The 3rd International Conference on Ambient Systems, Networks and Technologies. In: Computer Science, vol. 10, pp. 357–365.

Zhang, D., Adipat, B., Mowafi, Y., 2009. User-centered context-aware mobile applications–the next generation of personal mobile computing. Commun. Assoc. Inf. Syst. 24, 27–46.

Zhang, D., Huang, H., Lai, C.F., Liang, X., Zou, Q., Guo, M., 2013. Survey on context-awareness in ubiquitous media. Multimed. Tools Appl. 67, 179–211.

Zhang, D., Gu, T., Wang, X., 2005. Enabling context-aware smart home with semantic web technologies. Int. J. Hum. Welf. Robot. Syst. 6, 12–20.

CHAPTER 15

Visual Information-Based Activity Recognition and Fall Detection for Assisted Living and eHealthCare

Yixiao Yun, Irene Yu-Hua Gu
Department of Signals and Systems, Chalmers University of Technology, SE-412 96, Gothenburg, Sweden

15.1 INTRODUCTION

According to United Nations (2015), population aging is taking place in nearly all countries of the world, with overall a considerably high rate of growth. It is estimated that the global share of elderly people accounts 11.7% in 2013 and will continue to grow, reaching 21.1% by 2050 (United Nations, 2013). Many of them choose to live at home while others live in elderly care centers. There is a general lack of caring personnel, hence a great need emerges for automatic machine-based assistance.

Ambient assisted living is one of such systems that offer intelligent services supporting people's daily lives. An ambient assisted living system creates human-centric smart environments that are sensitive, adaptive and responsive to human needs, habits, gestures and emotions. The innovative interaction between human and surrounding environment makes ambient intelligence a suitable candidate for healthcare.

As shown in Figure 15.1, on creating a human–centric smart environment for assisted living, two of the most fundamental tasks are to (a) classify normal daily activities; and (b) detect abnormal activities. Examples of ambient assisted living for elderly include detecting anomalies like falls, robbery or fire at home, recognizing daily living patterns, and obtaining statistics of various daily activities over time. Among all activities, detecting falls is one of the basic topics attracting much attention, due to the associated risks (Mubashir et al., 2013), e.g. bone fracture, stroke, or even death. Triggering emergent help is desired, especially for persons who live alone.

Many different methods have been developed by exploiting different types of sensors, e.g., smartwatches (Kostopoulos et al., 2015), sound sensors, wearable motion devices (gyroscopes, speedometers, accelerometers),

Ambient Assisted Living and Enhanced Living Environments.
DOI: http://dx.doi.org/10.1016/B978-0-12-805195-5.00015-6

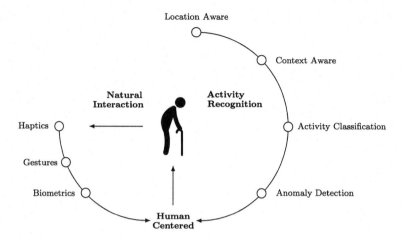

Figure 15.1 The key factors for ambient intelligence in assisted living.

visual, range and infrared (IR) cameras. Sound sensors could be used for collecting sound related to sudden falls, while smartwatches and other wearable devices measure the motion and can be used for fall detection as well. However, they are not always feasible, e.g., one does not always wear the device especially during showering, and the false alarm could be high if sound detection is used alone without combining other sensor information. Devices with imaging sensors (e.g., low-resolution RGB, depth, near or thermal IR) that offer real time analysis have drawn considerable attentions. In such cases, only analysis results, statistics and triggering information are stored without saving person's video information. Although some privacy concerns exist in video-based analytics, this issue can be mitigated by near real-time feature/information extraction from video followed discarding the original image data, or, by using very low resolution on depth or IR data that does not provide personal details where only the shape of a person is visible.

This chapter mainly focuses on describing visual-information based daily activity recognition and anomaly detection through using low-resolution visual sensors. The main reason for using visual-information based methods are due to: (i) vision plays a major role in recognition of activities which is a fundamental issue in a human-centric system; (ii) real-time visual sensor-based analysis may possibly offer high performance without intruding individual's privacy.

The chapter is organized as follows. First, we briefly review the current state-of-the-art methods on visual activity recognition for assisted living and healthcare. We then focus on three methods that are most promising, in particular, recognition of activities based on low-dimensional smooth manifolds.

15.2 EXISTING METHODS ON VISUAL ACTIVITY RECOGNITION FOR ASSISTED LIVING

For recognizing human activities from visual sensors, many different methods have been proposed, including both image-based and video-based methods. Some activity analyses only exploit *static cues* from still images or key image frames (Yun et al., 2013, 2014), while others use the *dynamics* of entire video (e.g. falls) (Yun and Gu, 2015; Yun et al., 2015). Surveys of image/video-based methods can be found in Guo and Lai (2014), Aggarwal and Ryoo (2011), Vishwakarma and Agrawal (2013), Cheng et al. (2015).

There are many different ways to categorize the methods used for visual analysis of human activities. In this section, we primarily categorize them according to three different application fields, including the recognition of activities for: (a) daily living at home environments; (b) healthcare, medical treatments and patient rehabilitations; (c) falls and other abnormal activities.

(a) *Recognition of Activities of Daily Living (ADL) at Home Environments*

The main purpose of such activity analysis is for life logging and the assessment of health conditions/functions. Context plays an important role for understanding ADL. One way of exploiting context is to use the location information, e.g., by dividing the living area into different functional regions and tracking the sequence of regions visited by the person for activity analysis (Duong et al., 2005). Alternatively, one could track the sequence of human-object interactions (e.g., enter the door, open fridge, put food into oven) as another type of contextual information to recognize an activity (Duong et al., 2006). It is also beneficial to jointly use the location, speed, shape and motion of the person for activity recognition (Zhou et al., 2008). Further, a variety of methods can be found by applying different models, e.g. using bag-of-words (BoW) model based on HOG and HOF features (Avgerinakis et al., 2013), or extracting features from local body parts under deformable part models (DPM) (Yan et al., 2014).

(b) *Recognition of Activities for Healthcare, Medical Treatments and Patient Rehabilitations*

A number of studies were conducted on visual analysis of behaviors of patients with stroke, Alzheimer's disease and other dysfunctions in hospitals or nursing homes, including medicine or food intake (Ghali et al., 2003; Gao et al., 2004), sit-to-stand motion (Goffredo et al., 2009) or the gait patterns (Leu et al., 2011; Li et al., 2011). The activity analysis often involves recognizing human–object interactions (Ghali et al., 2003), the interactions of human body parts (Gao et al., 2004), and human-to-human interactions (Liu et al., 2007). Studies were also conducted on nursing activities from caregivers to the elderly (e.g. hygiene, feeding, giving medicine, taking vital signs), through identification of related objects or tools (e.g., paper towels, pillbox, diaper, plate, cup, sphygmomanometer) (Martinez-Perez et al., 2010). Different techniques can be applied for such analysis, e.g., hidden Markov model (HMM) (Gao et al., 2004; Liu et al., 2007), dynamic time warping (DTW) (Tham et al., 2014), and a variety of classification methods (k-NN, LogitBoost, SVM) (Zhan et al., 2012).

(c) *Detection of Falls and Other Abnormal Activities*

Fall detection is a major issue in anomaly detection, due to its potential severe impact. Detection of falls is often realized through classification based on the one-against-all strategy. A variety of features related to falls were studied, including shape features based on extracted silhouettes (Rougier et al., 2011; Banerjee et al., 2013), curvature scale space (CSS) (Ma et al., 2014) and Riemannian manifolds (Yun and Gu, 2015), motion features based on optical flows (Yun et al., 2015), features based on target bounding box such as the aspect ratio (Fleck and Straßer, 2008; Charfi et al., 2013) and centroid position of the box (Yun et al., 2015; Charfi et al., 2013), and 3D modeling of human body (Auvinet et al., 2011; Banerjee et al., 2014; Stone and Skubic, 2015). Many different types of classifiers were employed for fall detection, such as SVM (Yun and Gu, 2015; Yun et al., 2015; Fleck and Straßer, 2008), AdaBoost (Charfi et al., 2013), Gaussian mixture model (GMM) (Rougier et al., 2011), Gustafson-Kessel (GK) clustering (Banerjee et al., 2013), extreme learning machine (ELM) (Ma et al., 2014) and decision trees (Stone and Skubic, 2015). Further, a range of camera types and settings were utilized, e.g., distributed cameras (Fleck and Straßer, 2008), IR (Banerjee et al., 2013), depth (Banerjee et al., 2013; Ma et al., 2014; Yun et al., 2015; Stone and Skubic, 2015), RGB cameras (Rougier et al., 2011; Banerjee et al., 2013; Yun and Gu, 2015; Fleck and Straßer, 2008; Charfi et al., 2013; Auvinet et al., 2011), in day and night scenes (Banerjee et al., 2013).

15.3 VISUAL ACTIVITY RECOGNITION USING MANIFOLD-BASED APPROACHES

Manifold-based methods are often employed for efficient low-dimensional representation of high-dimensional data meanwhile maintaining important properties of the data such as topology and geometry (Lee, 2012). Roughly speaking, a manifold can be considered as a set of low dimensional spaces embedded in a high dimensional space. An intuitive example of manifold is the earth which is globally a sphere in 3D space but locally flat in 2D maps.

Manifold is found useful in many vision tasks where (a) measured data naturally reside on nonlinear curved spaces, (b) data representation requires low-dimensional and efficient description or dimensionality reduction, or (c) non-Euclidean metrics better capture the non-linear relationship between data elements. Hence, manifolds may be exploited for efficiently characterizing the dynamic process of human activities in videos. In this section, we first summarize the essential mathematics required in the subsequent description of manifold-based analysis. We then describe three robust methods based on Riemannian manifolds for activity classification and fall detection.

15.3.1 Riemannian Geometry

In case of nonlinear manifolds that are not vector spaces, the Euclidean calculus and conventional statistics do not apply. A Riemannian manifold is a smooth manifold that is differentiable (Lee, 2012), where a set of metrics can be defined to measure the distance and angle on the manifold. Further, in the tangent space of points on Riemannian manifold, linear operations can be performed.

The geodesic is the shortest curve between two points on a manifold. The geodesic distance, the length of the geodesic, is used to measure the distance between two manifold points.

We give two examples of manifolds with the Riemannian geometry, namely the space of symmetric positive definite (SPD) matrices and the unit n-sphere, for the sake of mathematical convenience in the subsequent subsection.

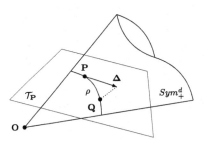

Figure 15.2 Example of Sym_+^d ($d = 2$) embedded in a 3D space $\mathbb{R}^3$. **O** is the origin. **P** and **Q** are manifold points, i.e., **P, Q** $\in Sym_+^d$. $\mathcal{T}_\mathbf{P}$ is the tangent space at **P**. $\mathbf{\Delta} \in \mathcal{T}_\mathbf{P}$ is the tangent vector whose projected point on the manifold is **Q**. The geodesic ρ is the shortest curve between **P** and **Q** on the manifold.

15.3.1.1 Space of Symmetric Positive Definite Matrices

The space of $d \times d$ SPD matrices (Sym_+^d) is an open convex cone whose strict interior is a Riemannian manifold (Lee, 2012). To compute the statistics on Sym_+^d, the affine–invariant metric (Pennec et al., 2006) and the log-Euclidean metric (Arsigny et al., 2008) are commonly used. These two metrics are mathematically equivalent, however, numerical results can slightly differ. Log-Euclidean metric is usually computationally more efficient (Arsigny et al., 2008).

Two operators, the exponential map and the logarithm map, are defined over differentiable manifolds to switch between the manifold and tangent space at a given point (Khan and Gu, 2014).

The exponential map $(\mathcal{T}_\mathbf{P} \mapsto Sym_+^d)$ is a function that maps a tangent vector $\mathbf{\Delta}$ (in the tangent space $\mathcal{T}_\mathbf{P}$ associated with a manifold point $\mathbf{P} \in Sym_+^d$) to its corresponding point $\mathbf{Q}$ on the manifold Sym_+^d (as shown in Figure 15.2). Under the log-Euclidean metric (Subbarao and Meer, 2009), it is given by

$$\exp_\mathbf{P}(\mathbf{\Delta}) = \exp(\log(\mathbf{P}) + \mathbf{\Delta}) = \mathbf{Q}, \tag{15.1}$$

where $\exp(\cdot)$ is the matrix exponential (Arsigny et al., 2008), and $\log(\cdot)$ is the principal logarithm of a matrix which is defined as the inverse of the matrix exponential (Arsigny et al., 2008).

The logarithmic map $(Sym_+^d \mapsto \mathcal{T}_\mathbf{P})$ is a function that maps a manifold point $\mathbf{Q} \in Sym_+^d$ to its corresponding tangent vector $\mathbf{\Delta}$ in the tangent space $\mathcal{T}_\mathbf{P}$ associated with another manifold point $\mathbf{P} \in Sym_+^d$ (as shown in Figure 15.2). Under the log-Euclidean metric (Subbarao and Meer, 2009), it is given by

$$\log_\mathbf{P}(\mathbf{Q}) = \log(\mathbf{Q}) - \log(\mathbf{P}) = \mathbf{\Delta}. \tag{15.2}$$

Geodesic is the shortest curve ρ between two manifold points $\mathbf{P}$, $\mathbf{Q}$ on Sym_+^d. The geodesic distance is the length of ρ given by

$$d(\mathbf{P}, \mathbf{Q}) = \| \log_\mathbf{P}(\mathbf{Q}) \| = \| \log(\mathbf{Q}) - \log(\mathbf{P}) \|, \tag{15.3}$$

where (15.3) is defined under the log-Euclidean metric (Subbarao and Meer, 2009), and $\| \cdot \|$ is the Frobenius norm.

The Karcher mean, also known as the Fráchet or Riemannian mean, is the intrinsic mean of a set of points on a Riemannian manifold. Given a set of manifold points $\{\mathbf{X}_i\}_{i=1}^N$, the Karcher mean $\mathbf{X}^*$ is defined as

$$\mathbf{X}^* = \arg\min_\mathbf{X} \sum_{i=1}^N w_i d^2(\mathbf{X}, \mathbf{X}_i), \tag{15.4}$$

where $w_i \in \mathbb{R}$ is the weight for each point, and $d(\cdot, \cdot)$ is the geodesic distance defined in (15.3). The minimization problem can be solved by iteratively mapping from manifold to tangent spaces and vice versa until convergence (Khan and Gu, 2014; Subbarao and Meer, 2009):

$$\mathbf{X}_{j+1} = \exp_{\mathbf{X}_j}\left(\sum_{i=1}^N w_i \log_{\mathbf{X}_j}(\mathbf{X}_i) / \sum_{i=1}^N w_i \right), \tag{15.5}$$

where $\exp_{\cdot}(\cdot)$ and $\log_{\cdot}(\cdot)$ are the pair of exponential and logarithm mapping functions defined in (15.1) and (15.2) under log-Euclidean metric.

The Riemannian geometry of Sym_+^d can be exploited when the extracted feature descriptors are covariance matrices, e.g., region covariance (Tuzel et al., 2006), due to the fact that SPD cone is exactly the set of non-singular covariance matrices. Since covariance matrices $\mathbf{C} \in Sym_+^d$, they may be viewed as connected points on a Riemannian manifold (Tuzel et al., 2008).

15.3.1.2 The Unit n-Sphere

The unit n-sphere, $\mathcal{S}^n$, is an n-dimensional sphere with a unit radius, centered at the origin of $(n + 1)$-dimensional Euclidean space. An example where $n = 2$ is illustrated in Figure 15.3. It can be considered as the simplest Riemannian manifold after the Euclidean space (Lovett, 2010). The geodesic distance between two manifold points $\mathbf{p}$, $\mathbf{q}$ on $\mathcal{S}^n$ is the great-circle distance:

$$\rho(\mathbf{p}, \mathbf{q}) = \arccos(\mathbf{p}^T\mathbf{q}), \tag{15.6}$$

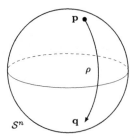

Figure 15.3 Example of an n-sphere $\mathcal{S}^n$ ($n = 2$) embedded in an $(n + 1)$-D space $\mathbb{R}^{n+1}$. **p** and **q** are manifold points, i.e., **p**, **q** $\in \mathcal{S}^n$. The geodesic ρ is the shortest curve between **p** and **q** on the manifold.

where $\arccos(\cdot)$ is the inverse cosine function (Jayasumana et al., 2013). The great-circle distance between two manifold points is unique.

The Riemannian geometry of unit n-sphere can be utilized when the extracted feature vectors are normalized by the ℓ_2 norm, e.g., HOG (Dadal and Triggs, 2005). The descriptors hence lie on a unit n-sphere $\mathcal{S}^n$, for some n.

15.3.2 Activity Recognition Methods by Exploiting Riemannian Manifolds

In this subsection, we describe three robust methods that exploit the geometry of Riemannian manifolds for video analysis of daily activities and falls. *Method-1* and *Method-2* both involve the space of SPD matrices for activity recognition. The similarities between these two methods are that both of them (a) employ log-Euclidean embedding of SPD matrices for mean computation; and (b) divide local spatial-temporal video volumes into 3D grids for incorporating structural information. The major differences between them are (i) that covariance matrices are computed from spatio-temporal domain in *Method-1* but from spatial domain in *Method-2*, and (ii) that local descriptors are encoded under different models. *Method-3* uses the unit n-sphere as a simple example for manifold-based fall detection. Different from the first two methods, *Method-3* extracts features as statistics of manifold points, instead of representing features as manifold points directly.

15.3.2.1 Method-1: *Action Recognition Using Log-Euclidean Bag of Words (LE-BoW)*

In *Method-1* (Faraki et al., 2014), the bag-of-words (BoW) model is used for activity recognition, based on covariance matrices of spatio-temporal features formed from histogram of optical flow (HOF). The BoW representation is usually obtained by first clustering a set of selected local descriptors (e.g., with k-means) to generate a visual vocabulary (or, codebook), followed by extracting a histogram by assigning each descriptor to its closest visual word (Fei-Fei and Perona, 2005).

The essence of this method is to extend the conventional BoW approach to its Riemannian version, where the local descriptors are covariance matrices. Since covariance matrices form a special type of Riemannian manifold, the space of SPD matrices, the underlying Riemannian geometry should be taken into account for the generation of the visual dictionary (codebook).

Specifically, given a set of N training samples $\mathbb{S} = \{\mathbf{X}_i\}_{i=1}^N$ from the underlying Sym_+^d manifold (where each point on the manifold corresponds to a covariance matrix), i.e., $\mathbf{X}_i \in Sym_+^d, \forall i = 1, \cdots, N$, a codebook is generated by applying k-means on the training data. Directly computing the arithmetic mean of clusters by neglecting the Riemannian geometry of SPD matrices may lead to undesirable outcome caused by swelling effect (Arsigny et al., 2008). On the other hand, replacing the arithmetic mean with the Karcher mean in (15.4) solves the problem intrinsically by exploiting Riemannian geometry but is computationally demanding, as it requires mapping back and forth between the manifold and its tangent spaces. Hence, an alternative approach is adopted extrinsically to simplify the problem by embedding the manifold into the space of $d \times d$ symmetric matrices, Sym^d. This is done by embedding Sym_+^d into its tangent space at identify matrix $\mathbf{I}$ through principal logarithm in (15.2):

$$\log_{\mathbf{I}}(\mathbf{X}) = \log(\mathbf{X}) - \log(\mathbf{I}) = \log(\mathbf{X}) = \mathbf{\Delta}, \qquad (15.7)$$

where $\mathbf{X} \in Sym_+^d$ and $\mathbf{\Delta} \in Sym^d$. Since symmetric matrices (or equivalently tangent spaces) form a vector space, Euclidean calculus and conventional statistics can be applied.

Each SPD matrix $\mathbf{X}$ and its log-Euclidean embedding $\log(\mathbf{X})$ have $l = d \times (d + 1)/2$ independent values respectively due to symmetry. Therefore, it can be uniquely represented as a vector of these independent values, i.e., only the upper triangular part of $\log(\mathbf{X})$

$$\mathbf{x} = \text{vec}(\log(\mathbf{X}))$$
$$= [\Delta_{1,1}, \sqrt{2}\Delta_{1,2}, \sqrt{2}\Delta_{1,3}, \cdots, \sqrt{2}\Delta_{1,d}, \Delta_{2,2}, \sqrt{2}\Delta_{2,3}, \cdots, \Delta_{d,d}]^T \in \mathbb{R}^l, \quad (15.8)$$

where $\Delta_{m,n}$ is the m-row and n-column entry of $\mathbf{\Delta}$. The off-diagonal entries are multiplied with $\sqrt{2}$ as they are counted twice during norm computation (Tuzel et al., 2008).

With all training samples being converted to log-Euclidean vector forms in (15.8), k clusters $C_1, C_2, \cdots, C_k$ are estimated with cluster centers $\{\mathbf{c}_j\}_{j=1}^k$ by applying conventional k-means clustering in the tangent space at identity matrix, as summarized in Algorithm 1.

Algorithm 1: Log-Euclidean k-means over Sym_+^d for learning the visual dictionary (Faraki et al., 2014).

Input :
- Training set $\mathbb{S} = \{\mathbf{X}_i\}_{i=1}^N$ from the underlying Sym_+^d manifold
- The number of iterations n_{Iter}

Output:
- Visual dictionary $\mathbb{D} = \{\mathbf{c}_j\}_{j=1}^k$, $\mathbf{D}_j \in \mathbb{R}^l$

Compute $\mathbb{S}' = \{\mathbf{x}_i\}_{i=1}^N$, log-Euclidean representation of $\mathbb{S}$ using (15.8);
Initialize the dictionary $\mathbb{D} = \{\mathbf{c}_j\}_{j=1}^k$ by selecting k samples from $\mathbb{S}'$ randomly;

for $t = 1 : n_{\text{Iter}}$ **do**

  Assign each point $\mathbf{x}_i$ to its nearest cluster in $\mathbb{D}$;

  Compute the average dispersion from cluster centers by

  $\epsilon = \dfrac{1}{N} \sum_{j=1}^k \sum_{\mathbf{x}_i \in C_j} \|\mathbf{x}_i - \mathbf{c}_j\|^2$;

  if ϵ *is less than a predefined threshold* **then**

   break the loop;

  else

   Recompute cluster centers $\{\mathbf{c}_j\}_{j=1}^k$ by $\mathbf{c}_j = \dfrac{1}{|C_j|} \sum_{\mathbf{x}_i \in C_j} \mathbf{x}_i$;

  end

end

Given an input video, a set of covariance matrices (local descriptors) $\mathbb{Q} = \{\mathbf{Q}_i\}_{i=1}^M$, $\mathbf{Q}_i \in Sym_+^d$, are extracted and converted to their log-Euclidean vector forms $\{\mathbf{q}_i\}_{i=1}^M$ using (15.8) such that $\mathbf{q}_i = \text{vec}(\log(\mathbf{Q}_i)) \in \mathbb{R}^l$. With the codebook $\mathbb{D} = \{\mathbf{c}_j\}_{j=1}^k$ learned from Algorithm 1, a histogram-based representation can be obtained for the video in three different ways, namely hard assignment (HA), spatio-temporal pyramids (STP), and sparse coding (SC). HA assigns vectors to their closest vocabulary word in the dictionary using Euclidean distance. STP encodes structure information into LE-BoW model. SC is employed due to the sparsity of histograms obtained by HA or STP in nature.

The computation of covariance matrices is based on overlapping spatio-temporal blocks extracted from a given video and the HOF features of densely extracted trajectories within each block. For more details on trajectory extraction, readers are referred to Faraki et al. (2014). For classification, a non-linear SVM with an RBF-χ^2 kernel is used.

15.3.2.2 Method-2: *Action Recognition Using Video Covariance Matrix Logarithm (VCML)*

In *Method-2* (Bilinski and Bremond, 2015), actions are characterized by VCML descriptors, based on a covariance matrix representation. The VCML descriptor models the relationship between different low-level features in each video frame, such as intensity and gradient. Local spatio-temporal video volumes are extracted by dense trajectories from each video, and the appearance information of these volumes are encoded by VCML descriptors.

The essence of this method is to encode local appearance and spatio-temporal structural information for characterizing actions in an efficient and effective way, without the requirement of human detection and event segmentation.

Given a single video frame at time t, let $\mathbf{f}_j$ be the d-dimensional feature vector for the j-th pixel inside it, where $j = 1, \cdots, N$, and N is the total number of pixels. The features can be, e.g., pixel coordinates, intensity, color, gradients, magnitudes and phase angles. The region is represented by a $d \times d$ covariance matrix

$$\mathbf{C}_t = \frac{1}{N-1} \sum_{j=1}^{N} (\mathbf{f}_j - \boldsymbol{\mu})(\mathbf{f}_j - \boldsymbol{\mu})^T, \tag{15.9}$$

where $\boldsymbol{\mu}$ is the mean feature vector. Since covariance matrices $\mathbf{C}_t \in Sym_+^d$, they may be viewed as connected points on a Riemannian manifold (Tuzel et al., 2008).

Similar to *Method-1*, covariance matrices are converted to log-Euclidean vectors by (15.8). Hence, each video frame is represented by a compact vector $\text{vec}(\log(\mathbf{C}_t))$ of size $l = d \times (d + 1)/2$, referred to as video frame descriptor.

Given a spatio-temporal video volume, it is treated as a cuboid, and divided into a 3D grid. Each cell of the grid is of size $g_x \times g_y \times g_t$, where $g_x \times g_y$ is the spatial size and g_t is the temporal size (number of frames). For each video frame in each cell of the grid, a separate video frame descriptor is computed. Then, each cell of the grid is described by a mean vector of all video frame descriptors from the cell:

$$\mathbf{v}_{\text{cell}} = \frac{1}{g_t} \sum_{t=1}^{g_t} \text{vec}(\log(\mathbf{C}_t)). \tag{15.10}$$

Finally, the video covariance matrix logarithm (VCML) descriptor $\mathbf{d}$ is defined as the concatenation of all descriptors from all cells of the grid:

$$\mathbf{d} = [\mathbf{v}_{\text{cell}_1}, \mathbf{v}_{\text{cell}_2}, \cdots, \mathbf{v}_{\text{cell}_m}]^T, \tag{15.11}$$

where m is the number of cells of the spatio-temporal grid.

A given video sequence can be represented by VCML descriptors of local spatio-temporal video volumes. Similar to *Method-1*, local spatio-temporal video volumes are extracted based on dense trajectories in a video sequence. To compute dense trajectories, a dense sampling is applied to extract interest points. Then, these interest points are tracked using a dense optical flow field. For details on the computation of dense trajectories (DT) and improve dense trajectories (IDT), readers are referred to Wang et al. (2011), Wang and Schmid (2013). Given detected trajectories, local spatio-temporal video volumes are extracted around them. In this way, a good coverage of a video sequence is provided and the extraction of meaningful features are ensured.

The extracted VCML descriptors of local spatio-temporal video volumes are further processed by Fisher vector encoding (Perronnin et al., 2010). Then, each video is represented by a $2DK$-dimensional Fisher vector, where D is the descriptor size and K is the number of Gaussians in the model. Linear SVM classifiers are employed for action classification using Fisher vectors, with one-against-all strategy for multi-class classification.

15.3.2.3 Method-3: Fall Detection through Shape Analysis Based on the Riemannian Manifold

In *Method-3* (Yun and Gu, 2015), manifold-based shape analysis is adopted in a single camera view for fall detection. The basic ideas of this method include: Instead of using features extracted from rigid bounding boxes (e.g.,

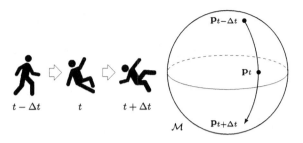

Figure 15.4 The intuitive illustration of converting the analysis of human shape dynamics to the study of velocity statistics of points (e.g., $\mathbf{p}_{t-\Delta t}$, $\mathbf{p}_t$, and $\mathbf{p}_{t+\Delta t}$) on a smooth manifold $\mathcal{M}$.

centroid, aspect ratio), the focus is shifted to the analysis of human shape inside the box. Since it is a broadly accepted intuition that a human shape deforms drastically in camera views while the person falls onto the ground, better features could be obtained by studying the rate of shape change in a certain time interval. A suitable metric is preferred for measuring the rate. Riemannian geometry fulfills this requirement, given the assumption that many image features including shape can be efficiently described by using Riemannian manifolds. By converting the analysis of human shape dynamics in an arbitrary camera view to the study of velocity statistics on a unified manifold for different camera views, it is expected that these features are less sensitive to view angles. This could lead to a simple and effective solution, without combining multiple cameras. An intuitive illustration of this method is given in Figure 15.4.

The shape descriptor is based on the histogram of oriented gradient (HOG) (Dadal and Triggs, 2005), where object shape is described by the distribution of intensity gradients through voting the dominant edge directions. Comparing to other shape representations (e.g., a set of boundary points) based on image segmentation algorithms such as GrabCut (Rother et al., 2004), the shape descriptor chosen by *Method-3* is less intuitive but also less computationally demanding. Besides, it is endowed with a manifold structure (Jayasumana et al., 2013) which suffices the purpose for manifold-based analysis.

Consider a target image region $\mathbf{R}$ defined by the bounding box at time t that is of size $w \times h$ and centered at (x_0, y_0), the region of interest (ROI) is obtained by modifying $\mathbf{R}$ to be of size $l \times l$ but still centered at (x_0, y_0), where $l = \max(w, h)$. Then, the ROI is normalized to $\lambda \times \lambda$, where λ is a predefined length that is usually smaller than l for computational efficiency.

$$\mathbf{R} : w \times h$$
$$l = \max(w, h)$$

ROI: $l \times l$

ROI: $\lambda \times \lambda$
$$\mathbf{R}' : (w \cdot \lambda/l) \times (h \cdot \lambda/l)$$

Figure 15.5 Size normalization for each ROI as the preprocessing step for shape description based on HOG features. The aspect ratio of the target person remains the same after size normalization.

In the normalized ROI, the corresponding tracked image region $\mathbf{R}'$ becomes of size $(w \cdot \lambda/l) \times (h \cdot \lambda/l)$. In this way, normalization of image size does not impact the aspect ratio of $\mathbf{R}$ inside ROI, as shown in Figure 15.5.

Given the normalized ROI, histograms of oriented gradient are formed, similarly to HOG (Dadal and Triggs, 2005). However, the difference here is that image gradients are only collected inside the tracked image region $\mathbf{R}'$. Thus, the background noise is effectively suppressed while keeping the aspect ratio of the foreground object. For a cell completely outside $\mathbf{R}'$, the histogram bins for that cell are assigned equal unit votes. Finally, the shape is described by a vector $\mathbf{p}_t \in \mathbb{R}^n$ concatenating all elements of ℓ_2 normalized histograms from all blocks. This vector $\mathbf{p}_t$ is a point residing on a unit n-sphere $\mathcal{S}^n$, i.e., $\mathbf{p}_t \in \mathcal{S}^n$.

For any input video segment, by representing the shape of a target person in each frames as connected points on a Riemannian manifold, the analysis of this dynamic feature are thus converted to the study of velocity statistics on that manifold. Intuitively, the more drastically the shape deforms, the more rapidly the corresponding manifold point moves.

To analyze the dynamics of points on the manifold in certain time intervals, a set of manifold points $\mathcal{Z}$ is collected:

$$\mathcal{Z} = \{\mathbf{p}_t\}_{t=1}^L, \tag{15.12}$$

where the manifold point $\mathbf{p}_t \in \mathcal{S}^n$, t is the frame index and L is the length of the video segment.

The instantaneous velocity v_t of a point moving on the manifold at time t is defined as

$$v_t = \frac{\rho(\mathbf{p}_{t-1}, \mathbf{p}_t)}{\Delta t}, \tag{15.13}$$

where $\mathbf{p}_{t-1}$ and $\mathbf{p}_t$ are the samples of $\mathbf{p}$ at time $(t-1)$ and t, $\rho(\mathbf{p}_{t-1}, \mathbf{p}_t)$ is the geodesic distance between them, and Δt is the time step. By assuming unit time step between consecutive frames (i.e., $\Delta t = 1$), the velocity simply becomes the geodesic distance between two neighboring points in $\mathcal{Z}$. The formula for computing geodesic distance $\rho(\mathbf{p}_{t-1}, \mathbf{p}_t)$ differs depending on the specific type of Riemannian manifold. *Method-3* uses (15.6) for computing geodesic distances on the manifold of unit n-sphere.

In this way, a vector containing piecewise velocities of moving points on the manifold can be constructed for each set of manifold points $\mathcal{Z}$ learned from the given video segment:

$$\mathbf{v} = [v_1, v_2, \cdots, v_{L-1}]^T. \tag{15.14}$$

The feature vector $\mathbf{f}$ describing the shape dynamics for the given video segment is a concatenation of simple statistics for the velocity of moving points in $\mathbf{v}$:

$$\mathbf{f} = [\ \mathbb{E}[\mathbf{v}],\ \text{Var}[\mathbf{v}],\ \max(\mathbf{v}),\ \min(\mathbf{v}),\ R(\mathbf{v}),\ \tau\ ]^T, \tag{15.15}$$

where each component features in (15.15) are defined as follows.
(a) the *mean* $\mathbb{E}[\mathbf{v}]$:

$$\mathbb{E}[\mathbf{v}] = \mu = \frac{1}{L-1} \sum_{t=1}^{L-1} v_t, \tag{15.16}$$

which measures the average speed;
(b) the *variance* $\text{Var}[\mathbf{v}]$:

$$\text{Var}[\mathbf{v}] = \mathbb{E}[(\mathbf{v} - \mathbb{E}[\mathbf{v}])^2] = \frac{1}{L-1} \sum_{t=1}^{L-1} (v_t - \mu)^2, \tag{15.17}$$

which measures the dispersion of speed;
(c) the *maximum* $\max(\mathbf{v})$:

$$\max(\mathbf{v}) = \{v_t \mid \forall k : v_k \le v_t\}, \tag{15.18}$$

and
(d) the *minimum* $\min(\mathbf{v})$:

$$\min(\mathbf{v}) = \{v_t \mid \forall k : v_k \ge v_t\}, \tag{15.19}$$

which evaluate the maximum and minimum speed;
(e) the *range* $R(\mathbf{v})$:

$$R(\mathbf{v}) = \max(\mathbf{v}) - \min(\mathbf{v}), \tag{15.20}$$

which describes the dynamic range in $\mathbf{v}$ (although it is correlated with other two feature components $\max(\mathbf{v})$ and $\min(\mathbf{v})$), providing a direct measure for the range of speed;

(f) the *Kendall rank correlation coefficient* τ (Kendall, 1938):

$$\tau = \frac{n_c - n_d}{\frac{1}{2}(L-1)(L-2)},$$
(15.21)

which estimates the likelihood of speed increase. In (15.21), n_c is the number of concordant pairs between $\mathbf{v}$ and $\hat{\mathbf{v}}$, and n_d is the number of discordant pairs between $\mathbf{v}$ and $\hat{\mathbf{v}}$, where $\hat{\mathbf{v}}$ is a vector that contains exactly the same elements of $\mathbf{v}$ but sorted in ascending order.

Fall detection is formulated as a binary classification problem that distinguishes the fall from other activities. That is, human falls are treated as the positive class, while all remaining activities are treated as the negative class. The training process of a binary SVM classifier (Burges, 1998) for fall detection via shape analysis based on Riemannian manifold is summarized in Algorithm 2.

Given a feature vector $\mathbf{f}$ in (15.15) which characterizes the shape deformation of a tracked person in a given video segment, its class label $c \in \{-1, +1\}$ is determined according to the decision rule as follows:

$$\hat{y} = \text{sgn}(\alpha),$$
(15.22)

where $\text{sgn}(\cdot)$ is a sign function, and α is the output margin of the trained SVM classifier by taking $\mathbf{f}$ as the input, where a fall is indicated as $\hat{y} = +1$.

15.4 EXPERIMENTAL RESULTS

In this section, some results from the previously described methods are shown, including comparisons with other closely related methods. Discussions are then given.

15.4.1 Publicly Available Datasets for Visual Activity Recognition

We include information on some publicly available datasets to facilitate the use of benchmark datasets for testing in the near future. A list of commonly used video datasets on activities of daily living and fall detection are described below.

15.4.1.1 Video Datasets on Activities of Daily Living

- *MSR Daily Activity 3D Dataset (MSRDailyActivity3D)* (Wang et al., 2012): The dataset contains 320 RGB-D videos that are captured by

---

Algorithm 2: Training SVM classifier for fall detection based on Riemannian manifold (Yun and Gu, 2015).

Input :

- Raw training dataset $\mathcal{Z} = \{z_i, y_i\}_{i=1}^{N}$, where z_i is the i-th video segment of length L_i, $y_i \in \{-1, +1\}$ is the corresponding class label

Output:

- The regularization coefficient C for SVM classifier
- The kernel parameter γ for RBF kernel

for $i = 1 : N$ *(each video segment)* **do**

    for $t = 1 : L_i$ *(each frame)* **do**

        Detect human as foreground object with a target bounding box;

        Form the target image region $\mathbf{R}_t$ defined by the target bounding box;

        Normalize ROI with preservation of aspect ratio;

        Compute shape descriptor $\mathbf{p}_t \in \mathcal{M}$ from normalized ROI, where $\mathcal{M}$ is the underlying Riemannian manifold;

    end

    Collect the set of manifold points $\{\mathbf{p}_t\}_{t=1}^{L_i}$;

    Compute the velocity vector $\mathbf{v}_i$ in (15.14), based on the velocity of manifold points defined in (15.13);

    Compute final feature vector $\mathbf{x}_i$ on velocity statistics according to (15.15);

end

Construct training set $\mathcal{X} = \{(\mathbf{x}_i, y_i)\}_{i=1}^{N}$;

Train a binary SVM classifier using $\mathcal{X}$ with cross-validation;

---

a Kinect device, with skeleton joint positions. There are 16 activities: drink, eat, read book, call cellphone, write on a paper, use laptop, use vacuum cleaner, cheer up, sit still, toss paper, play game, lie down on sofa, walk, play guitar, stand up, and sit down. Each activity is performed twice by 10 different persons, one in standing position, and the other in sitting position.

- *Cornell Activity Datasets (CAD)*: *CAD-60* (Sung et al., 2011) and *CAD-120* (Koppula et al., 2013) contain 60 and 120 RGB-D videos respectively that are captured by a Kinect device, with skeleton joint position (as shown in Figure 15.6). There are 12 activities in *CAD-60*:

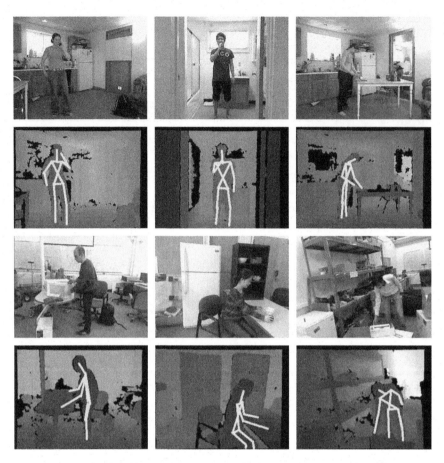

Figure 15.6 Example key frames from *CAD-60* (rows 1–2) (Sung et al., 2011) and *CAD-120* (rows 3–4) (Koppula et al., 2013). Permission for reproducing the above figure from http://pr.cs.cornell.edu/humanactivities/data.php has been obtained from the copyright owner (original author) of the datasets.

rinsing mouth, brushing teeth, wearing contact lens, talking on the phone, drinking water, opening pill container, cooking (chopping), cooking (stirring), talking on couch, relaxing on couch, writing on whiteboard, and working on computer. There are 10 activities in *CAD-120*: making cereal, taking medicine, stacking objects, unstacking objects, microwaving food, picking objects, cleaning objects, taking food, arranging objects, and having a meal. Activities in both datasets are performed by 4 different persons.

- *University of Rochester Activities of Daily Living Dataset (URADL)* (Messing et al., 2009): This dataset contains 150 RGB videos that are captured by an RGB camera. There are 10 activities: answering a phone, dialing a phone, looking up a phone number in a telephone directory, writing a phone number on a whiteboard, drinking a glass of water, eating snack chips, peeling a banana, eating a banana, chopping a banana, and eating food with silverware. Each activity is performed 3 times by 5 different persons.
- *RGBD-HuDaAct Dataset* (Ni et al., 2011): The dataset contains 1189 RGB-D videos that are captured by a Kinect device. There are 12 activities: make a phone call, mop the floor, enter the room, exit the room, go to bed, get up, eat meal, drink water, sit down, stand up, take off the jacket, and put on the jacket. Each activity is performed 2–4 times by 30 different persons.
- *ACT4² Dataset* (Cheng et al., 2012): The dataset contains 6844 RGB-D videos that are captured by 4 synchronized Kinect devices with different heights and viewpoints. There are 14 activities: Collapse, Drink, MakePhonecall, MopFloor, PickUp, PutOn, ReadBook, SitDown, SitUp, Stumble, TakeOff, ThrowAway, TwistOpen, and WipeClean. Each activity is performed several times by 24 different persons.
- Other video datasets on activities of daily living that are publicly available online include: *ICDSC 2009 Challenge: Smart Homes Dataset* (Aghajan and Vieeschouwer, 2009), *TST Intake Monitoring Database* (Cippitelli et al., 2014), *Eating and Drinking Activity Recognition Database (MOBISERV-AIIA Database)* (Iosifidis et al., 2015), etc. Moreover, some datasets that contains basic actions and sports activities are also commonly used for benchmarks, e.g., *KTH Action Database* (Schuldt et al., 2004), *UCF50 Action Recognition Dataset* (Reddy and Shah, 2012).

15.4.1.2 *Video Datasets on Fall Detection*

- *Multiple Cameras Fall Dataset* (Auvinet et al., 2010): This dataset contains 24 scenarios recorded by 8 IP video cameras. The first 22 first scenarios contain a fall and confounding events, the last 2 ones contain only confounding events. Some example key frames are shown in Figure 15.7.
- *UR Fall Detection Dataset* (Kwolek and Kepski, 2014): This dataset contains 30 fall scenarios measured by two Kinect sensors (parallel to the floor and ceiling mounted, respectively) and an accelerometer, as well

Figure 15.7 Example key frames from *Multiple Cameras Fall Dataset* (Auvinet et al., 2010). Upper row: falls. Lower row: other activities. Images are zoomed in for better inspection of the person. Permission for creating the above figure from *Multiple Cameras Fall Dataset* (Auvinet et al., 2010) has been obtained from the copyright owner (original author) of the dataset.

Figure 15.8 Example key frames from *UR Fall Detection Dataset* (Kwolek and Kepski, 2014). Upper row: falls. Lower row: other activities. Images are zoomed in for better inspection of the person. Permission for creating the above figure from *UR Fall Detection Dataset* (Kwolek and Kepski, 2014) has been obtained from the copyright owner (original author) of the dataset.

as 40 other activities (sitting down, crouching down, lying down) measured by one Kinect sensor parallel to the floor. Some example key frames are shown in Figure 15.8.

- Other video datasets on fall detection that are publicly available online include: *Le2i Fall Detection Dataset* (Charfi et al., 2012), *CIRL Fall Recognition Datasets* (Anderson et al., 2009), *TST Fall Detection Dataset* (Gasparrini et al., 2014), *SDUFall, EDF, OCCU* datasets (Zhang et al., 2015), etc.

Table 15.1 Precision and recall for *Method-1* (LE-BoW) (Faraki et al., 2014) with STP encoding on *URADL* (Messing et al., 2009) and *KTH* (Schuldt et al., 2004) datasets

Dataset	Activity	Precision (%)	Recall (%)
URADL	Answer phone	86.7	86.7
(Messing et al., 2009)	Chop banana	86.7	86.7
	Dial phone	86.7	86.7
	Drink water	100	100
	Eat banana	81.3	86.7
	Eat snack	86.7	86.7
	Lookup in phonebook	100	100
	Peel banana	92.9	86.7
	Use silverware	93.3	93.3
	Write on whiteboard	93.3	93.3
KTH Actions	Boxing	98.6	97.9
(Schuldt et al., 2004)	Hand-clapping	94.6	97.2
	Hand-waving	97.1	95.1
	Jogging	95.1	93.8
	Running	93.8	94.4
	Walking	95.2	95.8

15.4.2 Results and Comparisons

The three manifold-based methods described previously are evaluated on publicly available datasets and compared with state-of-the-art methods according to objective measures.

15.4.2.1 Evaluations of Method-1

Method-1 (LE-BoW) (Faraki et al., 2014) is evaluated on *URADL* (Messing et al., 2009) and *KTH* (Schuldt et al., 2004) datasets. Results are shown in Table 15.1 in terms of precision and recall. For each dataset, comparisons are made with three state-of-the-art methods according to the mean correct classification rate (CCR) in Table 15.2. It can observed in Table 15.2 that LE-BoW approach with STP and SC encoding schemes consistently outperforms state-of-the-art methods.

15.4.2.2 Evaluations of Method-2

Method-2 (VCML) (Bilinski and Bremond, 2015) is evaluated on *MSR-DailyActivity3D* (Wang et al., 2012), *URADL* (Messing et al., 2009) and

Table 15.2 Comparisons between *Method-1* (LE-BoW) (Faraki et al., 2014) and state-of-the-art methods on *URADL* (Messing et al., 2009) and *KTH* (Schuldt et al., 2004) datasets in terms of CCR

Dataset	Method	CCR (%)
URADL (Messing et al., 2009)	Laptev et al. (2008)	80
	Messing et al. (2009)	89
	Matikainen et al. (2010)	70
	Method-1: LE-BoW (HA) (Faraki et al., 2014)	90.0
	Method-1: LE-BoW (STP) (Faraki et al., 2014)	90.7
	Method-1: LE-BoW (SC) (Faraki et al., 2014)	**91.3**
KTH Actions (Schuldt et al., 2004)	Laptev et al. (2008)	91.8
	Gilbert et al. (2011)	94.5
	Wang et al. (2013)	95.3
	Method-1: LE-BoW (HA) (Faraki et al., 2014)	95.0
	Method-1: LE-BoW (STP) (Faraki et al., 2014)	95.7
	Method-1: LE-BoW (SC) (Faraki et al., 2014)	**97.4**

UCF50 (Reddy and Shah, 2012) datasets according to accuracy. Results are shown in Table 15.3, with comparisons to several state-of-the-art methods. It can be observed in Table 15.3 that VCML approach with IDT consistently outperforms state-of-the-art methods.

15.4.2.3 Evaluations of Method-3

Method-3 (Yun and Gu, 2015) is evaluated on two video datasets for fall detection, namely the *Multiple Cameras Fall Dataset* (Auvinet et al., 2010) and *UR Fall Detection Dataset* (Kwolek and Kepski, 2014). Comparisons are made with 6 existing methods in terms of sensitivity (true positive rate, TPR) and specificity (true negative rate, TNR) (Macmillan and Creelman, 2004), as shown in Table 15.4. It is worth mentioning that *Method-3* uses single-view videos without distinguishing different camera views. Therefore, it is evaluated differently from the mult-view or multi-modal methods that are compared to. For detailed information on the evaluation protocol of *Method-3*, readers are referred to Yun and Gu (2015). In Table 15.4,

Table 15.3 Comparisons between *Method-2* (VCML) (Bilinski and Bremond, 2015) and state-of-the-art methods on *MSRDailyActivity3D* (Wang et al., 2012), *URADL* (Messing et al., 2009) and *UCF50* (Reddy and Shah, 2012) datasets in terms of accuracy

Dataset	Method	Accuracy (%)
MSRDailyActivity3D (Wang et al., 2012)	Koperski et al. (2014)	72.0
	JPF: Wang et al. (2012)	78.0
	Orefej and Liu (2013)	80.0
	AE: Wang et al. (2012)	85.7
	DT: Wang et al. (2011)	76.2
	Method-2: **VCML (DT)** (Bilinski and Bremond, 2015)	78.1
	Method-2: **VCML (IDT)** (Bilinski and Bremond, 2015)	**85.9**
URADL (Messing et al., 2009)	Benabbas et al. (2010)	81.0
	Raptis and Soatto (2010)	82.7
	Messing et al. (2009)	89.0
	Bilinski and Bremond (2012)	93.3
	DT: Wang et al. (2011)	94.0
	Method-2: **VCML (DT)** (Bilinski and Bremond, 2015)	94.0
	Method-2: **VCML (IDT)** (Bilinski and Bremond, 2015)	**94.7**
UCF50 (Reddy and Shah, 2012)	Kantorov and Laptev (2014)	82.2
	Shi et al. (2013)	83.3
	Oneata et al. (2013)	90.0
	Wang and Schmid (2013)	91.2
	DT: Wang et al. (2011)	84.2
	Method-2: **VCML (DT)** (Bilinski and Bremond, 2015)	88.1
	Method-2: **VCML (IDT)** (Bilinski and Bremond, 2015)	**92.1**

Method-3 depicts high detection rates while maintaining small false alarms on both datasets. Although *Method-3* does not outperform other methods that are based on multiple camera calibration or multiple modality information, it still produces comparable results, especially considering the fact that only single arbitrary camera view is employed and that videos from different camera views are mixed in its test.

Table 15.4 Comparison between *Method-3* and state-of-the-art methods on two fall detection datasets (Auvinet et al., 2010; Kwolek and Kepski, 2014) in terms of sensitivity and specificity. *Arbitrary RGB View*: using single RGB camera from arbitrary view angles, without distinguishing different camera views

Dataset	Method	Sensor type	Sensitivity (%)	Specificity (%)
Multiple Cameras Fall Dataset (Auvinet et al., 2010)	Auvinet et al. (2011)	Multiple RGB views	80.6	**100**
	Rougier et al. (2011)	Multiple RGB views	95.4	95.8
	Hung and Saito (2012)	Multiple RGB views	95.8	**100**
	Ma et al. (2014)	RGB + depth	**99.93**	91.97
	Method-3 (Yun and Gu, 2015)	Arbitrary RGB view	91.30	91.67
UR Fall Detection Dataset (Kwolek and Kepski, 2014)	Kwolek and Kepski (2014)	Depth + accelerometer	**100**	**96.67**
	Bourke et al. (2007)	Accelerometer	**100**	90.00
	Method-3 (Yun and Gu, 2015)	Arbitrary RGB view	96.77	89.74

15.5 DISCUSSION

Despite the notable progress made and promising results reported in recent research and development, visual information-based activity recognition for assisted living and healthcare in a real-world deployment still remains a challenging issue and an open problem in various aspects that are summarized below.

Viewpoint Issue: Most existing algorithms are based on constrained viewpoints. For example, the target persons need to be in upright position or in front-view facing the camera. However, in real-world applications, videos are often captured from arbitrary camera viewpoints. Proposal solutions to this problem include the combination of multiple camera views and camera calibrations, at the cost of increased computational load or model complexity. Viewpoint invariant algorithms for monocular videos are more desirable.

Human Detection: Background subtraction is commonly used for detecting human as foreground moving objects. Challenges remain in developing a reliable background model that can handle inconsistent lighting conditions, occlusions and dynamic cluttered background, as well as systematically understanding the context for better segmentation of human objects.

Event Segmentation: Event segmentation defines the start and end of a video sequence containing an activity, thus having a strong impact on the recognition accuracy of activities. Common practice for segmenting video events are mainly based on the detection of abrupt change and motion, but it may become less effective in a daily living home environment where the transition between different activities is usually short and smooth.

Appearance Change: The appearance of a target person can change significantly over time due to many factors such as illumination variation, background clutter, partial occlusions, clothing, and objects being carried. Effective descriptors should be less sensitive to appearance change but still capture the most important characteristics of activities.

Contextual Information: Context plays important role for human behavior understanding. The same activity may have different behavior interpretations depending on the context in which it is performed. Besides the common contextual information such as time and place, the number of repetitions of an activity or the interaction between people or between person and objects can also be informative.

15.6 CONCLUSION

This chapter focuses on visual information-based activity recognition and fall detection for assisted living and healthcare. Three application domains are reviewed, including the recognition of activities for: (a) daily living at home environments; (b) healthcare, medical treatments and patient rehabilitations; (c) fall and other abnormal activities. Three robust methods for activity recognition and fall detection based on Riemannian manifolds are described in detail, with experimental results and comparisons with state-of-the-art methods, providing further support to the effectiveness of manifold-based visual analysis. Information on some publicly available datasets is also included to facilitate the use of benchmark datasets for testing in the near future. In spite of the good progress made for visual information-based assisted living and healthcare, discussions are given on several challenges that remain to be addressed for real-world practical deployment of effective assisted living systems under complex real scenarios.

REFERENCES

Aggarwal, J.K., Ryoo, M.S., 2011. Human activity analysis: a review. ACM Comput. Surv. 43 (3). Article 16.

Aghajan, H., Vieeschouwer, C.D., 2009. Smart homes dataset. In: ACM/IEEE International Conference on Distributed Smart Cameras (ICDSC) Challenge. [Dataset (available online)]: http://wsnl2.stanford.edu/icdsc09challenge/.

Anderson, D., Luke, R.H., Keller, J.M., Skubic, M., Rantz, M., Aud, M.A., 2009. Modeling human activity from voxel person using fuzzy logic. IEEE Trans. Fuzzy Syst. 17 (1), 39–49. [Dataset (available online)]: http://www.derektanderson.com/fallrecognition/datasets.html.

Arsigny, V., Fillard, P., Pennec, X., Ayache, N., 2008. Geometric means in a novel vector space structure on symmetric-positive definite matrices. SIAM J. Matrix Anal. Appl. 29 (1), 328–347.

Auvinet, E., Rougier, C., Meunier, J., St-Arnaud, A., Rousseau, J., 2010. Multiple cameras fall dataset. Technical report no. 1350. Department of Computer Science and Operations Research (DIRO), University of Montreal. [Dataset (available online)]: http://www.iro.umontreal.ca/~labimage/Dataset/.

Auvinet, E., Multon, F., Saint-Arnaud, A., Rousseau, J., Meunier, J., 2011. Fall detection with multiple cameras: an occlusion-resistant method based on 3-D silhouette vertical distribution. IEEE Trans. Inf. Technol. Biomed. 15 (2), 290–300.

Avgerinakis, K., Briassouli, A., Kompatsiaris, I., 2013. Recognition of activities of daily living for smart home environments. In: AAIE International Conference on Intelligent Environment.

Banerjee, T., Keller, J.M., Skubic, M., Stone, E., 2013. Day or night activity recognition from video using fuzzy clustering techniques. IEEE Trans. Fuzzy Syst. 22 (3), 483–493.

Banerjee, T., Skubic, M., Keller, J.M., Abbott, C., 2014. Sit-to-stand measurement for in-home monitoring using voxel analysis. IEEE J. Biomed. Health Inform. 18 (4), 1502–1509.

Benabbas, Y., Lablack, A., Ihaddadene, N., Djeraba, C., 2010. Action recognition using direction models of motion. In: IAPR International Conference on Pattern Recognition (ICPR).

Bilinski, P., Bremond, F., 2012. Contextual statistics of space-time ordered features for human action recognition. In: IEEE International Conference on Advanced Video- and Signal-Based Surveillance (AVSS).

Bilinski, P., Bremond, F., 2015. Video covariance matrix logarithm for human action recognition in videos. In: International Joint Conference on Artificial Intelligence (IJCAI).

Bourke, A.K., O'Brien, J.V., Lyons, G.M., 2007. Evaluation of a threshold-based tri-axial accelerometer fall detection algorithm. Gait Posture 26 (2), 194–199.

Burges, C.J.C., 1998. A tutorial on support vector machines for pattern recognition. Data Min. Knowl. Discov. 2 (2), 121–167.

Charfi, I., Miteran, J., Dubois, J., Atri, M., Tourki, R., 2012. Definition and performance evaluation of a robust SVM based fall detection solution. In: IEEE International Conference on Signal Image Technology and Internet Based Systems (SITIS). [Dataset (available online)]: http://le2i.cnrs.fr/Fall-detection-Dataset?lang=en.

Charfi, I., Miteran, J., Dubois, J., Atri, M., Tourki, R., 2013. Optimized spatio-temporal descriptors for real-time fall detection: comparison of support vector machine and Adaboost-based classification. J. Electron. Imaging 22 (4), 041106, 1–17.

Cheng, Z., Qin, L., Ye, Y., Huang, Q., Tian, Q., 2012. Human daily action analysis with multi-view and color-depth data. In: European Conference on Computer Vision (ECCV) Workshops and Demonstrations. [Dataset (available online)]: http://vipl.ict.ac.cn/rgbd-action-dataset.

Cheng, G., Wan, Y., Saudagar, A.N., Namuduri, K., Buckles, B.P., 2015. Advances in human action recognition: a survey. Preprint arXiv:1501.05964.

Cippitelli, E., Gasparrini, S., Santis, A.D., Montanini, L., Raffaeli, L., Gambi, E., Spinsante, S., 2014. Comparison of RGB-D mapping solutions for application to food intake monitoring. In: Italian Forum on Biosystems & Biorobotics Ambient Assisted Living. [Dataset (available online)]: http://www.tlc.dii.univpm.it/blog/databases4kinect#IDFood.

Dadal, N., Triggs, B., 2005. Histograms of oriented gradients for human detection. In: IEEE Conference on Computer Vision and Pattern Recognition (CVPR), vol. 1, pp. 886–893.

Duong, T.V., Bui, H.H., Phung, D.Q., Venkatesh, S., 2005. Activity recognition and abnormality detection with the switching hidden semi-Markov model. In: IEEE Conference on Computer Vision and Pattern Recognition (CVPR), vol. 1, pp. 838–845.

Duong, T.V., Phung, D.Q., Bui, H.H., Venkatesh, S., 2006. Human behavior recognition with generic exponential family duration modeling in the hidden semi-Markov model. In: International Conference on Pattern Recognition (ICPR), vol. 3, pp. 202–207.

Faraki, M., Palhang, M., Sanderson, C., 2014. Log-Euclidean bag of words for human action recognition. IET Comput. Vis. 9 (3), 331–339.

Fei-Fei, L., Perona, P., 2005. A Bayesian hierarchical model for learning natural scene categories. In: IEEE Conference on Computer Vision and Pattern Recognition (CVPR).

Fleck, S., Straßer, W., 2008. Smart camera based monitoring system and its application to assisted living. Proc. IEEE 96 (10), 1698–1714.

Gao, J., Hauptmann, A.G., Bharucha, A., Wactlar, H.D., 2004. Dining activity analysis using a hidden Markov model. In: IAPR International Conference on Pattern Recognition (ICPR), vol. 2, pp. 915–918.

Gasparrini, S., Cippitelli, E., Spinsante, S., Gambi, E., 2014. A depth-based fall detection system using a Kinect sensor. Sensors 14 (2), 2756–2775. [Dataset (available online)]: http://www.tlc.dii.univpm.it/blog/databases4kinect#IDFall.

Ghali, A., Cunningham, A.S., Pridmore, T.P., 2003. Object and event recognition for stroke rehabilitation. In: SPIE International Conference on Visual Communications and Image Processing, pp. 980–989.

Gilbert, A., Illingworth, J., Bowden, R., 2011. Action recognition using mined hierarchical compound features. IEEE Trans. Pattern Anal. Mach. Intell. 33 (5), 883–897.

Goffredo, M., Schmid, M., Conforto, S., Carli, M., Neri, A., D'Alessio, T., 2009. Markerless human motion analysis in Gauss–Laguerre transform domain: an application to sit-to-stand in young and elderly people. IEEE Trans. Inf. Technol. Biomed. 13 (2), 207–216.

Guo, G., Lai, A., 2014. A survey on still images based human action recognition. Pattern Recognit. 47, 3343–3361.

Hung, D.H., Saito, H., 2012. Fall detection with two cameras based on occupied area. In: Japan–Korea Joint Workshop on Frontiers in Computer Vision (FCV), pp. 33–39.

Iosifidis, A., Marami, E., Tefas, A., Pitas, I., Lyroudia, K., 2015. The MOBISERV-AIIA eating and drinking multi-view database for vision-based assisted living. J. Inf. Hiding Multimed. Signal Process. 6 (2), 254–273. [Dataset (available online)]: http://www.aiia.csd.auth.gr/MOBISERV-AIIA/.

Jayasumana, S., Hartley, R., Salzmann, M., Li, H., Harandi, M., 2013. Combining multiple manifold-valued descriptors for improved object recognition. In: IEEE International Conference on Digital Image Computing: Techniques and Applications (DICTA), pp. 1–6.

Kantorov, V., Laptev, I., 2014. Efficient feature extraction, encoding and classification for action recognition. In: IEEE Conference on Computer Vision and Pattern Recognition (CVPR).

Kendall, M., 1938. A new measure of rank correlation. Biometrika 30 (1/2), 81–93.

Khan, Z.H., Gu, H.I.Y., 2014. Online domain-shift learning and object tracking based on nonlinear dynamic models and particle filters on Riemannian manifolds. Comput. Vis. Image Underst. 125, 97–114.

Koperski, M., Bilinski, P., Bremond, F., 2014. 3D trajectories for action recognition. In: IEEE International Conference on Image Processing (ICIP).

Koppula, H.S., Gupta, R., Saxena, A., 2013. Learning human activities and object affordances from RGB-D videos. Int. J. Robot. Res. 32 (8), 951–970. [Dataset (available online)]: http://pr.cs.cornell.edu/humanactivities/data.php.

Kostopoulos, P., Nunes, T., Salvi, K., Deriaz, M., Torrent, J., 2015. F2D: a fall detection system tested with real data from daily life of elderly people. In: IEEE International Conference on E-Health Networking, Application and Services (Healthcom).

Kwolek, B., Kepski, M., 2014. Human fall detection on embedded platform using depth maps and wireless accelerometer. Comput. Methods Programs Biomed. 117 (3), 489–501. [Dataset (available online)]: http://fenix.univ.rzeszow.pl/~mkepski/ds/uf.html.

Laptev, I., Marszalek, M., Schmid, C., Rozenfeld, B., 2008. Learning realistic human actions from movies. In: IEEE Conference on Computer Vision and Pattern Recognition (CVPR).

Lee, J.M., 2012. Introduction to Smooth Manifolds. Springer.

Leu, A., Ristic-Durrant, D., Graser, A., 2011. A robust markerless vision-based human gait analysis system. In: IEEE International Symposium on Applied Computational Intelligence and Informatics (SACI), pp. 415–420.

Li, Y., Miaou, S., Hung, C.K., Sese, J.T., 2011. A gait analysis system using two cameras with orthogonal view. In: IEEE International Conference on Multimedia Technology (ICMT), pp. 2841–2844.

Liu, C., Chung, P.C., Chung, Y.N., Thonnat, M., 2007. Understanding of human behaviors from videos in nursing care monitoring systems. In: Broadband Multimedia Sensor Networks in Healthcare Applications. J. High Speed Netw. 16 (1), 91–103.

Lovett, S.T., 2010. Differential Geometry of Manifolds. A K Peters/CRC Press.

Ma, X., Wang, H., Xue, B., Zhou, M., Ji, B., Li, Y., 2014. Depth-based human fall detection via shape features and improved extreme learning machine. IEEE J. Biomed. Health Inform. 18 (6), 1915–1922.

Macmillan, N.A., Creelman, C.D., 2004. Detection Theory: A User's Guide. Taylor & Francis.

Martinez-Perez, F.E., Gonzalez-Fraga, J.A., Tentori, M., 2010. Artifacts' roaming beats recognition for estimating care activities in a nursing home. In: EAI International Conference on Pervasive Computing Technologies for Healthcare (PervasiveHealth).

Matikainen, P., Hebert, M., Sukthankar, R., 2010. Representing pairwise spatial and temporal relations for action recognition. In: European Conference on Computer Vision (ECCV).

Messing, R., Pal, C., Kautz, H., 2009. Activity recognition using the velocity histories of tracked keypoints. In: IEEE International Conference on Computer Vision (ICCV). [Dataset (available online)]: http://www.cs.rochester.edu/~rmessing/uradl/.

Mubashir, M., Shao, L., Seed, L., 2013. A survey on fall detection: principles and approaches. Neurocomputing 100, 144–152.

Ni, B., Wang, G., Moulin, P., 2011. RGBD-HuDaAct: a color-depth video database for human daily activity recognition. In: IEEE International Conference on Computer Vision Workshops (ICCVW). [Dataset (available online)]: http://adsc.illinois.edu/sites/default/files/files/ADSC-RGBD-dataset-download-instructions.pdf.

Oneata, D., Verbeek, J., Schmid, C., 2013. Action and event recognition with Fisher vectors and a compact feature set. In: IEEE International Conference on Computer Vision (ICCV).

Orefej, O., Liu, Z., 2013. Hon4d: histogram of oriented 4D normals for activity recognition from depth sequences. In: IEEE Conference on Computer Vision and Pattern Recognition (CVPR).

Pennec, X., Fillard, P., Ayache, N., 2006. A Riemannian framework for tensor computing. Int. J. Comput. Vis. 66 (1), 41–66.

Perronnin, F., Sanchez, J., Mensink, T., 2010. Improving the Fisher kernel for large-scale image classification. In: European Conference on Computer Vision (ECCV).

Raptis, M., Soatto, S., 2010. Tracklet descriptors for action modeling and video analysis. In: European Conference on Computer Vision (ECCV).

Reddy, K.K., Shah, M., 2012. Recognizing 50 human action categories of web videos. Mach. Vis. Appl. 24 (5), 971–981. [Dataset (available online)]: http://crcv.ucf.edu/data/UCF50.php.

Rother, C., Kolmogorov, V., Blake, A., 2004. GrabCut: interactive foreground extraction using iterated graph cuts. ACM Trans. Graph. 23, 309–314.

Rougier, C., Meunier, J., Saint-Arnaud, A., Rousseau, J., 2011. Robust video surveillance for fall detection based on human shape deformation. IEEE Trans. Circuits Syst. Video Technol. 21 (5), 611–622.

Schuldt, C., Laptev, I., Caputo, B., 2004. Recognizing human actions: a local SVM approach. In: IAPR International Conference on Pattern Recognition (ICPR). [Dataset (available online)]: http://www.nada.kth.se/cvap/actions/.

Shi, F., Petriu, E., Laganiere, R., 2013. Sampling strategies for real-time action recognition. In: IEEE Conference on Computer Vision and Pattern Recognition (CVPR).

Stone, E.E., Skubic, M., 2015. Fall detection in homes of older adults using the Microsoft Kinect. IEEE J. Biomed. Health Inform. 19 (1), 290–301.

Subbarao, R., Meer, P., 2009. Nonlinear mean shift over Riemannian manifolds. Int. J. Comput. Vis. 84 (1), 1–20.

Sung, J., Ponce, C., Selman, B., Saxena, A., 2011. Human activity detection from RGBD images. In: AAAI Workshop on Pattern, Activity and Intent Recognition (PAIR). [Dataset (available online)]: http://pr.cs.cornell.edu/humanactivities/data.php.

Tham, J.S., Chang, Y.C., Fauzi, M.F.A., 2014. Automatic identification of drinking activities at home using depth data from RGB-D camera. In: IEEE International Conference on Control, Automation and Information Sciences (ICCAIS).

Tuzel, O., Porikli, F., Meer, P., 2006. Region covariance: a fast descriptor for detection and classification. In: European Conference on Computer Vision (ECCV), pp. 589–600.

Tuzel, O., Porikli, F., Meer, P., 2008. Pedestrian detection via classification on Riemannian manifolds. IEEE Trans. Pattern Anal. Mach. Intell. 30 (10), 1713–1727.

United Nations, Department of Economic and Social Affairs, Population Division. World population prospects: the 2015 revision, key findings and advance tables. United Nations publication 2015; ESA/P/WP.241.

United Nations, Department of Economic and Social Affairs, Population Division. World population ageing 2013. United Nations publication 2013; ST/ESA/SER.A/348.

Vishwakarma, S., Agrawal, A., 2013. A survey on activity recognition and behavior understanding in video surveillance. Vis. Comput. 29, 983–1009.

Wang, H., Schmid, C., 2013. Action recognition with improved trajectories. In: IEEE International Conference on Computer Vision (ICCV).

Wang, H., Klaser, A., Schmid, C., Liu, C.L., 2011. Action recognition by dense trajectories. In: IEEE Conference on Computer Vision and Pattern Recognition (CVPR).

Wang, J., Liu, Z., Wu, Y., Yuan, J., 2012. Mining actionlet ensemble for action recognition with depth cameras. In: IEEE Conference on Computer Vision and Pattern Recognition (CVPR). [Dataset (available online)]: http://research.microsoft.com/en-us/um/people/zliu/actionrecorsrc/.

Wang, H., Kläser, A., Schmid, C., Liu, C.L., 2013. Dense trajectories and motion boundary descriptors for action recognition. Int. J. Comput. Vis. 103 (1), 60–79.

Yan, Y., Ricci, E., Rostamzadeh, N., Sebe, N., 2014. It's all about habits — exploiting multi-task clustering for activities of daily living analysis. In: IEEE International Conference on Image Processing (ICIP).

Yun, Y., Gu, I.Y.H., 2015. Human fall detection via shape analysis on Riemannian manifolds with applications to elderly care. In: IEEE International Conference on Image Processing (ICIP).

Yun, Y., Gu, I.Y.H., Aghajan, H., 2013. Riemannian manifold-based support vector machine for human activity classification in images. In: IEEE International Conference on Image Processing (ICIP).

Yun, Y., Fu, K., Gu, I.Y.H., Aghajan, H., Yang, J., 2014. Human activity recognition in images using SVMs and geodesics on smooth manifolds. In: ACM/IEEE International Conference on Distributed Smart Cameras (ICDSC).

Yun, Y., Innocenti, C., Nero, G., Lindén, H., Gu, I.Y.H., 2015. Fall detection in RGB-D videos for elderly care. In: IEEE International Conference on E-Health Networking, Application and Services (Healthcom).

Zhan, K., Ramos, F., Faux, S., 2012. Activity recognition from a wearable camera. In: IEEE International Conference on Control, Automation, Robotics & Vision (ICARCV).

Zhang, Z., Conly, C., Athitsos, V., 2015. A survey on vision-based fall detection. In: ACM International Conference on PErvasive Technologies Related to Assistive Environments. Article 46.

Zhou, Z., Chen, X., Chung, Y.C., He, Z., Han, T.X., Keller, J.M., 2008. Activity analysis, summarization, and visualization for indoor human activity monitoring. IEEE Trans. Circuits Syst. Video Technol. 18 (11), 1489–1498.

CHAPTER 16

End-Users Testing of Enhanced Living Environment Platform and Services

Rossitza I. Goleva*, Nuno M. Garcia†,
Constandinos X. Mavromoustakis‡, Ciprian Dobre§,
George Mastorakis¶, Rumen Stainov‖

*Technical University of Sofia, Kl. Ohridski blvd 8, Faculty of Telecommunications, Department of Communication Networks, 1756 Sofia, Bulgaria
†Instituto de Telecomunicações, Universidade da Beira Interior, R Marquês d'Ávila e Bolama, 6200-001 Covilhã, Portugal
‡Department of Computer Science, University of Nicosia, 46 Makedonitissa Avenue, 1700 Nicosia, Cyprus
§University of Politehnica of Bucharest, Bucharest, Romania
¶Technological Educational Institute of Crete, Heraklion, Crete, Greece
‖Applied Computer Science Department, University of Applied Sciences, Fulda, Germany

16.1 INTRODUCTION

The Enhanced Living Environment (ELE) platform is an open distributed system for cloud computing that provides customized services for the society and is intended to be of great importance for elderly and disabled people (http://www-cps.hb.dfki.de/research/baall; http://robotik. dfki-bremen.de/en/research/robot-systems/aila.html). The science groups are using and applying different approaches, interfaces, background technologies, applications (Loshkovska and Koceski, 2016; Agüero et al., 2015). The need of ELE platform and testing is essential due to the fact that many people nowadays face partial disabilities or just difficulties and they need the support of information and communication technologies and services to continue to work and live without the need of personal assistance. On the other hand, active people in the society need support of the customized services that will help them to organize their life in a healthy way, i.e. by performing regular sport activities, by keeping track on the important health and physical parameters, by warning about the necessity of support to children, relatives, parents and so on. Testing, verification and validation are considered continuous problem because the platform is a live heterogeneous distributed system as defined in chapter 8.

Ambient Assisted Living and Enhanced Living Environments.
DOI: http://dx.doi.org/10.1016/B978-0-12-805195-5.00016-8

In some countries, telecom operators and ISPs could provide the Ambient Assisted Living as a Service (AALaaS) and ELE as a Service (ELEaaS). In other cases, the services could be provided locally using client/server application, or free cloud computing, or as a customized existing platform. In all of the cases, testing could answer the question whether the service is available, working and what the level of output and overall performance is. Telecom operators, ISPs, local application providers need to have general level test. Service developers and analysts need validation methods answering the question whether the correct service is designed. Details in design by means of correct functionality are a matter of verification algorithms.

End-user testing, verification and validation of the AAL/ELE services and platform parts are very hot topics. The dynamic nature of the network and platform itself, ad hoc solutions, heterogeneity of the proposals, necessity of interoperability and mobility of the data are a prerequisite for bugs and continuous changes in the services and functionality (http://www.aal-europe.eu). Many authors like Calvaresi et al. (2015) proposed testing at abstract level based on ontology. It is usually related to the functionality of the platform.

In this chapter, we propose testing at different levels and platform domains based mostly on the technical characteristics of the solution. The presented use-case scenarios are not specifically connected to the specific testing, verification of validation algorithm and are not pretending to be complete. Furthermore, the terms testing, verification and validation are not distinguished clearly in the chapter due to the lack of space.

16.2 STATE OF THE ART AND LIVING LABS EXPERIENCE

There are existing solutions for testing already proposed by Grgurić et al. (2013) that are related to the platform UniversAAL (see details in chapter 8). Al-Fuqaha et al. (2015) published a survey on Internet of Things with very limited tests at the end of the work.

Some of the living labs are very specific and intended to solve problems of people with wheel chairs like in Krieg-Brückner et al. (2015). The authors define formal language for ontology description and test the services in the living environment (Autexier and Hutter, 2015). Part of the solutions is related to the video services like in Mandel and Autexier (2016). Other part of the services is related to daily activities like cooking, dressing, washing etc. There is also research on heterogeneous smart environments like in Autexier et al. (2013) and Goleva et al. (2015a, 2015b). Metzen et al. (2014) presented specific case of robot that helps disabled people.

Depari et al. (2014) develop testing facilities for real-time sensor network. General guidelines for testing and verification could be seen in http://www.aal-europe.eu/get-involved/i-am-a-user-2/. Calvaresi et al. (2015) published an idea for testing of virtual environment services. Al-Fuqaha et al. (2015) defined general rules for 5G platform testing.

Bernardo et al. (2014) propose solution for quality of service and performance analyses that is end-to-end in sensor networks. Vijayakumar et al. (2015a, 2015b) propose big data analyses in a cloud computing platform. Authors also take into account end-user behaviour. Calheiros et al. (2011) proposed a solution for resource evaluation of cloud services. Banzai et al. (2010) worked on distributed network testing with stress on cloud computing. Subashini and Kavitha (2011) proposed tests on security and risk. Xu (2012) worked on engineering issues in cloud computing including testing. Hashem et al. (2015) presented algorithms for big data analyses in the mobile clouds.

Joshi et al. (2012) proposed security solution towards DDoS attacks. Dinh et al. (2013) worked on mobile cloud computing tests. Jadeja and Modi (2012) worked on Platform as a Service (PaaS) testing. In their book Rittinghouse and Ransome (2016) also propose security tests in the cloud. Complete test solutions in the cloud are demonstrated in Riungu-Kalliosaari et al. (2012). Attention to security for small business cloud is presented in Chang et al. (2016).

Cloud computing service ranking is proposed in Garg et al. (2013). Garrison et al. (2012) presented idea on deployment and testing of distributed platform. Small and medium business cloud computing implementation is shown in Gupta et al. (2013). Deployment of cloud computing for small enterprises is also discussed in Oliveira et al. (2014). Sanaei et al. (2014) proposed solutions for mobile cloud computing deployment. Wei et al. (2014) explain data verification models for distributed models. Navimipour et al. (2015) demonstrated automatic verification of the cloud solutions. Liu et al. (2015) presented external big data verification in mobile computing.

16.3 AALAAS AND ELEAAS PLATFORM

AAPELE platform for AAL/ELE services is defined in chapter 8 and shown in a simplified way on Figure 16.1. The main idea of the platform is to fol-

Sensors, Gateways, Home Servers   Server Farms, Data Centers

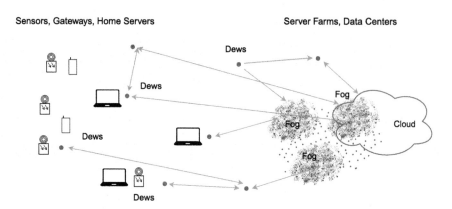

Figure 16.1 AALaaS and ELEaaS platform.

low the definitions and structure from communication networks, i.e. having access, edge and core parts. Access part is related to the sensor networks, Wi-Fi, Bluetooth, LAN networks used to allow end-users to access the platform. This part is specific with technology diversity. New technologies appear on the market almost every three years. Access part corresponds to the dew-computing infrastructure and ends at the home, school, hospital, car, tram, shop servers. Edge part of the network consists of fog computing infrastructure. These are servers, data centers, network infrastructure facilities that are primary or secondary computing level for the primary end-users. In case when dew-computing level is not supported for given domain fog level has the necessary functionality and capacity to support it. Fog computing servers are gathering and analyzing the data from primary users and dews. They save, backup, restore, search, and provide statistical services to all stakeholders and primary and secondary users. Part of the functionality of the cloud could be also part of the fog infrastructure. This is especially specific when the services are supported only in given domain of subnet.

The platform will be transparent to the data transmitted, cultural peculiarities, and specific needs of the end-customer. There is a need to profile the end-user behaviour taking into consideration the most important data to be transmitted, the dynamics of the data, mobility of the data, requirements to the data processing, data presentation, data storage, traffic patterns, Quality of Service and Quality of Experience levels (Goleva et al., 2015a, 2015b, 2016a, 2016b).

Testing should also support other organizations (public or private) that are responsible for the people with specific needs. It should be available to the doctors, nurses, supporting personnel, insurance companies, relatives, caregivers, non-profit organizations of elder or ill people, public organizations that work under support of government etc. The ultimate beneficiaries of the platform are the European citizens who will benefit of better AALaaS and ELEaaS solutions (Goleva et al., 2015a, 2015b).

The AAL forum defines end-users classifying them in three groups — primary users, secondary users, tertiary users (http://www.aal-europe.eu). The primary users are patients, active aging people, people that are a subject of monitoring. AAL/ELE platform is also gathering other information from static and mobile sensors and those sensors are considered also primary objects for testing in terms of machine-to machine (M2M) communication.

Secondary end-users are nurses, medical doctors, family members, caregivers that are directly related to the primary users. M2M communication secondary users are home robots, cleaning machines, home automation servers etc.

The tertiary users are usually people related to the primary and secondary users like internet providers, insurance company, utility company, hospital administration etc. In the context of M2M the tertiary users are servers in the fog and cloud that support AALaaS and ELEaaS (Stainov et al., 2016).

As seen from the comments above we need to define in addition to the definition of the AAL forum primary, secondary and tertiary objects subjects that support ELE services for the virtualized personal networks. Objects are related to the intelligent devices like robots, servers, applications that support the services. Subjects are related to the people who are owners of the service, who support the service and who provide the service.

16.4 STAKEHOLDERS AS TESTERS

The actors of the platform are defined on Figures 16.2 and 16.3. Patients of different type could have different type of access to the platform. For example, active patients will have access to raw data, alarms, prescriptions, historical data, feedbacks, consultations, and notifications as shown on Figure 16.2. On Figure 16.3, active patients could also navigate the service. Patients with dementia of other physical and psychological disability could have restricted access to the functions. Their caregivers and family members

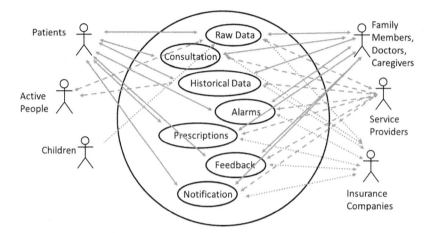

Figure 16.2 Primary and secondary AALaaS and ELEaaS users.

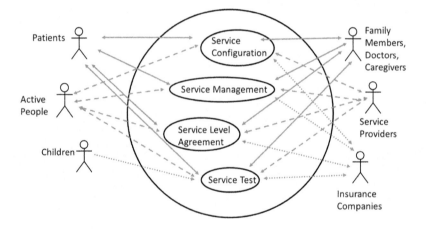

Figure 16.3 Management services.

will manage these functions instead of them. This is the case with children on the figures.

Active people do not need functions like consultations, feedback, alarms. They could have access to the raw and historical data only and activate other functions one by one on demand.

Family members, doctors, caregivers, service providers, system administrators, insurance companies, hospital administration need to have full access to the functions. Part of the actors will work with the data content. Others will work with service management. They need to have access not only as end-users but also on behalf of other end-users. This functionality is not shown on the figures for simplicity.

16.5 PLATFORM AND APPLICATION TESTING

The use-case scenarios should cover all levels of the hierarchy of the platform as well as protocols and interfaces. We highlight here only channel/network layers as well as application layer. Platform testing starts with sensor networks as part of the access plane. On Figure 16.4 we show how raw data is collected from sensors. Home servers represent dew level, regional servers and data centers represent fog level and big data centers and server farms represent cloud–computing level. The cloud level is not shown on the figure. The data on the figure is forwarded by intermediate sensor. It is not always the case in sensor networks. The data is generated on timer. Only one occurrence is shown on the figure. The data is acknowledged at all levels.

Raw data could be collected also on demand. It is demonstrated on Figure 16.5. In this case, the home server schedules the raw data requests and could manage scheduling based on the configuration, green energy availability, demands from the doctors and caregivers, context of the data etc.

On Figure 16.6 there is a case without acknowledgements. Data could be sent many times, i.e. between 1 and 5 times depending on the type of sensor and context. EnOcean technology is working without acknowledgements and with data repetitions. Sensors that are pure sources of raw data could sleep in some sensor technologies and harvest the energy. Sensor networks could be clustered or coordinator less.

Figure 16.7 presents an idea of historical data requests. It is in the form of client/server or peer-to-peer (including peer port, Stainov et al., 2014) mode between fog and cloud levels. At dew level, the diversity of the services might require only client/server in many cases. When there is a service for notification, prescription, feedback, alarms the data could start from the cloud without requests from the end-user device and be triggered by other applications in the fog or cloud level.

Figure 16.8 demonstrates multihoming, multipathing and load balancing through three different interfaces with historical data requests from home server to the fog and cloud servers. The three channels could use peer port to backup each other. Home server could backup at fog level. Multiple requests from caregivers, doctors, nurses, system administrators,

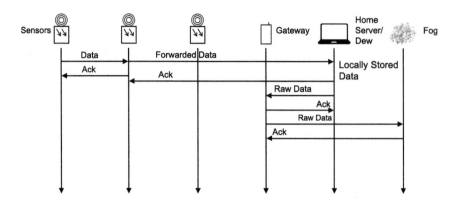

Figure 16.4 Sensors send raw data on timer.

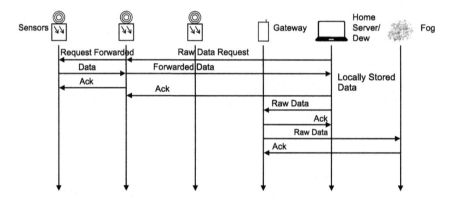

Figure 16.5 Raw data requests generated by the home server.

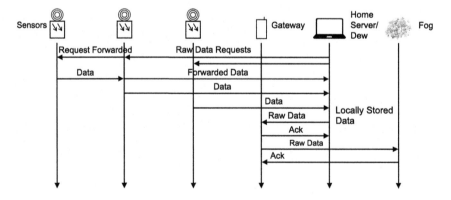

Figure 16.6 Raw data gathering without acknowledgements.

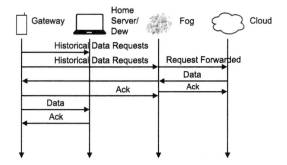

Figure 16.7 Historical data requests from end-user application.

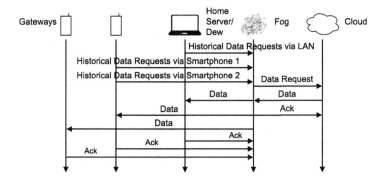

Figure 16.8 Miltihoming, multipathing and load balancing via different interfaces.

insurance companies use-case scenario are described on Figure 16.9. The multiplication of the requests is not shown on the figure for simplicity. Tests on protocols, specific interfaces and data formats are not part of the chapter because of the lack of space. Ontology based testing could support interoperability of the data models. For example, part of the data could be incompatible to the other. There might be a data overlapping between users that could be solved using algorithms for big data analyses (Hashem et al., 2015).

Companies like insurance and service providers may need additional statistical analyses of the end-user data. Big data analyses are also invaluable there.

Performance tests are related to the number of end-users using the system at the same time and traffic patterns to the server farms and data centers (Goleva et al., 2015a, 2015b). There is a proposal how and when to turn on and off virtual machines and physical servers in cases of high or low traf-

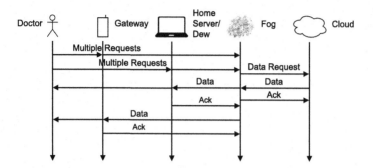

Figure 16.9 Multiple requests from the doctors and caregivers.

fic. The work is presenting also green technologies algorithms for future smart cities. The output of the testing could be data necessary for objective Quality of Service (QoS) and subjective Quality of Experience (QoE). This is especially true for the data necessary for network configuration and management.

Testing is considered a continuous process. This is due to the fact that the platform and services are developed continuously. Due to the big requirements for security and reliability of the platform and services (Goleva et al., 2015a, 2015b) the tests are developed and performed also all the time (Figures 16.2 and 16.3).

16.6 CONCLUSION AND FUTURE WORK

This chapter presents the basic use-case scenarios for end–user testing, verification and validation of the AALaaS and ELEaaS open distributed platform. The platform has pure access, edge and core parts. Access part testing is based on sensor network. It could be applicable to many other technologies. The dew/fog/cloud based computing and testing is client/server and peer-to-peer using peer port algorithms. We demonstrated the flow of data from any single actor and sensor in the platform towards the clouds and back. Performance tests are not part of this chapter due to the lack of space. Preliminary analyses on performance tests could be seen in the cited references. The same is valid for traffic patterns, QoS and QoE.

Our future work plans continue with performance test, standardization of the protocols, interfaces, and data formats within the distributed platform as well as definition of the way of peer port implementation and testing at dew/fog and cloud level.

ACKNOWLEDGEMENT

Our thanks to ICT COST Action IC1303: Algorithms, Architectures and Platforms for Enhanced Living Environments (AAPELE), ICT COST Action IC1406: High-Performance Modelling and Simulation for Big Data Applications (cHiPSet) and TD COST Action TD1405: European Network for the Joint Evaluation of Connected Health Technologies (ENJECT).

REFERENCES

http://robotik.dfki-bremen.de/en/research/robot-systems/aila.html. AILA, mobile dual-arm-manipulation, University of Bremen, Germany (accessed June, 2016).

http://www.aal-europe.eu/get-involved/i-am-a-user-2/ (accessed June, 2016).

http://www-cps.hb.dfki.de/research/baall, BAALL: The Bremen Ambient Assisted Living Lab, University of Bremen, Germany (accessed June, 2016).

Agüero, R., Zinner, T., Goleva, R., Timm-Giel, A., Tran-Gia, P. (Eds.), 2015. Mobile Networks and Management. Springer International Publishing. ISBN 9783319162911.

Al-Fuqaha, A., Guizani, M., Mohammadi, M., Aledhari, M., Ayyash, M., 2015. Internet of things: a survey on enabling technologies, protocols, and applications. IEEE Commun. Surv. Tutor. 17 (4), 2347–2376. http://dx.doi.org/10.1109/COMST.2015.2444095. Fourthquarter 2015.

Autexier, S., Hutter, D., 2015. SHIP — A logic-based language and tool to program smart environments. In: LOPSTR 2015, pp. 313–328.

Autexier, S., Hutter, D., Stahl, C., 2013. An implementation, execution and simulation platform for processes in heterogeneous smart environments. AmI 2013, 3–18.

Banzai, T., Koizumi, H., Kanbayashi, R., Imada, T., Hanawa, T., Sato, M., 2010. D-cloud: design of a software testing environment for reliable distributed systems using cloud computing technology. In: Proceedings of the 2010 10th IEEE/ACM International Conference on Cluster, Cloud and Grid Computing (CCGRID '10). IEEE Computer Society, Washington, DC, USA, pp. 631–636.

Bernardo, V., Curado, M., Braun, T., 2014. An IEEE 802.11 energy efficient mechanism for continuous media applications. Sustain. Comput. Inf. Syst. (ISSN 2210-5379) 4 (2), 106–117. http://dx.doi.org/10.1016/j.suscom.2014.04.001.

Calheiros, R.N., Ranjan, R., Beloglazov, A., De Rose, C.A.F., Buyya, R., 2011. CloudSim: a toolkit for modeling and simulation of cloud computing environments and evaluation of resource provisioning algorithms. Softw. Pract. Exp. (ISSN 1097-024X) 41 (1), 23–50. http://dx.doi.org/10.1002/spe.995.

Calvaresi, P., et al., 2015. Using a virtual environment to test a mobile App for the ambient assisted living. In: AIME 2015.

Chang, V., Kuo, Y.H., Ramachandran, M., 2016. Cloud computing adoption framework: a security framework for business clouds. Future Gener. Comput. Syst. (ISSN 0167-739X) 57, 24–41. http://dx.doi.org/10.1016/j.future.2015.09.031. http://www.sciencedirect.com/science/article/pii/S0167739X15003118.

Depari, A., Ferrari, P., Flammini, A., Rinaldi, S., Rizzi, M., Sisinni, E., 2014. Development and evaluation of a WSN for real-time structural health monitoring and testing. Proc. Eng. 87, 680–683. Available at: http://linkinghub.elsevier.com/retrieve/pii/S1877705814027374.

Dinh, H.T., Lee, C., Niyato, D., Wang, P., 2013. A survey of mobile cloud computing: architecture, applications, and approaches. In: Wireless Communications and Mobile Computing. John Wiley & Sons, Ltd., pp. 1587–1611.

Garg, S.K., Versteeg, S., Buyya, R., 2013. A framework for ranking of cloud computing services. Future Gener. Comput. Syst. (ISSN 0167-739X) 29 (4), 1012–1023. http://dx.doi.org/10.1016/j.future.2012.06.006, http://www.sciencedirect.com/science/article/pii/S0167739X12001422.

Garrison, G., Kim, S., Wakefield, R.L., 2012. Success factors for deploying cloud computing. Commun. ACM 55 (9), 62–68. http://dx.doi.org/10.1145/2330667.2330685.

Goleva, R., Atamian, D., Mirtchev, S., Dimitrova, D., Grigorova, L., Rangelov, R., Ivanova, A., 2015a. Traffic analyses and measurements: technological dependability. In: Mastorakis, G., Mavromoustakis, C., Pallis, E. (Eds.), Resource Management of Mobile Cloud Computing Networks and Environments. Information Science Reference, Hershey, PA, pp. 122–173.

Goleva, R., Stainov, R., Savov, A., Draganov, P., 2015b. Reliable platform for enhanced living environment. In: First COST Action IC1303 AAPELE Workshop Element 2014, in Conjunction with MONAMI 2014 Conference. Wurzburg, 24 Sept., 2014. Springer International Publishing, pp. 315–328.

Goleva, R., Stainov, R., Savov, A., Draganov, P., Nikolov, N., Dimitrova, D., Chorbev, I., 2016a. Automated ambient open platform for enhanced living environment. In: Loshkovska, S., Koceski, S. (Eds.), ICT Innovations 2015. ELEMENT 2015 Workshop, Ohrid, FyROM, 1 Oct. 2015. In: Advances in Intelligent Systems and Computing. Springer International Publishing, Switzerland, pp. 255–264.

Goleva, R., Stainov, R., Wagenknecht-Dimitrova, D., Mirtchev, S., Atamian, D., Mavromoustakis, C.X., Mastorakis, G., Dobre, C., Savov, A., Draganov, P., 2016b. Data and traffic models in 5G network. In: Mavromoustakis, C.X., Mastorakis, G., Batalla, J.M. (Eds.), Internet of Things (IoT) in 5G Mobile Technologies, 2016. ISBN 978-3-319-30913-2. Springer International Publishing, Cham, pp. 485–499.

Grgurić, Mošmondor, M., Kušek, M., Stocklöw, C., Salvi, D., 2013. Introducing gesture interaction in the ambient assisted living platform universAAL. In: 2013 12th International Conference on Telecommunications (ConTEL). Zagreb, pp. 215–222.

Gupta, P., Seetharaman, A., Raj, J.R., 2013. The usage and adoption of cloud computing by small and medium businesses. Int. J. Inf. Manag. (ISSN 0268-4012) 33 (5), 861–874. http://dx.doi.org/10.1016/j.ijinfomgt.2013.07.001. http://www.sciencedirect.com/science/article/pii/S026840121300087X.

Hashem, I.A.T., Yaqoob, I., Anuar, N.B., Mokhtar, S., Gani, A., Khan, S.U., 2015. The rise of "big data" on cloud computing: review and open research issues. Inf. Syst. (ISSN 0306-4379) 47, 98–115. http://dx.doi.org/10.1016/j.is.2014.07.006. http://www.sciencedirect.com/science/article/pii/S0306437914001288.

Jadeja, Y., Modi, K., 2012. Cloud computing — concepts, architecture and challenges. In: 2012 International Conference on Computing, Electronics and Electrical Technologies (ICCEET). Kumaracoil, pp. 877–880.

Joshi, B., Vijayan, A.S., Joshi, B.K., 2012. Securing cloud computing environment against DDoS attacks. In: 2012 International Conference on Computer Communication and Informatics (ICCCI). Coimbatore, pp. 1–5.

Krieg-Brückner, B., Autexier, S., Nokam, S.G., 2015. Formal modelling for cooking assistance. In: Software, Services and Systems, pp. 355–376.

Liu, C., Yang, C., Zhang, X., Chen, J., 2015. External integrity verification for outsourced big data in cloud and IoT: a big picture. Future Gener. Comput. Syst. (ISSN 0167-739X) 49, 58–67. http://dx.doi.org/10.1016/j.future.2014.08.007. http://www.sciencedirect.com/science/article/pii/S0167739X14001551.

Loshkovska, S., Koceski, S. (Eds.), 2016. ICT Innovations 2015. Advances in Intelligent Systems and Computing. Springer International Publishing, Switzerland.

Mandel, C., Autexier, S., 2016. People tracking in ambient assisted living environments using low-cost thermal image cameras. In: ICOST 2016, pp. 14–26.

Metzen, J.H., Fabisch, A., Senger, L., de Gea Fernandez, J., Kirchner, E.A., 2014. Towards learning of generic skills for robotic manipulation. KI 28 (1), 15–20.

Navimipour, N.J., Navin, A.H., Rahmani, A.M., Hosseinzadeh, M., 2015. Behavioral modeling and automated verification of a Cloud-based framework to share the knowledge and skills of human resources. Comput. Ind. (ISSN 0166-3615) 68, 65–77. http://dx.doi.org/10.1016/j.compind.2014.12.007.

Oliveira, T., Thomas, M., Espadanal, M., 2014. Assessing the determinants of cloud computing adoption: an analysis of the manufacturing and services sectors. Inf. Manag. (ISSN 0378-7206) 51 (5), 497–510. http://dx.doi.org/10.1016/j.im.2014.03.006. http://www.sciencedirect.com/science/article/pii/S0378720614000391.

Pierdicca, A., Clementi, F., Isidori, D., Concettoni, E., Cristalli, C., Lenci, S., 2016. Numerical model upgrading of a historical masonry palace monitored with a wireless sensor network. Int. J. Mason. Res. Innov.

Rittinghouse, J.W., Ransome, J.F., 2016. Cloud Computing: Implementation, Management, and Security. CRC Press, 340 pages.

Riungu-Kalliosaari, L., Taipale, O., Smolander, K., 2012. Testing in the Cloud: exploring the practice. IEEE Softw. 29 (2), 46–51. http://dx.doi.org/10.1109/MS.2011.132.

Sanaei, Z., Abolfazli, S., Gani, A., Buyya, R., 2014. Heterogeneity in mobile cloud computing: taxonomy and open challenges. IEEE Commun. Surv. Tutor. 16 (1), 369–392. http://dx.doi.org/10.1109/SURV.2013.050113.00090. First Quarter 2014.

Stainov, R., Goleva, R., Demirova, M., 2014. Reliable transmission over disruptive cloud using peer port. In: 10th Annual International Conference on Computer Science and Education in Computer Science 2014 (CSECS 2014). 4–7 July 2014, Albena, Bulgaria, pp. 153–166. ISSN 1313-8624.

Stainov, R., Goleva, R., Mirtchev, S., Atamian, D., Mirchev, M., Savov, A., Draganov, P., 2016. AALaaS intelligent backhauls for P2P communication in 5G mobile networks. In: BlackSeaCom 2016. June 6–9, Varna, Bulgaria.

Subashini, S., Kavitha, V., 2011. A survey on security issues in service delivery models of cloud computing. J. Netw. Comput. Appl. (ISSN 1084-8045) 34 (1), 1–11. http://dx.doi.org/10.1016/j.jnca.2010.07.006. http://www.sciencedirect.com/science/article/pii/S1084804510001281.

Vijayakumar, V., Neelanarayanan, V., Panackal, J.J., Pillai, A.S., 2015a. Adaptive utility-based anonymization model: performance evaluation on big data sets. In: Big Data, Cloud and Computing Challenges. Proc. Comput. Sci. (ISSN 1877-0509) 50, 347–352. http://dx.doi.org/10.1016/j.procs.2015.04.037.

Vijayakumar, V., Neelanarayanan, V., Ragunathan, T., Battula, S.K., Jorika, V., Mounika, Ch., Sruthi, A.U., Vani, M.D., 2015b. Advertisement posting based on consumer behaviour. In: Big Data, Cloud and Computing Challenges. Proc. Comput. Sci. (ISSN 1877-0509) 50, 329–334. http://dx.doi.org/10.1016/j.procs.2015.04.040.

Wei, L., Zhu, H., Cao, Z., Dong, X., Jia, W., Chen, Y., Vasilakos, A.V., 2014. Security and privacy for storage and computation in cloud computing. Inf. Sci. (ISSN 0020-0255) 258 (10), 371–386. http://dx.doi.org/10.1016/j.ins.2013.04.028. http://www.sciencedirect.com/science/article/pii/S0020025513003320.

Xu, Xun, 2012. From cloud computing to cloud manufacturing. Robot. Comput.-Integr. Manuf. (ISSN 0736-5845) 28 (1), 75–86. http://dx.doi.org/10.1016/j.rcim.2011.07.002. http://www.sciencedirect.com/science/article/pii/S0736584511000949.

CHAPTER 17

M2M Communications and Their Role in AAL

Radosveta I. Sokullu, Abdullah Balcı
Ege University, Department of Electrical & Electronics Engineering, Turkey

17.1 INTRODUCTION

Machine-to-Machine (M2M) communications can be defined as the technology that enables the communication between various devices (e.g. like portable computers devices, smartphones, WPAN health devices, or less sophisticated devices like smart sensors, embedded controllers, actuators) and allows them to perform a variety of actions without or with only limited human intervention (Wu et al., 2011). Because of the extremely large number of devices possibly involved as well as the different communication pattern, the implementation of M2M poses many new challenges related to network architecture, data transmission, protocol design, system integration, power and spectrum efficiency which make it a very hot research topic. Cellular network operators are considering the infrastructure to implement M2M communications over existing and emerging network technologies (3G, LTE, LTE-A) in order to generate new services in many areas like e-Health, Traffic Management, Intelligent Buildings etc. (Wu et al., 2011; Kartsakli et al., 2014; Borgia, 2013; Atzori et al., 2010; Chen, 2012; Taleb and Kunz, 2012; Kim et al., 2014; Clayman and Galis, 2011). This chapter covers topics related to M2M communications and the role they play and are expected to play in Ambient Assistant Living (AAL), more specifically eHealth and assisted living for the elderly. The M2M technology, the major standards for M2M architectures and services are described, and existing solutions in the area of eHealth and AAL are summarized. With the proliferation of Wireless Sensor Networks (WSN), applications based on WSNs have become a commodity. Many of these systems are related to providing eHealth services — e.g. remote monitoring of the physical state of elderly people or patients with chronical diseases and reporting it over some connected network infrastructure (LAN, wireless, cellular) to care givers and clinical personnel. However, those sys-

Ambient Assisted Living and Enhanced Living Environments.
DOI: http://dx.doi.org/10.1016/B978-0-12-805195-5.00017-X

tems represent only a faint glimpse into the future of integrated solutions that can be built utilizing M2M. The possibilities that M2M communications bring to bridge different network technologies (e.g. wired, mobile and fixed wireless, and cellular networks) with a range of contextual applications like for example social networks and ontology based systems unprecedented both in scale and in context.

The chapter is based on the following structure: first we define M2M communications, their characteristics, applications and existing standards. Then we elaborate on how M2M can serve the AAL paradigm. This is a hot research area and many scientists examine the different possibilities to integrate patient medical data, collected from wireless sensors, into a meaningful structured eHealth system which will be scalable, proactive and easy to use, unrestricted by the ability, knowledge or physical condition of the elderly person. A number of interesting implementations and architectural solutions have been suggested and are summarized in this chapter. To be practically applicable eHealth solution requires the integration of entities based on different standards, related to specific levels of the ICT infrastructure: e.g. the ETSI and 3GPP standards for M2M communications which cover communication issues over existing Wide Area Networks (WAN); the Health Level 7 (HL7) standard which covers procedures for communication with health platforms; the Electronic Health Records (openEHR) standard which deals with data semantics and organization of raw clinical data. Since the book aims to address the more general public these platforms and standards are examined and summarized balancing in-depth technical information with the large scope of related issues involved. Finally we conclude by pointing out some potential problems in the development of a unified approach to incorporating M2M technology in the AAL platforms and outline open research issues in the area.

17.2 M2M COMMUNICATIONS AND ARCHITECTURES

In its simplest form the definition of M2M can be given as "the technology which will enable the coordinated communication between machines and devices without human intervention, allowing them to perform actions beneficial to humans". Two terms are used interchangeable by the major standardization bodies — MTC (Machine Type Communications) adopted by 3GPP and M2M (Machine-to-Machine Communications) adopted by IEEE and ETSI. In this chapter, we use preferably the term M2M, but use also MTC when required especially in referring to the 3GPP standards (Wu et al., 2011; Kartsakli et al., 2014). In its core, M2M might be considered as a technology rooted in previous industrial automation systems like SCADA (Industrial supervisory control and data acquisition) which were introduced much earlier. Some authors look at M2M as an extended, more sophisticated version of those systems. However, many more think that this is not a comprehensive enough definition of M2M technologies. SCADA systems, which were very popular in the 1980's are predominantly based on separate proprietary solutions and their newer implementations can more precisely be described as "extended embedded systems". This concept is quite restricted for describing M2M communications, which with its potential to connect billions of very different devices over various network structures is a revolutionary technology of the future, that some visionaries call "The Embedded Mobile Internet of the Future". Apart from these enthusiastic characterizations, the ETSI Technical Committee on Machine-to-Machine Communications (ETSI TC M2M) has provided a formal exact definition: "Machine-to-Machine communication is the communication between two or more entities that do not necessarily need any direct human intervention" (Borgia, 2013; Atzori et al., 2010; Chen, 2012). Another interesting definition can be found in a study conducted by Intel (Wu et al., 2011), which defines M2M as "a future technology where billions to trillions of everyday objects scattered in the surrounding environment are connected and managed through a range of devices, communication networks, and cloud-based servers". The expectations that M2M technology can really open the door for another information revolution like the introduction of the computers and the Internet is also based on predictions of 25% compound annual growth (CAGR) for M2M communications, driven by very strong technological and economic factors (Taleb and Kunz, 2012). First of all, today's advanced wireless and cellular networks can provide broadband services at much lower costs than in the past and can also provide a lot of the features required for M2M communications. On the other hand, a major economic drive is anchored in the fact that revenue from voice services is continually deteriorating which compels operators to look for and adopt new, different revenue creating services.

M2M technology contains 3 major building blocks: the devices (these are expected to be very low-cost and low-power devices in huge numbers as well as lesser numbers of higher-end devices, which will serve as aggregation points, control modules, gateways or intelligent interfaces); the scalable network (independent on the specific network infrastructure used the network has to ensure reliable interconnection of millions and billions of heterogeneous devices with very diverse application requirements); the services and device management (probably cloud-based) (it is expected that centralized management will be required because of the complexity of the involved applications, in addition to some distributed functionality) (Kim et al., 2014).

However, the focus of this technology is not only on the essential building blocks, it is bringing them together which will be a key for turning M2M communications into a reality. There are a number of challenges that such inter-operation is bound to overcome: *manageability* of collected information in terms of providing suitable solutions for the optimal distribution (as a clue one might consider one of today's popular application — the large surveillance networks, in which the number of video feeds is compared to the number of operators); *backward compatibility* — the network plug-and-play capabilities have to be ensured for both legacy and new devices; the *gateways/aggregation* points have to be designed taking into consideration new and emerging value-added services to really enable the explosive growth of short-range smart sensor technologies; *security issues* on a scale never seen or imagined before; last but not least the issues of *standardization*, because unlike the current vertical, highly proprietary communication systems solutions, M2M communications should be based on horizontal design principles and require new architectures and service platforms (Clayman and Galis, 2011). In the following section we will concentrate specifically on the standardization efforts related to M2M architectures.

17.2.1 M2M Architectures

Based on today's existing wide area and local area networks there are multiple connection options available for M2M applications. The first mandatory step in order to provide smooth interoperability and avoid fragmentation among solutions and services is that a standard end-to-end architecture for M2M is proposed. That is why in recent years, standardization activities have been going on at an increasing pace, powered by a number of international standardization bodies. Among them we can first mention the 3GPP (Third Generation Partnership Project) (3GPP, 2010a), IEEE and ETSI (European Telecommunications Standards Institute) (M2M, 2011; M2M, 2012; M2M, 2013), based a large worldwide participation as well

Table 17.1 ETSI standards

ETSI document reference number	Specification name
TS 102 689	M2M Service Requirements
TS 102 690	M2M Network Architecture
TS 102 921	M2M Interfaces

as the newer Open Mobile Alliance (OMA) (openmobilealliance) and oneM2M (onem2m); some powerful more region oriented ones as TIA TR-50 Engineering Committee (The Telecommunications Industry Association is the United States counterpart to ETSI) (tiaonline), the CCSA TC10 (The China Communications Standards Association) and (ccsa) The Global ICT Standardization Forum for India (GISFI) (gisfi). The efforts of 3GPP and IEEE are focused particularly on issues related to wireless cellular networks support for M2M, while ETSI addresses the M2M high-level architecture, components, and interactions between the application, network and device domain (M2M, 2011, 2012, 2013). In the following ETSI and 3GPP approaches are discussed in more detail.

17.2.1.1 ETSI Architecture

So far ETSI has published a number of standards related to M2M. The major ones are presented in Table 17.1.

The ETSI technical specification (M2M, 2011) defines the M2M architecture presented in Figure 17.1 below. The lowest domain (**Device and Gateway Domain**) consists of M2M devices, which can be fixed (for example in applications like factory control, metering devices, fixed sensors etc.) or mobile (fleet management sensors, wearable sensor devices etc.) with heterogeneous structure and features, transmitting under different patterns (one-way communication, burst transmissions etc.) which will allow operators to provide some degree of optimization based on grouping especially for charging and controlling purposes. To ensure backward compatibility this domain provides network access for legacy devices, connectivity between M2M devices and M2M gateways through the M2M Area Network as well as services (through the M2M Gateway) to legacy devices hidden from the Network Domain. As examples of M2M Area Networks we can point out existing Wireless Personal Area Network (WPAN) technologies (IEEE 802.15.1, ZigBee, Bluetooth) or more control oriented networks as PLC, M-BUS, Wireless M-BUS and KNX.

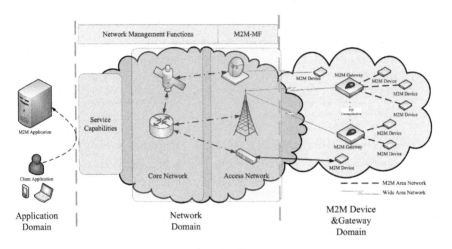

Figure 17.1 M2M high level architecture as specified by ETSI.

The **Network Domain** consists of two sub-entities: the Access Network (AN) and the Core Network (CN). The AN function is to provide connectivity between "M2M Devices and Gateway Domain" and the Core Network. The CN on the other hand, is responsible for IP connectivity and interconnection with other networks, roaming and network control. As examples of AN we can point xDSL, HFC, satellite, UTRAN, eUTRAN, W-LAN and WiMAX, while examples of CN include core networks like 3GPP CNs, ETSI TISPAN CN and 3GPP2 CN.

The last domain is defined as the **Application Domain** and it integrates the M2M Application Servers (AS) under the control of a mobile network operator or a third party.

The Management Plane is specified across the Network Domain and incorporates Network Management Functions (NMF) and M2M Management Functions (M2M-MF). NMF include functions for managing the AN and CN (resources, supervision, fault detection and management, etc.); M2M-MF denotes the group of functions related to managing the M2M Service Capabilities (SC).

A more detailed illustration which includes the different possibilities for underlying network technologies is presented in Figure 17.2.

Connections in the network can be established as follows: either an M2M Device connects to the M2M server directly through a WAN (cellular, 3G/4G etc.) or an M2M device connects through a M2M Gateway, or an aggregation point. The first one is known as "Direct Connectivity"

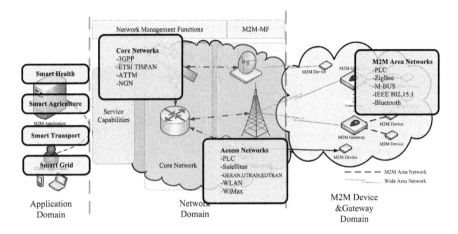

Figure 17.2 ETSI M2M architecture and existing network technologies.

and the connecting devices are responsible for carrying out procedures like network registration, device authentication and authorization, resource allocation and management. The second type of connectivity is known as "Gateway as a Network Proxy". The role of M2M Gateway or aggregation point can be fulfilled by any device sophisticated enough to collect and process data from simple, low-cost and low-power M2M devices. Under this definition it is possible to create a hierarchical structure that will include extremely large numbers of low-cost-low-power devices, where the above mentioned procedures of network registration, authentication, provisioning and management are performed by the gateway.

For the M2M Service Architecture ETSI has adopted the so called Representational State Transfer Model (REST) (M2M, 2012). According to that each logical or physical entity is represented as a "resource"; each resource can have different states and these states can be manipulated in a certain way. As an example, a sensor according to the REST model can be described formally as a resource which can be configured or read. It is very important to note that resources are uniquely addressable and also can be accesses as web links (similar to web browsers) by using some well-established protocols as HTTP. On the other hand, each resource is an M2M applications and network entities can exchange information in a well prescribed way.

The M2M Service Capability Layer (SCL) consists of a number of different Service Capability Entities (SCE). It is defined on top of the connectivity layers and resides in all network servers, gateways and M2M

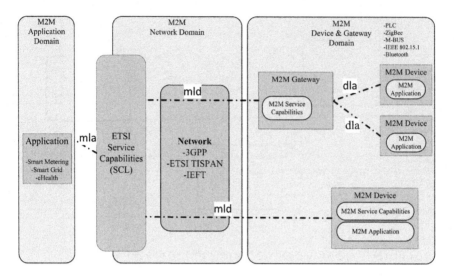

Figure 17.3 M2M architecture and interfaces (ETSI).

devices (low power low capacity devices will host lighter version of the SCL). It is responsible for registration, access right, security, authentication, and data transfer (data containers) as well as for managing devices (activities like joining, leaving, notification, error reporting etc.). The API for the applications is based on the REST principles to allow for scalability and binding with protocols like HTTP and CoAP.

Furthermore, ETSI has defined three major interfaces: the *"mla"* interface — between the application and the SC in the M2M Core Network; the *"dla"* interface which is defined between the application and the SC in the M2M Device/M2M Gateway; and finally the *"mld"* interface between SC in the M2M Core Network and the SC in the M2M Device/M2M Gateway — i.e. Service Capability Client (SCC) — Service Capability Server (SCS). The above described interfaces are illustrated in Figure 17.3.

A recent technical report addressed specifically the issues of M2M application scenarios in the area of eHealth and Aging, which are an integral part of AAL (M2M, 2013). In general the application of M2M in the realm of eHealth relies on a variety of sensors which collect information about the well-being of the individual and his immediate environment and send it to a backend server at regular intervals or when triggered by an event. The amount of data to be send, as well as the specific time intervals or trigger times depend on the patient's case or the particular disease. In many cases the system is required to configure disease management devices (e.g. adjust

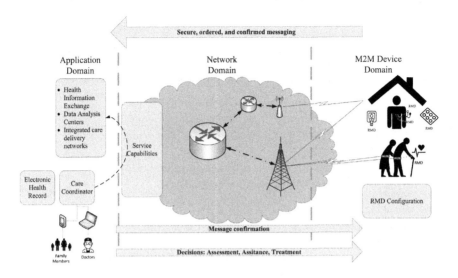

Figure 17.4 Healthcare remote patient monitoring scenario.

the reporting period) and/or verify correct operation (check the correct functioning of the sensors as well as verify connectivity).

Several scenarios have been defined — remote patient monitoring, patient-provider secure messaging, measurement of very low voltage body signals and telecare between home and remote center. As an example an illustration version of the general "Remote M2M Device (RMD) Patient Monitoring" scenario is presented in Figure 17.4.

For the communication process related to this scenario the following phases can be defined: ***RMD Initialization Phase, RMD Registration and Configuration Phase, Data Retrieval Phase and Data Delivery Phase.*** Details about these phases are presented in Table 17.2.

The main stakeholders that would possibly be involved in these types of eHealth scenarios are defined as follows:

The Patient: this is the person, using the RMD and for whom vital signs measurements are taken. He/she can be found in a range of different environments like for example at home or at work, in a hospital, traveling, residing in some public assisted living facilities, etc.

Care Coordinator: this term can refer to either a person or computer applications that continuously or periodically monitor the information received from the patient's device(s). One of the functions of the Care Coordinator is to intervene or inform clinicians if required (e.g. when measurements fall outside of a predetermined range or alerts have been raised).

Table 17.2 Establishing M2M communication — major phases for the RMD scenario

Initialization phase	RMD may be attached to a laptop or another device
	RMD performs initial power up sequence
	(If required secure start-up procedure should be executed)
Registration and configuration phase	Patient/provider execute authentication procedure
	Remote registration and Configuration (if required)
	RMD registers with appropriate SCE (direct/indirect -M2M-GW)
	Appropriate SCE provides name and address mapping
	Appropriate SCE extracts network address
	Appropriate SCE provides network selection
Data retrieval phase	RMD periodically/event-triggered wakes up
	RMD prepares message to send
	RMD check connectivity — if connected sends message
	— if not connected stores message
	(Life-critical applications have to be always connected)
	RMD sends message either direct or through an M2M-GW
Data delivery phase	Appropriate SCE extracts name and address from M2M message
	Appropriate SCE transports message between M2M Device/M2M-GW and
	M2M Application in the Network
	Appropriate SCE stores message copy and delivery status report, error report etc.
	M2M application (M2M Device/Network M2M Application) receives message

Clinician: this term refers to all (physicians, nurses, assistants, psychologists or any medical personnel) whose responsibility is to make decisions and carry out interventions if required by state of the monitored patient.

Remote Monitoring Device (RMD): this is the electronic device, denoted as "M2M Device", which can be equipped with a wireless or wearable sensor, special user interface and/or actuator. It should be able to collect information from the patient and/or his immediate surroundings; should be able to communicating the raw or aggregated information to the appropriate M2M SC entity or M2M application through the M2M network. The connection should be a two-way connection so the device may receive commands from the M2M SC entity or provide information through a given interface (screen, sound, light etc.) to the Patient. As mentioned before this type of devices are usually very low power, low capacity so they require very low complexity communication protocols to allow them to connect to the network via the M2M Gateway (M2M-GW).

M2M Service Capability Entity (SCE): the Network Entity that provides M2M communication services to the M2M application entities. These applications may support functional capabilities specific to health information exchange activities. Additionally, the M2M SCE communicates with the Remote Monitoring Device to collect data or send commands.

M2M Application Entity: this is a term used to define all high level system elements (stakeholders) which are not covered in the scope of M2M. As such we can mention health information services, data centers, care providers and care providing organizations, record banks or other health system related public organizations.

Electronic Health Record (EHR): comprises the medical data, held in a specific digital format and usually maintained by the responsible health care system (Electronic Health Record — EHR). It also refers to the formal or more informal medical records kept by the patients themselves, their families and relatives.

The efforts of ETSI in the area of M2M standardization are on-going. Figure 17.5 below gives an overview of these efforts. In July 2012 a new alliance oneM2M, was formed (Founding partners — ETSI, ATIS, ARIB, TIA, CCSA, TTC, TTA) with the aim to accelerate and facilitate the work on M2M standardization activities (onem2m). One M2M efforts are specifically focused on service capabilities related to M2M communications. The main idea is to create standardized end-to-end services, i.e. derive open standard interfaces considering the charging aspects, the security and privacy issues. Another focal point is the addressing, identification and naming structure which is quite critical given the range of devices' and applications' characteristics and their huge numbers.

17.2.1.2 3GPP Architecture

The 3GPP is the second most important standardization organization which is deeply involved in M2M communications. Its efforts are parallel and complementing the standardization activities of ETSI. As a matter of fact, ETSI is also an active member of the 3G Partnership Project. While there are slight variations in the notation of the entities, the most important difference between the two standards is that 3GPP focus is on ensuring high capacity by creating a hierarchical architecture based on mobile and cellular infrastructure. This subsection will present more details on the M2M

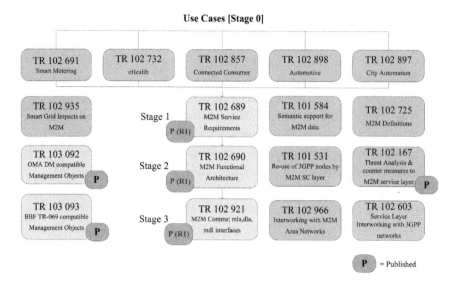

Figure 17.5 ETSI M2M standardization documents.

architecture suggested by 3GPP (3GPP, 2013, 2010b, 2010c; Ghavimi and Chen, 2015).

First of all, parallel to the term "M2M communications", adopted by ETSI, 3GPP defines the "MTC communications" — i.e. Machine-Type Communication (MTC). MTC is a form of data communication which involves one or more entities that do not necessarily need human interaction" (3GPP, 2013). It is clear that there is a slight difference between the two, whether the later (MTC communications) is a broader term that also covers interactions between machines and humans (i.e., M2H communications or H2M communications).

The aim of 3GPP is to optimize the LTE network to support MTC traffic and applications. For this purpose, it defines the so-called 3GPP MTC architecture depicted in Figure 17.6. Similarly to the ETSI architecture, three major domains are distinguished: the MTC Device Domain, the Communication Network Domain, and the MTC Application Domain. The major difference, compared to the ETSI specification is the fact that the communication network is a 3GPP mobile network, assuming LTE-A as an underlying technology with multi-tier connectivity.

Starting from the lowest architectural level (MTC Device Domain) comprising various devices (**user equipment** (UE) and MTC devices), the **evolved Node B** (eNB) and the **Home eNB** (HeNB), are defined. Their role in the network is to fulfill the functions of an LTE Base station. Above them in the hierarchy, the Communication Network Domain, (LTE/LTE-A network) is subdivided into the Radio Access Network (RAN), comprising base stations (BSs) — eNBs or HeNBs and the Evolved Packet Core (EPC) — known as the Core Network (CN).

The RAN provides an air interface with the existing UEs and MTC devices. Each of the eNBs can connect to other eNBs through the X2 interface and the EPC through the S1 interface. The HeNBs, which can be viewed as eNBs for indoor coverage improvement, can be connected to the EPC directly or via a gateway (if a large number of HeNBs is involved).

The EPC is a flat all-IP Core Network, responsible for the overall control of mobile devices and establishment of Internet Protocol (IP) packet flows. It can be accessed through 3GPP radio access (e.g. WCDMA, HSPA, and LTE/LTE-A) and non-3GPP radio access (e.g., WiMAX and WLAN). This access flexibility of the EPC is very desirable for network operators since it allows a straight forward approach for upgrading and modernization of their core data networks to support a wide variety of access types over a common core.

One of the main entities of the EPC is the **Serving Gateway** (S-GW). The S-GWs handle primary operations inherent to the 3GPP network like routing of incoming and outgoing network packets, handoff and mobility issues within the network. They are a major connection point between the radio access part of the 3GPP network and the core part, the EPC. The S-GW is connected to the eNB through S1-U interface and to the P-GW through S5 interface. Each UE/MTC device is associated to a unique S-GW, which can be hosting several functions.

Another important core element of the 3GPP network is the **Packet Date Network Gateway** (P-GW), the "anchor" of mobility, the gateway between 3GPP and non-3GPP networks. Its main function is to provide connectivity for user entity (UE) traffic form and to other networks (WiMAX, WLAN, EvDO etc.) as well as handle issues of policy enforcement, charging, packet filtration etc. The P-GW provides also a secure connection between UEs/MTC devices by using Internet protocol security (IPSec) tunnels between UEs/MTC devices connected to an untrusted, non-3GPP access network with the EPC.

A primary control entity in the EPC is the Mobility Management Entity (MME). Besides being responsible for managing mobility, roaming, handover and security functions like authentication, authorization and NAS signaling, it is also responsible for choosing the Serving Gateway (S-GW)

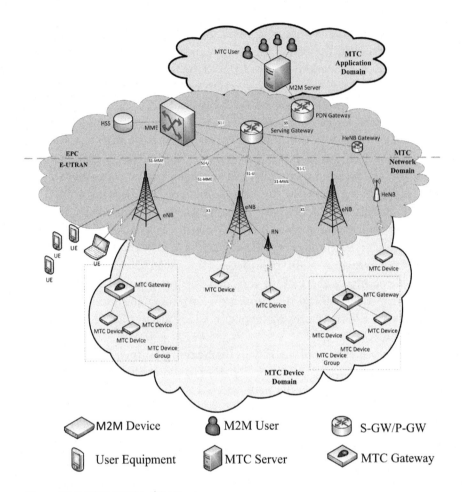

Figure 17.6 3GPP M2M architecture.

and Packet Data Network Gateway (P-GW) for an UE/M2M device at its initial stage of attachment. The S1-MME interface connects the eNBs with the. EPC-MME.

There are three types of transmissions among MTC devices as specified in (onem2m);

- **Direct transmission:** an MTC device can send and receive information to and from the eNB directly. The MTC device itself performs the control processes in the network domain.
- **Transmission via MTC Gateway:** The MTC Gateway collects the data from the connected devices and provides the connection to the

network domain through the LTE RAN and allows MTC to send and receive data through the network. It connects to the MTC devices in a one-to-many and many-to-one fashion, i.e. one MTC Gateway can serve various MTC devices, and at the same time an MTC device can be connected to more than one MTC Gateway.

• **Peer-to-peer transmission:** M2M devices can send/receive data to/from other M2M device directly.

The above described communication models can pertain to two quite different operator involvement approaches. The first one, the case where MTC users access and control the MTC devices through one or more MTC servers, mandates that the MTC server and the APIs used are provided and specified by the operator. Since the physical location of the server is not relevant to the access modality, the operators can position them in their domains or outside, and the server access procedures will remain unchanged for the user. On the other hand, the case when MTC devices communicate directly in a P2P mode (peer-to-peer), the users are totally independent of the operator that the MTC devices are connected to.

3GPP efforts are focused on optimizing the transmission through the network with the purpose of providing seamless connectivity and extending the coverage. Macrocells were introduced to provide coverage and support of higher mobility. A recent approach to increase the coverage, especially in densely populated areas and close to the borders of the macrocells, is the introduction of picocells (with eNBs) and femtocells (with HeNBs). They allow bringing the link connectivity closer to the end devices and hold high promises for increasing the capacity and reliability. This is an important issue for low-power low-capacity devices, like most of the MTC devices are expected to be.

17.2.2 Characteristics of M2M Applications

The realm of possible M2M applications is far from defined. It is quite obvious that there will be great differences among the applications and it is very difficult to put them under single denominator. However there are some characteristics that will be common to a lot of the applications and we try to summarize them below:

• Simultaneous or nearly simultaneous mass device transmission request/access to a single base station — this springs from the expectations that M2M communications will connect millions of devices related to numerous applications, which is quite a different situation compared to existing Human-to-Human (H2H) or (human-to-machine) H2M type applications. The most important point is that a much larger number of devices will try to access the network from a given location/area at the same time, i.e. device that will try to connect to a single base station. Since most of the initial network procedures used today are random access based, where device compete for the resources, this will constitute a major problem.

- Traffic pattern — the traffic that M2M devices create will have characteristics very different from H2H or H2M communications. For a large number of applications frequent transmission of small packages will be sufficient. This means that the traffic pattern will be characterized by infrequent, small bursts transmissions (SMS like) which however might have strict delay requirements. Other applications might require continuous time controlled traffic, which again is very different from most current H2H traffic patterns.

- High reliability transmissions independent of environment conditions like channel quality, mobility, network connectivity issues — compared to existing H2H communications, M2M applications pose much more stringent restrictions for reliability which are rooted in the nature of applications they are expected to support. Consider for example a scenario where critical physiological data about an individual is transmitted and it has to trigger an alarm and/or due response from medical personnel. Loss of information or disruption in transmission due to variable wireless channel conditions can be tolerated in H2H where the connection will be resumed later, but will be devastating for the M2M application.

- Extremely low power consumption for devices that act together — in a lot of cases the devices involved in M2M communications might not be replaced for years. This mandates that the power consumption should be near zero or that methods similar to the ones used in WSNs, benefiting from possibilities for high redundancy and energy scavenging, should be applied in order to extend the operation life of the system even if single devices fail.

- Low/No mobility — regarding mobility issues M2M applications in general can be separated into those that require/rely on high mobility and those that do not have any mobility (or very low mobility). Many researchers expect that a larger number of applications will fall into the second group.

- Time tolerance — similarly to the mobility characteristic, M2M applications can be divided into two large groups considering delay tolerance: the first group encompasses applications that require extremely low latency and while the second refers to the so called "time tolerant" applications. For the first group "extremely low latency" refers to the end-to-end latency which includes both transmission delay and network access delay and mandates that both should be minimized. The time tolerance is very important in many emergency related application scenarios (e.g., healthcare, surveillance).
- Addressing extremely large number of devices and providing group control based on predefined criteria — as mentioned before the number of devices expected to be involved in M2M communications is in the order of billions, excluding the existing devices today. Considering an all IP network the addressing and address management is a major research challenge.
- Priority issues like enhanced priority options for alarms etc. — M2M applications are very diverse and the network should be able to distinguish between their requirements in order to provide relevant QoS.
- Security, monitoring and authentication issues — it is obvious that many applications will be related to sensitive information and adequate methods for authentication, privacy and security should be devised. A specific restriction here is the fact that most of the devices involved will be extremely limited in power and computational capabilities so proposed methods should rely on extreme lightweight stacks and procedures. This is really a great challenge considering the complexity of the security methods existing for H2H communications.

17.3 M2M AS AN ENABLING TECHNOLOGY FOR AAL — STATE OF THE ART

17.3.1 The Role of M2M as an Enabling Technology for eHealthcare Applications

So far, we have discussed different network architectural aspects and general characteristics of M2M communications. Actually, M2M as an emerging communication technology is part of a larger vision of a future world, the so called physical-cyber world. In recent years, there has been a myriad of concepts that are quite close or similar to M2M communications. No doubt, one of the most popular ones is the Internet of Things (IoT) (PPR ITU-T

Y.2060, 2012) which claims IP connectivity between all devices. There are also authors who support the term Smart Device Communications (TIA TR-50, 2010) due to its connection to smart environments, or similarly Machine-Oriented Communication (MOC) (PPR ITU-T Y.2061, 2012). Ubiquitous Sensor Networks (USNs) on the other hand is mostly supported in the WSNs research community (ITU-T Y.2221, 2010).

No matter what the specific details of these concepts are, the realization of the general vision has to go through an inevitable evolution of the network and services' infrastructure. The approach used in different large systems is known as "silo" or "stove-pipe" because of its vertical design: each application is built on its proprietary ICT infrastructure, dedicated devices and protocols (Borgia, 2013; Atzori et al., 2010; Chen, 2012). Currently existing applications, even if very similar, do not generally share features for managing network operations and services, which obviously results in unnecessary redundancy, low efficiency of the ecosystem as a whole, increased costs and nearly impossible inter-operation. For realizing the vision of the physical–cyber world it is imperative that the vertical, "silo" approach to system design is replaced by a more flexible, horizontal approach. Many authors confirm the idea that a common operational platform for organizing and supervising network operation and services, which will abstract across a diverse range of data sources, will allow higher degree of interoperability between various kinds of applications.

As discussed by Borgia (2013), "applications will no longer work in isolation, but will share infrastructure, environment and network elements, and a common service platform will orchestrate on their behalf". These ideas are illustrated in Figure 17.7 and Figure 17.8.

Healthcare is one of the sectors to greatly benefit from developments and substantiation of the physical–cyber world. Numerous applications developed with the introduction of WSNs have opened the way for the advancement in sensing devices and today there is a proliferation of solutions for real-time monitoring of vital functions and parameters (e.g. temperature, blood pressure, heart rate, sugar level, cholesterol level, etc.) (Delmastro, 2012; Abbate et al., 2012; Triantafyllidis et al., 2013; Jara et al., 2013; Tseng and Ke, 2012). Standards like ZigBee, Bluetooth, WirelessHART, ISA100 etc. have been developed which further promoted the adoption of WPAN (Wireless Personal Area Networks) and BAN (Body Area Networks) based on wearable devices as a means for remote monitoring, diagnosis and control of patients and elderly people. Less

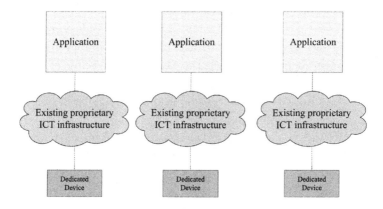

Figure 17.7 The "silo" system concept.

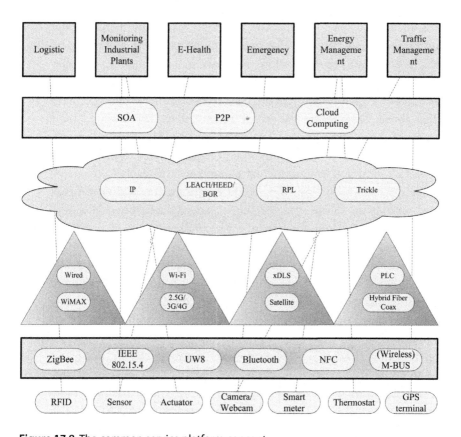

Figure 17.8 The common service platform concept.

mainstream applications related to the tracking and identification of medical supplies (e.g. smart labels) and equipment (e.g. tracking of lost or stolen equipment) were also developed to facilitate medical inventory (Borgia, 2013). Even though such systems help increase the well-being of many people, especially patients with chronical diseases and the elderly the problem is that they are isolated systems and serve only a single well defined purpose and/or organization or institution.

On the other hand, there are a number of network services and applications related to healthcare and independent living which, even though technologically possible have not been realized, because they require a much higher degree of interoperation between existing vertical systems. Recently the special term "e-inclusion", has emerged, referring to the improvement in the life standards of citizens and supporting independent living. As part of the European Commission Joint Program for AAL, the e-inclusion focuses on specific categories of people, primarily the physically disabled or the aging citizens. The goal is to establish the technological and societal conditions that will allow the above mentioned groups longer, healthier life and more fulfilling participation in the society. These highly complex and extensive systems are the next level from the systems we have described above. Despite the variety of aspects, they will cover, in their core many of them will be based on WSNs and M2M communications. Monitoring the physical status (physiological parameters) together with other psychological and environmental parameters will evolve into comprehensive applications for emulating medical consultation, taking complex decisions, setting up alarms and reacting to specific situations in a very prompt manner not waiting for human involvement (Abbate et al., 2012). Applications based on such integrated systems will allow diagnosing diseases like dementia, Alzheimer and Parkinson at a very early stage and suggesting possible hospitalization. Service integration with home entertainment systems, using PC and TV screens, will provide a new channel to stimulate people to exercise, provide surveillance, guide them in searching for objects in the house (Dias et al., 2012).

On the other hand, since elderly people move much less and prefer to stay at home thus getting naturally isolated from society, social networks will allow them to connect, communicate and participate more actively in various discussion groups based on their interests (AALJP, 2012; AAL-2011-4-099, 2011). Key elements in this respect will be highly simplified, multi-modal interfaces which will allow the systems to acquire substantial information about human behavior and predict human actions. Another aspect of assisted living, which can highly benefit from the underlying M2M technology, is mobilizing the elderly by providing them with an easy and secure way to move in the neighborhood or use public transportation. Aggregating and evaluating data from personal mobile devices (equipped with e.g. position sensors, orientation sensors, movement obstacle detection sensors, video cameras) and data from sensors positioned in the surroundings, the physical–cyber system can reconstruct perceptions of the environment. The information collected can be used to adapt certain actuators in the surroundings (open doors, activate services etc.) or can be later verbalized to individuals by synthesized voice. Similar applications can also highly benefit visually impaired people and extend their ability to move in the city (AAL-2011-4-099, 2011; Manduchi and Coughlan, 2012). Wellbeing and lifestyle services are another very important field directed at improving people's quality of life by capturing capture users' reactions to the environment.

17.3.2 The Concept of Ambient Assisted Living

In this section we dwell on the concept of Ambient Assisted Living and discuss its relation to other relevant research areas.

In the last decade, the large amount of research related to WSNs, smart environments, robotics, artificial intelligence, content aware computing and multi-agent systems, etc. have precipitated in the definition of a new paradigm, the Ambient Intelligence (AmI). The Ambient Intelligence paradigm (AmI) is defined in Augusto (2007) as "a digital environment that proactively, but sensibly, supports people in daily lives". AmI goes far beyond what we know as ubiquitous (or pervasive) computing by specifically focusing on the users, their experience and expectations from the devices surrounding them. The AmI paradigm presents a great shift in viewpoint as compared to the way ICT technology was considered only a decade ago. This new paradigm puts the intelligence not in a single device, operated and controlled by a human, but in the environment — the physical-cyber environment. The philosophy of AmI is that the surrounding environment should adapt to the inhabitant(s) and not vice versa (Grguric, 2012). The developments in this area are greatly facilitated by the ever increasing number of intelligent networked devices and their continuous miniaturization secured by the progress in the semiconductor industry. The Information Society Technologies Advisory Group (ISTAG) (ISTAG, 1999) defines five key technological requirements for AmI. AmI systems should be built on *unobtrusive hardware*, creating **high density, dynamic, distributed device networks** which rely on **novel interfaces** with natural human feel, and provide **seamless connectivity** through mobile or fixed communication infrastructure, with very high degree of *fault tolerance, dependability and security*.

AmI will expend people's capabilities by augmenting the digital environment and enabling innovative human-machine interactions. Assisted living technologies based on AmI are called Ambient Assisted Living (AAL) tools. The term "Ambient Assisted Living" was adopted by the general public with the identically named Framework Program 6 funded Support Action (FP6-SA) that developed scope, procedures and legal basis of the AAL Joint Program (AALIANCE, 2010). The scope of AAL is detailed with the "European Action Plan on Ageing Well in the Information Society" (Ageing, 2016) according to which the idea of AAL is "to extend the time older people can live in their homes by increasing their autonomy and self-confidence"; "to promote healthier way of life and functional capability of elderly"; "to prevent social isolation and maintaining social network around the individual" and also, last but not least "to offer common (software and hardware) platforms that will fulfill expectations and enable easy development and deployment of AAL solutions", increasing the efficiency of resources in our aging society. AAL tools are envisaged to serve many different aspects of people's life: from monitoring, improving and curing health conditions, medication management, controlling of medical treatments (Qudah et al., 2010; Khan et al., 2010) to providing safety (Eklund et al., 2005; Aghajan et al., 2007; Fleck and Strasser, 2008), mobility (Pollack et al., 2003; Dubowsky et al., 2000) and also opportunities for including the elderly more actively into society and social life (Mynatt et al., 2001; Vetere et al., 2009).

The research and development activities in AAL mandate that new advancements coming from various fields should be considered on a common ground, using a multidisciplinary approach: enabling technologies range from distributed embedded-network sensors and actuators to M2M communications and highly sophisticated reasoning engines and anticipatory human-computer interfaces; current research spreads over a wide range of topics from robotics, artificial intelligence, smart environments, multi-agent systems to human activity recognition and behavior understanding. Figure 17.9 below gives an idea about some of the most important research areas relevant to AAL.

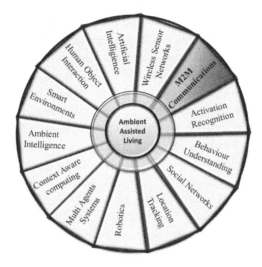

Figure 17.9 Research areas relevant to AAL.

The dominant characteristics of the underlying technological solutions for realizing the paradigm of AAL can be summarized as (ADEPA):

- Adaptive (to user and user's environment)
- Distributed (throughout environment)
- Embedded (non-invasive, invisible if possible)
- Personalized (depending on the user's needs)
- Anticipatory (anticipating user's desires).

A very informative review of AAL is presented in Rashidi and Mihailidis (2013) which provides a summary of recent advancements in various technological areas closely related to AAL including smart homes, assistive robotics, e-textile, and mobile and wearable sensors (Rashidi and Mihailidis, 2013; Monekosso et al., 2015). In this chapter we extend the list by adding one more emerging technology that has a great potential to further the development of AAL — the M2M technology. The rest of this chapter focuses especially on research in which M2M is used as an enabling technology for AAL systems.

17.3.3 M2M Based Applications for AAL — State of the Art

An in-depth overview of the role that ICT technology is playing in the life of the elderly, covering technological, sociological and economical aspects is presented in Augusto (2007). The author affirms that M2M communications together with cloud computing are the most powerful and promising technologies of the next generation communication networks. However, according to him the problems ahead will lay not with the technology itself, as a physical implementation but with the amount of information collected, specifically its interpretation and further utilization. Another important issue is the new way we define IT and our interaction with it. Technology in its bare form, as devices and equipment used by humans, will be pushed far in the background. The interaction of humans will be based on devices hidden in the environment (device "disappearing physically") or devices that are not perceived as computational objects but rather through the changes they create in terms of sound, movement, light, smell or taste (devices "disappearing mentally"). Today's devices, which can be characterized by a single type (mode) of interface, will be replaced by multi-modal interfaces that have the ability to simultaneously affect several of the seven human senses. Interaction through such multi-modal devices and systems is especially appropriate for elderly people since they often suffer from a specific reduced capability (sensory, physical and intellectual) (Alm et al., 2001; Chellouche et al., 2013).

In Chung (2012) a multi-modal sensing M2M healthcare monitoring system is presented, which is specifically tailored to homecare or countryside situations where the elderly do not have any resident caregivers. The context aware sensing allows, besides monitoring day-to-day activities, also for providing insights into the development of chronic diseases or diseases where noticeable changes occur over longer periods and are difficult to catch using a simple, single parameter monitoring system. The authors propose the design of a multi-modal system, Multi-Modal Sensing u-Healthcare System (MSUS), which is schematically presented in Figure 17.10. The systems is composed of intelligent M2M devices (wireless sensor nodes for patient monitoring and tracking); a base station (smartphone), which functions as a local gateway or aggregation point, a central server and a terminal PC or PDA. Since it is designed to track the condition of elderly patients over long periods of time it allows collection of data during various activities like working, walking, running, sleeping and also integrates both indoor and outdoor location tracking. Experiments have shown that using multimodality and context aware data evaluation increases the accuracy in early symptoms detection. As an example, in the case of monitoring arrhythmic heart disease, altered ECG signals can be attributed to a number of factors other than an intrinsic cardiac condition, e.g. physical or mental stress. The authors believe that MSUS can

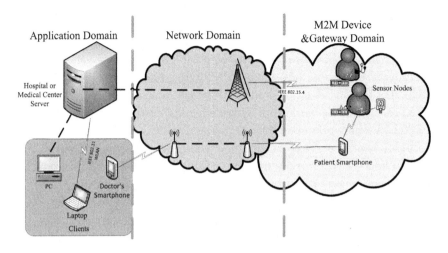

Figure 17.10 Multi-modal sensing u-Healthcare system (MSUS).

be successfully used to provide further insight into the natural cause and progression of a specific old age related diseases.

To achieve context aware data collection the authors use simultaneously different types of sensors: physiological biosensors (ECG, SpO2 and blood pressure, heart rate sensors) in conjunction with other types of sensors (e.g. accelerometer sensors, ultrasonic transceivers for location tracking). A novel design of an integrated sensor is proposed and implemented, the Integrated Fusion Sensor (IFUS) which contains conductive fabric and an accelerometer sensor and is used for measuring ECG signals and relative speed/acceleration. The data collected by the IFUS is transmitted over using an IEEE 802.15.4-based radio protocol to the base station. An interesting detail in the design of the system is that it allows the simultaneous processing of two different types of signals — the so called waveform-dependent (ECG signal) and the waveform-independent (blood pressure, heart rate, oxygen content). These two types have very different sampling and transmission (in terms of packet loss and QoS) characteristics — while one needs high rate sampling the others might require to be transmitted only when above a certain threshold. To provide context-aware information an indoor location tracking module is used, based on ceiling-mounted reference beacons which periodically publish location information using RF and ultrasonic signals. The design of the MSUS also involves very precise indoor positioning — between 7 ~ 15 cm as compared to other approaches like radar, active bat and active Badge, whose tracking accuracy is in the range of several meters.

An important component of the proposed MSUS is the Personal Mobile Healthcare Diagnosis Sub-system (PMHDS). The above described subsystem, IEEE802.15.4-enabled medical devices and a web server are connected to the outside world through a merged infrastructure consisting of IEEE802.15.4 networks and CDMA mobile networks. The PMHDS is constructed to handle received data, respond to requests to and from cellular phones, and host the server monitoring program for real-time monitoring, analysis and management of abnormal data.

The MSUS system integrates medical applications vital for elderly people living alone in a context-aware setting with emerging communication technologies like mobile networks and M2M. This experimental work provides valuable insight into the advantages and challenges that integrating M2M in future more complex applications of AAL might bring.

In Jung et al. (2013) propose a prototype wireless M2M healthcare system that combines mobile and IPv6 techniques with a WSN to monitor the health condition of an individual and provide a range of healthcare services. A low power, embedded wearable sensor collects real-time data and is connected, over low power WPAN through an M2M Gateway to the Internet or external IP-enabled networks. Recorded biomedical signals are displayed on an Android mobile device. The block diagram of this wireless M2M healthcare solution, using mobile devices in a global network is presented in Figure 17.11.

The core design comprises M2M End Devices (photoplethysmogram (PPG) and the M2M Node), hosting a lightweight TinyOS operating system which can collect data and transmit signals through the M2M Gateway. The function of the M2M gateway is to allocate IP address to the M2M node and perform address translation from global addresses to 16-bit short addresses or IEEE EUI64-bit extended addresses. The 6LoWPAN protocol stack is implemented on top of the IEEE 802.15.4 layer in the M2M nodes, which allows extending the IP based WSN environment with IPv6, so external hosts can directly communicate with M2M nodes with assigned global IPv6 addresses. The system architecture of the wireless M2M healthcare prototype is given in the Figure 17.11.

The measured biomedical signals are captured at the M2M node, further processed by the M2M Gateway, sent to the server PC (conceptually an M2M Application Server), which performs the monitoring, processing (extraction of useful data, address verification) and storing of the received

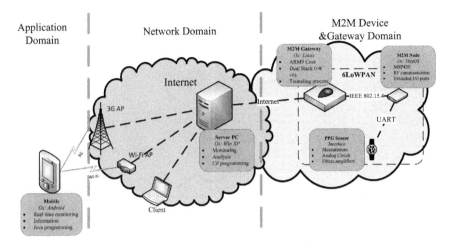

Figure 17.11 M2M Android based solution for healthcare applications.

data packets. Furthermore, it also sends the received data to a mobile Android platform (Samsung Galaxy S) running at 1 GHz ARM processor. Thus, the suggested configuration can be used over both wired and wireless networks, to connect to mobile devices and perform measurements collection and testing. Another major advantage is that once configured it can operate independently of the user, so data can be collected and transmitted even if the patient is unconscious. The communication between the mobile and the server is based on a query type model. The mobile device can display biomedical signals graphically in real-time. Analysis of the collected and saved data is carried out at the server — heart rate variability (HRV) analysis in time and frequency — and detected variations can be used to determine psychological and physiological stress and fatigue. Time domain analysis allows extracting stable/unstable HRV signals for evaluating whether the stressed and unstressed state is physiologically based, while frequency domain analysis allows evaluating the likelihood of emotions associated with stress.

Real-life test have been performed using a Samsung Galaxy S model to monitor the biomedical signals, the IPv6 address of the M2M node, the HR, and the blood oxygen saturation. The authors also point out that the popularity of mobile App Stores has opened another effective means that will ensure a wide spread of high quality healthcare applications delivered globally with minimum user effort. The authors conclude that with the

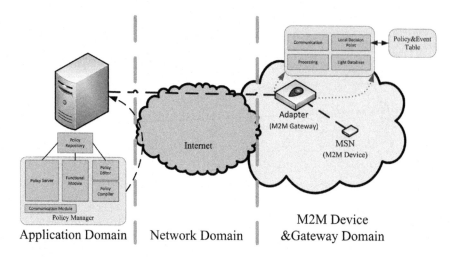

Figure 17.12 Intelligent M2M healthcare sensor network.

evolution of network integration and embedded devices (M2M End Devices) a universal healthcare system is within reach.

A different patient monitoring solution, which adds intelligence to the M2M Gateway, is presented in Shin et al. (2012). The authors propose to augment the end components of the network with the ability to make distributed decisions using policy-selection-based automatic management software where the nodes can learn from the environment and adapt to it. The system comprises three major modules: the Mobile Sensor Agent (MSA), the Adaptor and the Policy Manager (PM). Depending on the condition of the patient and in accordance with the policy determined by the PM manager the MSA can make local decisions regarding the communication process with the base station, the processing interval and the frequency of the vital signs measurements.

The MSA incorporates a number of sensors (for measuring blood pressure; heart beat and blood oxygen levels), a communication unit and a processing unit. The Adaptor (conceptually the M2M Gateway) contains, besides communication and processing modules, a local decision point and a light database. The local decision point chooses the policy to be sent to the MSA from the existing ones in the light policy repository. In case that a local decision cannot be made the Adaptor sends a request to the manager. The general architecture of the system is presented in Figure 17.12.

The proposed mobile agent based solution demonstrates a simple way of introducing intelligence into the M2M Gateway. It is an example of a healthcare solution which can, within certain limits and predefined policies operate without external human intervention. This work is a step towards more complicated, more encompassing and more intelligent M2M based AAL solutions.

Another very innovative approach of using M2M for improving the quality of life of elderly people is presented in Bhowmik et al. (2015). In their work, Bhowmik et al., discuss the use of M2M technologies to support AAL by incorporating them with Social Networks (SN). SNs, such as Facebook, Twitter and Google+, are continually growing in popularity and allow all individuals to remain connected with family and friends, create profiles and highly enhance the level of social interactions. However, getting involved in this new virtual world is not a straightforward option for the older generation. On the other hand, current applications depend on user initiative and action to upload and update information manually. That is where M2M can come to rescue — the suggested approach allows creating the content for the SN in an automated fashion, without human intervention. In this respect mobile phones, which have already found their place in everybody's daily life, can play a crucial role — they can be used as M2M Gateways to connect various M2M devices with SNs. For example, the GPS (Global Positioning System) sensors mounted in the mobile phone can be used to automatically update the location information of the user or a wearable sensor placed on the body of an elderly person can automatically (through the mobile phone) post information about his physical condition to his COI (Community of Interests) like family and friends. These and similar applications necessitate that the M2M architecture supports common features of SNs like authentication, posting, retrieving and at the same time should be able to support highly resource restricted M2M devices. Other requirements include scalability and application domain independence.

In their work, Bhowmik et al. propose architecture for M2M enabled Social networks (Bhowmik et al., 2015). It is fully in line with the ETSI standards described in the previous section. An overview of the suggested architecture is presented in Figure 17.13. The main functions required for connecting M2M devices to the SN are concentrated in the M2M Gateway, that's why the indirect communication model is adopted in this case. The data is collected by the M2M devices, sent to the Gateway for processing and then through the Network to the Application domain where the SN servers are located. The SN server receives and analyzes to determine the occurrence of an event. Once an event is detected it saves data in the database and the SN server propagates it onto the SN. In the Network Domain on the other hand, the Access Network and the Internet ensures the connectivity of the SN server with the end-users and the M2M Gateway. The authors have specifically selected the Gateway-based architecture because the M2M devices (wearable sensors) to be connected are very resource constrained, with limited processing capabilities and are unable to directly connect to the SN server. On the one hand, the M2M Gateway supports lightweight communication with the M2M devices, on the other hand it communicates with the SN server. It also supports joining departure procedures (self-organizing) and both synchronous and asynchronous communications because the system has to send notifications when changes occur.

To fulfill the functionalities described above Bhowmik and his colleagues propose four different categories of entities ("nodes") integrated in the M2M Gateway, defined depending on their functionalities: the ones interacting with the mobile nodes (Sink Entry Point — SEP), the ones interacting with the SN (Publisher Point — PP), storage ones (Data Storage — S) and the so-called super-peers for each of the above mentioned groups (SSEP, SPP and SDM — Super Data Management) (Figure 17.14).

The SEP nodes take on the role of entry points to the M2M Gateway, while the PPs are responsible for publishing the information on the SN. A SEP node can send requests to the M2M End Devices to request data and the S nodes store the data for the M2M Gateway. The so called super nodes on the other hand — the Super Publisher Point (SPP), the Super Sink Entry Point (SSEP) and Super Data Management (SDM) — connect to several respective nodes (PPs, SEPs and S nodes) and can communication among themselves (see Figure 17.14). The Super SEP nodes decides whether information should be stored, filtered, processed or sent to the SN, while the SDM replies to data requests from the other gateway nodes. To allow for both synchronous and asynchronous communications between M2M End Devices, the M2M Gateway peer nodes and the SN the standard, lightweight Constrained Application Protocol (CoAP) is used (Pereira et al., 2014).

Finally, the SN Server contains the following major components: the Request Handler, the M2M Adapter and the Data Analyzer. The Request Handler is responsible for handling user requests and forwarding them to the appropriate handler; the M2M adapter maps M2M End Devices with user profiles from the SN; the Content Adaptation Manager translates M2M data to SN-sharable information.

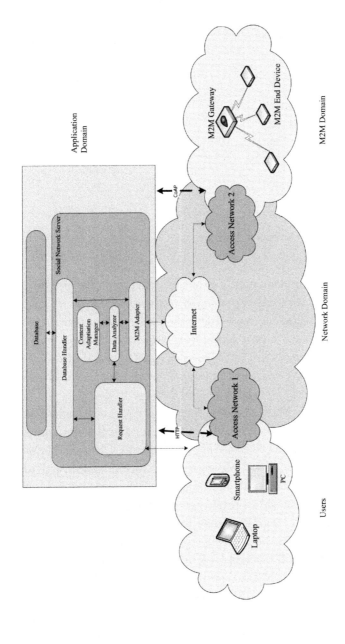

Figure 17.13 Architecture for social enabled M2M networks.

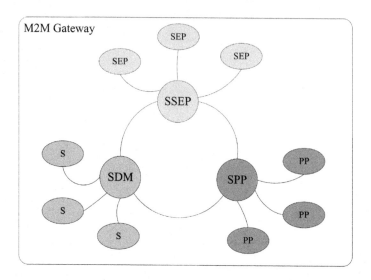

Figure 17.14 M2M gateway functional structure.

The authors have implemented a proof of concept prototype for the following scenario: *"John is wearing an accelerometer sensor that can sense rapid movements. The sensor periodically sends data to the SN server through the gateway on his smart phone. When a sudden movement is registered the gateway analyzes the data and forwards it to the SN. The SN server analyzes the sensory data in more detail. If the analyzer detects a fall event, it requests content adaptation information from the SN and posts about the event in the SN."*

In the prototype, the M2M Gateway functionality is implemented on a laptop, while another computer hosts the SN Server and the two connect using a Wi-Fi network. The SHIMMER Platinum Dev Kit accelerometer sensor collects the sensory data registering falls during random activities (walking, sitting and sudden rapid movements). The prototype is implemented on top of an open-source implementation of SNs (Zing).

As discussed before a major challenge for realizing more sophisticated eHealthcare systems lays in the difficulty to integrate various systems and stakeholders in a single interoperable solution. The examples discussed so far, even though including M2M communications were integrating only two of the stakeholders involved in healthcare — the patient and the caregiver/doctor. However, in order to advance to a really integrated e-Healthcare system it is required that both the involvement and the trust of all stakeholders (patients, doctors and other medical personnel, families and caregivers, health care providers, regulators as well as related electronic health records) is secured.

An interesting work in this respect can be found in Shelby et al. (2014). Pereira and his colleagues suggest a novel eHealthcare system which integrates components from ICT infrastructure with upcoming standards in the medical sector, like HL7 (Health Level 7) for realizing interconnection with health platforms and the openEHR standard which defines the data semantics, storing and accessibility of medical data. This is one of the first and also very promising frameworks that enable integration of monitoring devices using M2M communications based on the ETSI M2M architecture model with the de-facto standard for medical records, the openEHR (Electronic Health Records) system and the standardized procedures for exchanging medical records known as HL7. To bridge the gap existing between M2M communication standards and the health standards the authors propose that the manager and the server entities in the ETSI model are based on the openEHR and HL7. Besides interoperability among the subsystems, which is the main design goal, the authors point out the following design requirements: *"wearability"* regarding the sensor devices; *connectivity*, which depending on the specific case might vary from delay tolerant, one-a-day connection to *"always connected"* solution; *privacy* which is of paramount importance since personal medical records are involved; and last but not least *reliability* and *fault tolerance* of the system and the networking process.

To illustrate the proposed solution the authors consider the following scenario: *"Mrs. Maria is 73 years old, living by herself on the third floor of a building in the outskirts of a small town. She is wearing a set of sensors, a specific kit related to her health condition, which is continuously collecting physiologic data because she has heart complications and is at risk of developing Alzheimer disease. The collected data contains information about her heart rate variability and R-R interval, breathing rate, posture, activity level, geographical position and the number of steps she takes. In the mornings, the system uploads the data, using a mobile phone, to her primary care unit, where it is stored in her EHR, indexed by her national health identification number. The system analyzes the data and searches for patterns that indicate problematic conditions. On a specific day, the system detects the occurrence of several mild arrhythmias. An alarm is immediately sent to her primary care unit and the nurses on call follow up by phoning her. It is recommended that she rests the rest of the day and that a doctor visits her in later during the day. The visit prompts a change in medication. The doctor advises that she should rest the next day, but after that she could resume her daily walks. During the following days the monitored values return to normal and so does Mrs. Maria's walks, as indicated by the system."* (Shelby et al., 2014)

In the scenario described above the patient's phone functions as the M2M Gateway, and collects information from the M2M End Devices (sensors) using an M2M application. The application connects to a Network Domain SCL that can manage it, provide access control and connect to other M2M applications and services. As for the transmission links the connection between M2M End Devices and the M2M Gateway uses Bluetooth, while the connection between the M2M Gateway and the M2M Network Application (M2M-NA) SCL uses 3G cellular technology.

In order to create proper clinical documents from the collected raw data Pereira and his colleagues make use of the openEHR (openEHR, 2013; Atalag et al., 2015), a non-proprietary architecture used for creating electronic medical records. The openEHR allows capturing and storing clinical data in a structured way, independent of specific software. In the considered scenario, it is used for the creation of the relevant clinical documents, by processing the raw data received from the M2M device at the M2M-NA, so that the templates when accessed by the primary care unit will contain Mrs. Maria's relevant clinical information properly structured. On the other hand, the HL7 (2007) messages are used to communicate with the openEHR repository and to receive and update the clinical information of Mrs. Maria e.g. the M2M-NA produces. The HL7 is a message-based ANSI standard for the exchange, management and integration of electronic information in the clinical domain. It covers the data exchange procedures, the timing of data exchanges and the handling of communication errors.

The general view of the proposed interoperable framework is given in Figure 17.15.

We can track the path of raw data starting from Mrs. Maria's wearable sensors (M2M device), through Mrs. Maria's smartphone (M2M Gateway) — Bluetooth connection — then through the 3G/4G cellular network to the NSCL (Network service Capability Layer), where the raw data is processed and sent to all content entities that are interested in it (e.g. the Processing Application — NA and the openEHR). Since the openEHR should receive the information in a structured form the task of providing the required data format is allocated to several Processing Applications (PA). The PAs are very specific Network Applications responsible for processing the raw data received from the M2M device into a HL7 structured messages and sending it back to the NSCL.

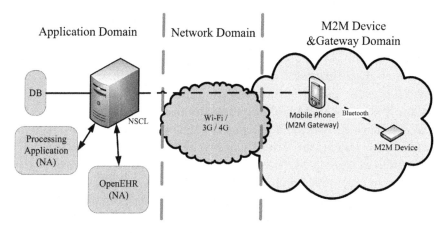

Figure 17.15 Inter-operable eHealthcare system architecture integrating data and network standards.

The communication process discussed so far is triggered by events at the patient side. So it is reasonable to adopt an event-based, publish-subscribe model of operation. In the considered scenario PA is a "subscriber" that notifies the "broker" (NA) for its interest in the events "published" by the M2M device of Mrs. Maria (the publisher). Once this exchange of messages is completed the PA will in its turn publish the properly structured information (HL7 format) in the NSCL so the openEHR can subscribe to them. All the procedures and resource mappings follow the ETSI RESTful resource architecture described before and use CRUD (Create, Read, Update, Delete) operations to manipulate the resources.

It is also possible to have the communication flow in the opposite direction — starting from the openEHR, i.e. the openEHR can post, delete or update information in the NSCL. An example of this is the situation when the clinical staff informs Mrs. Maria that she needs to rest for one day. In this case the openEHR system has to register itself as an Application resource and another PA will subscribe to it. The appropriate PA will make the required changes and the information will be published onto the NSCL to be received the M2M end device through the M2M Gateway.

This interesting and comprehensive work provides a solution which besides demonstrating interoperability is both scalable and highly reliable. It also allows for fast reactions in case of critical conditions. The proposed framework is one of the first complete solutions to using M2M communications in AAL.

Discussing interoperability in the network and application domain of the M2M architectures we should not overlook interoperability issues in the semantic domain, i.e. the semantic interoperability. As the authors of Chellouche et al. (2013) point out there is still lack of semantic in the data sharing process. Taking into consideration the diversity of the stakeholders involved in the M2M healthcare systems, from patients, physicians, hospitals, insurance agencies, and pharmaceutical companies, each with its own information system, the lack of semantic mapping of information is a major hurdle in turning M2M based Healthcare to reality. Chellouche et al. propose an innovative ontology-based framework for pervasive M2M Healthcare which will allow paving the road towards more comprehensive and more intelligent systems that can automatically take complex decisions and rapid actions based on the vital information of the patient's status and his environment. Chellouche and his colleagues propose the design of an ontology-based data model with explicit semantics that incorporates a variety of information on the patient's health and environmental context. Furthermore they propose a reasoning based middleware that can support different task involved in the data management.

Ontology, as defined in Gruber (1991) is a "formal explicit specification of a shared conceptualization of a domain of interest". It allows us to formally describe the semantics of some context information in terms of concepts and roles. In the setting of M2M communications ontology can provide possibilities to model unstructured information sensed through different M2M End Devices and collected in the network in a unified, structured way. This will enable healthcare services to be integrated with other relevant services like home applications controlling services, environment controlling services, home entertainment etc. Even though such services do not seem to affect the patient's condition directly, they are part of the AAL environment and can considerably contribute to generally improving the quality of assisted living.

In order to create the data model the authors first identify the relevant context information. "Context" is defined (Dey, 2001) as: "any information that can be used to characterize the situation of an entity. The entity is a general label which can refer to a person, place or object that is considered relevant to the user — application interaction, including also the user and the application themselves". The suggested ontology is based on the Web Ontology Language (OWL) which is designated as the world standard and there are numerous applications integrated with it. At a conceptual level it consists of owl:ClassElements, related by owl:ObjectProperty and characterized by owl:DataProperty. In the case under consideration these concepts are used to model the following generic entities defined by the authors: Person, Device, Service, SensedParm. Each one is further detailed by properties e.g. person can be a doctor, a patient etc., each one with a different profile: general profile — name, age, location, activity, availability etc.; medical profile — contains the medical data of the patient; the social profile — contains persons to contact, different member groups etc. Similarly the service entity is extended with the service profile in terms of inputs, outputs, preconditions and effects (IOPE). Devices can be sensors or other devices that run M2M applications. Events are classified in three types: discovery events that trigger service discovery, invocation events that trigger activation of events, and adaptation events that trigger changes in services. Based on these definitions, higher level context can be derived by reasoning from the low-level context, triggering context-aware decisions and actions. The services on the other hand are advertised in the network by the so called Service Platform in terms of IOPEs, which allows automatic and personalized service discovery and invocation. Defined on top of these are the rules specific to the medical environment that will allow context aware evaluation of the data collected from the M2M Device. For example the case of hyperglycemia in a pregnant woman cannot only be determined by physiological sensor measurements only; it is the context information that influences the interpretation of the measured value.

Taking in consideration the ETSI M2M architecture the authors suggest that these functionalities are implemented in the SCL of home M2M Gateway. Besides ease of implementation and flexibility this solution is also advantages in terms of scalability, because the system complexity is a function of the ontology size (in terms of individuals). Furthermore a smartphone can play the role of an M2M Gateway hosting a light version of the SCL. Figure 17.16 below provides an overview of the functional architecture of the proposed system. For the implementation of this ontology-based model the authors propose a middleware which creates a bridge between the consumers and the data provided. The raw data is collected from different sensors, the integrated into the knowledge system from which a new level context is produced and formatted adequately is fed to the service platform to take decisions and/or perform actions. The main functional components of the middleware are: the Context Integrator, The Healthcare and Environmental Manager, the Data Manager and the Interface Engine.

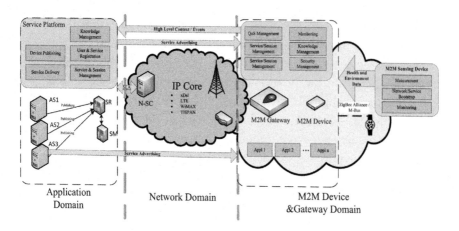

Figure 17.16 Ontology-based pervasive M2M healthcare architecture.

The work can be further extended by the development of an overall M2M Gateway and an interface for automatically converting user defined rules into semantic web language rules (SWRL).

Another interesting suggestion, a system for monitoring environmental risk for the elderly, is discussed here because even though not directly related to medical care, it can be considered as an important element of an ambient assisted environment (Tseng and Ke, 2012). It is a robotic Mobile Agent (MA) designed to transverse a target environment acting as a fire detector, fire alarm device and an extinguisher and can be used for individual homes or elderly care facilities. The initial prototype contains a wireless module, sensors and a mobile robotic cart on which a microprocessor, motor controller and a five-link articulated robotic arm are mounted. The low cost system composed of off-the-shelf elements can be adapted to other common household applications that will assist elderly people. The cart moves along a predefined path and upon detection of fire (abnormal levels detected by IR–flame sensors/CO) leaves the predefined path and moves toward the fire. Guided by signal strength measurements (for choosing the optimal path towards the destination) it reaches the fire location and gives command for emptying water to extinguish the fire. The prototype can connect to a PC or the network through a wireless connection. Further development of this work can include a number of such devices (operated in separate sections of an elderly care facility) which can be monitored and controlled through the network. In that scenario, each device will play the role of an M2M End Device, while the PC or a microcontroller will take over the functions of an M2M Gateway.

Another very important aspect of introducing ICT technology in the life of the elderly population is to evaluate how they react to it and what their acceptance level is. A study of this kind has been performed in Taiwan over a population group of around 50 individuals (Jianan Elderly Nursing Home Facility) and the results and evaluations are presented in Tseng et al. (2013). The study is motivated by the fact that the current generation of elderly people has quite low IT literacy and technology adoption. Furthermore, such systems are designed keeping closely in mind the needs of the elderly but there is no feedback about how much, depending on the adoption process, the actual users (the elderly themselves) think these systems meet their actual needs. The authors base their research on the intelligent framework they have designed and a novel friendliness user interface.

The system as installed in the facility consists of three layers: base layer containing the user RFID tags; second layer, containing the functional base and finally a database, with two types of interfaces — the client-based interface and the web-based one. The web interface allows logging into the system through an HTTP application, while the client side requires logging in with an RFID card. The RFID card is also used for other services in the nursing home like identity authentication, as a door key, and as an electronic wallet, as well as for storing medical data. The system architecture incorporates four main modules: the Communication Module (CM), the Membership Management Module (MMM), the Information Integration Module for Health Education of the Elderly (IIM-HEE) and the Personal Health Management Module (PHMM). The RFID tag and the sensors for measuring vital physiological parameters the M2M End Device which through a CareGiver Device/PC (conceptually equivalent to an M2M Gateway) can connect to a GSM network and to an Application Server or Database Server (conceptually the SCL). A second option is that an end device connects to the network through a Wi-Fi connection: chips installed in medical devices at the facility allow direct transmission of medical data to a PC connected to the Internet. Once the system is installed the researches continue with the second phase of exploring the factors that affect the elderly in their acceptance and usage of the intelligent monitoring systems. This process of exploring the acceptance includes the following procedures:

1. Introduction: explanation of the experiment that will be performed

2. Distribution of RFID cards (to be used with the platform) to the elderly patients in the care facility

3. Taking blood pressure measurements recording the results using a simplified scale of "sunny day", "cloudy day" and "rainy day" (representing the current physical condition of the user)

4. Evaluation and distribution of questionnaires that should be filled by the patients.

5. Conclusion: Each participating elderly is awarded a small gift.

The evaluation of the results is based on the "Integrated Technology Acceptance Model" (ITAM) developed by Venkatesh and Davis (2000) and includes five variables: performance expectation, endeavor expectation, social influence, facilitating condition and bandwagon effect. *Performance expectation* is defined as the degree that users expect to improve their performance by using the given system; the *endeavor expectation* is defined as the degree of effort the individual puts in using the system; the *social influence* is defines the effect of the significant others on the user in operating the system; *facilitating condition* measures the degree that the system assists users; the *bandwagon effect* is the majority influence which makes individuals follow public thinking and behavior; finally *user intention* refers to the subjective probability that an individual is likely to engage in certain behavior and *user behavior* refers to the fact that the user actually operates on the system. The assessment is done based on a 5-point Likert Scale, where 1 = strongly disagree, 5 = strongly agree. The designed questionnaire has been validated by external experts as well and data analysis is performed through the SPSS 14.0 statistical software. The major results of the study can be summarized as follows:

- Results for the performance expectation show that users' intention is positive and significant, which signifies that the tested healthcare system is helpful to the elderly
- Elderly showed that with minor additional endeavor and easy learning step procedures they can accept the system
- The acceptance of the elderly is highly influenced by the opinion of others around, especially peers
- The more provided technological assistance the higher the intensity of the elderly using the system
- Behavior intention does not affect the actual behavior of the elderly. This means that users who have intention to use the system do not necessarily use it. This translates to the conclusion that, even if the elderly have the willingness to use the system they still do not have the

habit to actually and stably use it. The researches suggest that this might partially be due to mobility constraints.

This study, even though limited in scale allows us to be optimistic that given the proper user interfaces and required step-by-step guidance elderly people will feel comfortable with intelligent healthcare systems. It is very important to keep in mind that this acceptance rate is highly influenced not only by technological factors but also by sociological and management factors. Such systems should be designed with requirements for easy learning steps, should be positioned taking into consideration the mobility restrictions of the elderly and also keeping in mind the bandwagon effect that the more peers accept the system the more probable and easy it is for new elderly individuals to join in using it.

17.4 CONCLUSION

In this chapter we have discussed the concepts of Machine-to-Machine technology and role it can play in the materializing of AAL. Even though its roots lay in much older, well establish systems like SCADA and industrial embedded networks, M2M is an emerging technology of the future, one of the enabling technologies of the Embedded Mobile Internet of the Future. Standardization work is on-going and in the first section we have discussed in detail the efforts of the 2 major global standardization bodies involved in the development of M2M — the ETSI and the 3GPP Alliance. Details on the proposed M2M architecture standards, requirements definition and possible application scenarios have been presented. Furthermore the concept of AAL and its relation to M2M communications have been discussed. Despite the fact that there are very strong technological and economical drives for the wide adoption of M2M communications, the lack of interoperability hinders the progress. The final part of the chapter summarized some of the latest and most interesting research works which focus on using M2M in AAL — from simple, isolated solutions to a comprehensive, interoperable framework for eHealth based on M2M, EHR and HL7. Finally an interesting research is described which evaluates the feedback from the user. The study investigates how the elderly adopt and accept this technology provide some interesting conclusions that point out that not only technological factors (suitable device design and adequate simple interfaces) but also the opinion the peers and the management procedures influence the acceptance rate.

List of acronyms with explanation

3GPP The 3rd Generation Partnership Project

6LoWPAN Ipv6 over Low Power Wireless Personal Area Networks

AAL Ambient Assisted Living

AML Ambient Intelligence

AN Access Network

CCSA The China Communications Standards Association

CN Core Network

CoAP Constrained Application Protocol

EHR Electronic Health Recorder

eNB Evolved Node B

EPC Evolved Packet Core

ETSI European Telecommunications Standards Institute

eUTRAN Evolved UMTS Terrestrial Radio Access Network

GISFI The Global ICT Standardization Forum for India

HeNB Home Evolved Node B

HFC Hybrid Fiber-Coaxial

HL7 Health Level 7

HSS Home Subscriber System

HTTP Hyper-Text Transfer Protocol

ICT Information Communication Technology

IEEE The Institute of Electrical and Electronics Engineers

IoT Internet of Things

LAN Local Area Network

LTE Long Term Evolution

M2M Machine to Machine

M2M-MF M2M Management Functions

MME Mobility Management Entity

MTC Machine Type Communication

NMF Network Management Functions

OMA Open Mobile Alliance

P-GW Packet Data Network Gateway

RAN Radio Access Network

RMD Remote Monitoring Device

RN Relay Node

S-GN Serving Gateway

SC Service Capabilities

SCADA Supervisory Control and Data Acquisition

SCC Service Capability Client

SCL Service Capability Layer
SCS Service Capability Server
SN Social Network
UTRAN UMTS Terrestrial Radio Access Network
W-LAN Wireless Local Area Network
WAN Wide Area Network
WBAN Wireless Body Area Network
Wi-Fi Wireless Fidelity
WiMAX Worldwide Interoperability for Microwave Access
WPAN Wireless Personal Area Network
WSN Wireless Sensor Networks
xDSL Digital Subscriber Line

Glossary of terms with explanations

M2M Communication Emerging technology that enables the communication between different devices and allows them to perform a variety of actions without or with only limited human intervention.

Network Architecture Layout of communication network that consist of the hardware and software components, their functional organization and configuration, their operational principles and procedures.

Ambient Intelligence Digital environment that is sensible and responsible supporting people in their life.

Gateway Intermediary device to manage the other devices, which are related with it, and aggregate data from device to send the base station.

Application Server Software framework in a distributed network that handles all application operations between users and databases.

Ambient Assisted Living AAL is a concept that combine the daily life activations with the information and communication technology to improve and increase the quality of life for people in a daily life.

Internet of Things System of computing devices, mechanical machines, objects, animals or people that provide a communication between them without any human intervention.

Cyber Physical Systems Physical systems whose operations are monitored, controlled and coordinated by a communication core.

End Device End device is a simplest type of device in a network which transmits or receives a data but cannot route the data.

Communication Protocol System of rules that allow the devices in a network to transmit information.

REFERENCES

3GPP, 2010a. Service Requirements for Machine-Type Communications, TS 22.368 V10.1.0, June 2010.

3GPP, 2010b. TR 22.888 v1.0.0 — System Improvement for Machine-Type Communications. technical report.

3GPP, 2010c. 3GPP TS 36.300 v10.0.0 — Evolved Universal Terrestrial Radio Access (E-UTRA) and Evolved Universal Terrestrial Radio Access Network (EUTRAN), June 2010.

3GPP, 2013. TS 22.368. Service requirements for Machine-Type Communications (MTC); Stage 1 (Release 11), Mar, 2013.

AAL-2011-4-099, 2011. ALICE — Assistance for Better Mobility and Improved Cognition of Elderly. http://alice-project.eu.

AALIANCE, 2010. AALIANCE project, 7h Framework Programme of the European Union, Grant Agreement No. 85562, 2008–2010, http://www.aaliance.eu, deliverable 2.5: AAL Strategic Research Agenda, March 2010.

AALJP, 2012. Ambient Assisted Living Joint Programme — Call2, ALICE — Advanced Lifestyle Improvement System and New Communication Experience, May 2012. http://aal-alice.eu.

Abbate, S., Avvenuti, M., Bonatesta, F., Cola, G., Corsini, P., Vecchio, A., 2012. A smartphone-based fall detection system. Pervasive Mob. Comput. 8 (6), 883–899.

Ageing, 2016. The Ageing Well in the Information Society Action Plan web site: http://ec.europa.eu/information_society/activities/einclusion/policy/ageing/action_plan/ [Accessed: May 2016].

Aghajan, H., Augusto, J.C., Wu, C., McCullagh, P., Walkden, J.-A., 2007. Distributed vision-based accident management for assisted living. In: Proc. Int. Conf. Smart Homes Health Telemat, pp. 196–205.

Alm, N., Arnott, J.L., Dobinson, L., Massie, P., Hewines, I., 2001. Cognitive prostheses for elderly people. In: IEEE International Conference on Systems, Man, and Cybernetics, vol. 2. Tucson, AZ, pp. 806–810.

Atalag, K., Beale, T., Chen, R., Gornik, T., Heard, S., McNicoll, I., 2015. openEHR: a semantically-enabled, vendor-independent, health computing platform. openEHR. [Online]. Available: www.openehr.org/resources/white_paper_docs/openEHR_vendor_independent_platform.pdf.

Atzori, L., Iera, A., Morabito, G., 2010. The Internet of Things: a survey. Comput. Netw. 54 (15), 2787–2805.

Augusto, J., 2007. Ambient intelligence: the confluence of pervasive computing and artificial intelligence. In: Schuster, A. (Ed.), Intelligent Computing Everywhere. Springer, pp. 213–234.

Bhowmik, A.K., Khendek, F., Hormati, M., Glitho, R., 2015. An architecture for M2M enabled social networks. In: 2015 14th Annual Mediterranean, Ad Hoc Networking Workshop (MED-HOC-NET), pp. 1–8.

Borgia, E., 2013. The Internet of Things vision: key features, applications and open issues. Comput. Commun. 54, 1–31.

Chellouche, S.A., Chalouf, M.A., Lemlouma, T., 2013. Ontology-based pervasive M2M healthcare environment. In: First International Symposium on Future Information and Communication Technologies for Ubiquitous Healthcare (Ubi-Health Tech 2013), pp. 1–5.

Chen, K., 2012. Challenges and opportunities of Internet of Things. In: 2012 17th Asia and South Pacific Design Automation Conference (ASP-DAC). Sydney.

Chung, W.-Y., 2012. Multi-modal sensing M2M healthcare service, WSN. KSII Trans. Int. Inf. Syst. 6 (4).

Clayman, S., Galis, A., 2011. INOX: a managed service platform for inter-connected smart objects. In: ACM IoTSP 2011. Tokyo, Japan.

Delmastro, F., 2012. Pervasive communications in healthcare. Comput. Commun. 35, 1284–1295.

Dey, A.K., 2001. Understanding and using context. In: ACM Personal and Ubiquitous Computing, V.5., N0.1. Springer-Verlag, Atlanta, USA, pp. 4–7.

Dias, A., Gorzelniak, L., Jrres, R.A., Fischer, R., Hartvigsen, G., Horsch, A., 2012. Assessing physical activity in the daily life of cystic fibrosis patients. Pervasive Mob. Comput. 8 (6), 837–844.

Dubowsky, S., Genot, F., Godding, S., Kozono, H., Skwersky, A., Yu, H., Yu, L.S., 2000. PAMM — a robotic aid to the elderly for mobility assistance and monitoring for the elderly. In: Proc. Robot. Autom. Conf., vol. 1, pp. 570–576.

Eklund, J., Hansen, T., Sprinkle, J., Sastry, S., 2005. Information technology for assisted living at home: building a wireless infrastructure for assisted living. In: Proc. Eng. Med. Biol. Soc, pp. 3931–3934.

Fleck, S., Strasser, W., 2008. Smart camera based monitoring system and its application to assisted living. Proc. IEEE 96 (10), 1698–1714.

Ghavimi, F., Chen, H.H., 2015. M2M communications in 3GPP LTE/LTE-A networks: architectures, service requirements, challenges, and applications. IEEE Commun. Surv. Tutor. 17 (2).

Grguric, A., 2012. ICT Towards Elderly Independent Living. Research and Development Centre, Ericsson Nikola Tesla.

Gruber, T.R., 1991. The role of common ontology in achieving sharable reusable knowledge bases. In: Principles of Knowledge Representation and Reasoning, Proc. of the Second International Conference. Cambridge, MA, pp. 601–602.

Health Level Seven International, 2007. What is HL7? [Online]. Available: http://www.hl7.org/about/.

http://ccsa.org.cn/english/tc.php?tcid=tc10.

http://www.gisfi.org/index.php.

http://www.openmobilealliance.org/.

http://www.onem2m.org/.

http://www.tiaonline.org/standards/procedures/manuals/scope.cfm#TR50.

ISTAG, 1999. Information Society Technologies Advisory Group (ISTAG): orientations for work programme 2000 and beyond, pp. 3–4 (17 September 1999).

ITU-T Y.2221, 2010. Requirements for support of ubiquitous sensor network (USN) applications and services in the NGN environment. Jan. 2010.

Jara, A.J., Lopez, P., Fernandez, D., Zamora, M.A., 2013. Communication protocol for enabling continuous monitoring of elderly people through near field communications. Interact. Comput. 26 (2), 145–168.

Jung, S.J., Myllyla, R., Chung, M.Y., 2013. Wireless machine-to-machine healthcare solution using android mobile devices in global networks. IEEE Sens. J. 13 (5), 1419–1424.

Kartsakli, E., Lalos, A.S., Antonopoulos, A., Tennina, S., Di Renzo, M., Alonso, L., 2014. End-to-end communication challenges in M2M systems for mHealth applications. In: 2014 IEEE 19th International Workshop on Computer Aided Modeling and Design of Communication Links and Networks (CAMAD), pp. 353–359.

Khan, D.U., Siek, K.A., Meyers, J., Haverhals, L.M., Cali, S., Ross, S.E., 2010. Designing a personal health application for older adults to manage medications. In: Proc. Int. Health Inf. Symp, pp. 849–858.

Kim, J., Lee, J., Kim, J., Yun, J., 2014. M2M service platforms: survey, issues, and enabling technologies. IEEE Commun. Surv. Tutor. 16 (1), 61–76.

M2M, 2011. Machine-to-Machine communications (M2M) — Functional architecture. Technical specification, ETSI TS 102 690 V1.1.1 (2011-10).

M2M, 2012. Machine-to-Machine communications (M2M) — Service requirements. Technical specification, ETSI TS 102 689 V1.1.1 (2012-8).

M2M, 2013. Machine-to-Machine communications (M2M); Use cases of M2M applications for eHealth. ETSI TR 102 732 V1.1.1 (2013-09).

Manduchi, R., Coughlan, J., 2012. (Computer) vision without sight. Commun. ACM 55 (1), 96–104.

Monekosso, D., Revuelta, F.F., Remagnino, P., 2015. Ambient assisted living. IEEE Intell. Syst. 30 (4), 2–6.

Mynatt, E.D., Rowan, J., Craighill, S., Jacobs, A., 2001. Digital family portraits: supporting peace of mind for extended family members. In: Proc. CHI, pp. 333–340.

openEHR, 2013. What is openEHR. [Online]. Available: http://www.openehr.org/what_is_openehr.

Pereira, C., Frade, S., Brandao, P., Correia, R., Aguiar, A., 2014. Integrating data and network standards into an interoperable e Health solution. In: 2014 IEEE 16th International Conference on e-Health Networking, Applications and Services (Healthcom), pp. 99–104.

Pollack, M.E., Brown, L., Colbry, D., McCarthy, C.E., Orosz, C., Peintner, B., Ramakrishnan, S., Tsamardinos, I., 2003. Autominder: an intelligent cognitive orthotic system for people with memory impairment. Robot. Auton. Syst. 44 (3–4), 273–282.

Pre-published Recommendation ITU-T Y.2060, 2012. Overview of Internet of Things. Jun. 2012.

Pre-published Recommendation ITU-T Y.2061, 2012. Requirements for support of machine oriented communication applications in the NGN environment. Jun. 2012.

Qudah, I., Leijdekkers, P., Gay, V., 2010. Using mobile phones to improve medication compliance and awareness for cardiac patients. In: Proc. Int. Conf. Pervas. Technol. Related Assisted Environ, pp. 1–7.

Rashidi, R., Mihailidis, A., 2013. A survey on ambient-assisted living tools for older adults. IEEE J. Biomed. Health Inform. 17 (3), 2168–2194.

Shelby, Z., Hartke, K., Bormann, C., Frank, B., 2014. The Constrained Application Protocol (CoAP), RFC7252.

Shin, S.H., Kamal, R., Haw, R., Moon, S.I., Hong, C.S., Choi, M.J., 2012. Intelligent M2M network using healthcare sensors. In: 2012 14th Asia-Pacific, Network Operations and Management Symposium (APNOMS), pp. 1–4.

Taleb, T., Kunz, A., 2012. Machine type communications in 3GPP networks: potential, challenges, and solutions. IEEE Commun. Mag. 50 (3), 178–184.

TIA TR-50, 2010. Smart Device Communications. Feb. 2010.

Triantafyllidis, A.K., Koutkias, V.G., Chouvarda, I., Maglaveras, N., 2013. A pervasive health system integrating patient monitoring, status logging, and social sharing. IEEE J. Biomed. Health Inform. 17 (1).

Tseng, P.C., Ke, C.H., 2012. Monitoring the environmental risk for the elderly. In: 2012 International Symposium on Intelligent Signal Processing and Communications Systems (ISPACS), pp. 233–238.

Tseng, K.C., Hsu, C.L., Chuang, Y.H., 2013. Designing an intelligent health monitoring system and exploring user acceptance for the elderly. J. Med. Syst. 37, 9967.

Venkatesh, V., Davis, F.D., 2000. A theoretical extension of the technology acceptance model: four longitudinal filed studies. Manag. Sci. 46, 186–204.

Vetere, F., Davis, H., Gibbs, M., Howard, S., 2009. The magic box and collage: responding to the challenge of distributed intergenerational play. Int. J. Hum.-Comput. Stud. 67 (2), 165–178.

Wu, G., Talwar, S., Johnsson, K., 2011. M2M: from mobile to embedded Internet. Commun. Mag. (April), 36–43.

Zing. [Online]. Available: https://code.google.com/p/zing/.

INDEX

Printed in the United States
By Bookmasters